An African Green Revolution

Keijiro Otsuka • Donald F. Larson
Editors

An African Green Revolution

Finding Ways to Boost Productivity on Small Farms

Springer

Editors
Keijiro Otsuka
National Graduate Institute
for Policy Studies
7-2 2-1 Roppongi
Minato, Tokyo 106-8677
Japan

Donald F. Larson
The World Bank
Research Group
1818 H Street Northwest
Washington, DC 20433
USA

ISBN 978-94-007-5759-2 ISBN 978-94-007-5760-8 (eBook)
DOI 10.1007/978-94-007-5760-8
Springer Dordrecht Heidelberg New York London

Library of Congress Control Number: 2012951120

Printed on acid-free paper

Springer is part of Springer Science+Business Media (www.springer.com)

Preface

During the last 40 years, improved technologies have fueled an on-going revolution in agriculture in most developing countries, but not in Africa. Since the 1970s, grain yields have more than doubled in China, India, many other Asian countries, and all of South America, while yields in Africa grew by a third. Still, it would be wrong to believe that agriculture in Africa did not grow. Turning to a different metric, cereal production more than doubled in Africa and, decade over decade, grew slightly faster in Africa than in Asia and just slightly slower than in South America. Even so, how African agriculture grew – by bringing more land under cultivation and introducing slow improvements in yields – has had consequences. Low and stagnant productivity has brought with it low income growth for those whose livelihood depends on agriculture. And this is especially significant for Sub-Saharan Africa, where agriculture accounts for nearly 70% of employment and more than a third of the population lives on less than a dollar a day.

There are urgent reasons to find a different way for agriculture to grow in Africa. Over the next 40 years, Africa's population will likely grow by 93% – nearly triple the rate of increase for the world as a whole. Moreover, while rural populations are expected to decline significantly in Asia and South America, rural populations are expected to grow in Africa. Already, land for expanding agricultural production has become scarce and land frontiers are closing across much of the continent. What's more, disease and poor infrastructure prevent some potentially productive lands in Africa from being farmed while overuse and poor land management practices have led to declining soil fertility in others places. In addition, should productivity continue to stall and Africa be unable to contribute more to global food supplies, there is risk that food prices will rise internationally and hinder global efforts to reduce poverty.

At the same time, the current situation in Africa is not so different from conditions in Asia before a Green Revolution began there. In Asia then as in Africa now, large portions of the population lived in rural areas and depended on agriculture for their livelihoods; many farmed small parcels of land and were poor; grain yields were low and stagnant; and famine was a lingering reality. Today, after decades of sustained growth in agricultural productivity, rural poverty has been greatly reduced and food security greatly improved. For this reason, many African leaders have modeled their own efforts to improve agricultural productivity on lessons from Asia.

What lessons then can we draw from the Asian experience that are relevant for contemporary Africa? In the chapters of this volume, authors explore how Asia's Green Revolution began and how initial conditions then compare to Africa now. They explore various aspects of the Asian model, with its focus on smallholder agriculture, its emphasis on modern seed varieties and input-intensive methods, and its reliance on private markets. Some chapters examine in detail important institutions, including fertilizer markets, extension systems, and irrigation schemes, and key crops, maize and rice. Collectively, the chapters suggest that the Asian policy approach is well suited for Africa as a general framework.

However, the studies also point to key modifications needed to account for the great heterogeneity of conditions across Africa. In Asia, the dominant roles played by rice and wheat in farming systems and in the diets of the poor meant that a narrow set of technological breakthroughs had an outsized impact on farm productivity and incomes. In a related way, agro-climatic conditions within and across countries were often similar, which meant that a given technology that worked well in one place could work well in many places, thereby lowering the cost of achieving overall productivity growth. This is not to imply that productivity gains came easily or that country-based research did not play a role in adapting proven technologies to local conditions. Yet advancing increased productivity for staple grains was and remains a priority for many countries in Asia and gains there have been hard earned.

Still, Africa faces additional challenges. In Africa, diets are diverse, agro-climatic conditions vary greatly, and there are significant differences in population and infrastructure densities and the strength and efficacy of public and private institutions. Consequently, a mosaic of initial conditions prevails for the continent as a whole. Because of this, a portfolio of technologies is required to get the same impact as breakthroughs in rice and wheat brought to Asia. Differences in local conditions also imply greater experimentation and greater risks, even for technologies that are proven successful at field stations. It also put greater demands on extension services and farmers, who must sort out the effects of these differences on the economic viability of competing approaches to farming. And it puts a premium on markets for outputs, labor, land, and fertilizers, which must work across a larger set of core commodities.

So then, will all of these handicaps forestall an African Green Revolution? The chapters of this volume provide several reasons to believe they will not. To start, there are places in Africa today where productivity gains have been significant and where farmers obtain yields that rival their most productive counterparts in Asia. Moreover, the chapters' authors list new advances and find scope for transferring and adapting what has proved successful for smallholders in Asia and in Africa across the continent. And importantly, the twenty-first century has seen a recommitment by African governments and the development community to bring about a Green Revolution in African agriculture. It is our sincere hope that, by putting into perspective key aspects of both the Asian and African experiences, this volume contributes to that purpose and contributes to the design of an effective strategy to realize a Green Revolution in Africa.

This book is collaboration and there are many people to thank. It was Gershon Feder, the respected former economist at the World Bank, who first suggested a volume to explore the relevance of Asia's Green Revolution for Africa, while learning from the Asian experience, during the International Conference of Agricultural Economists in 2006 on the Australian Gold Coast. We would like to thank him for turning our attention to the exceedingly important issue. We owe much to his insights and advice. No less important were Peter Hazell and Derek Byerlee, two leading agricultural economists with rich experience in agricultural development in Asia and Africa, both of whom have served as economists at the research centers of the Consultative Group of International Agricultural Research and the World Bank. They directly contributed to the project leading to this book publication as authors and as advisers. Without their dedicated contributions, this project would not have been completed so successfully. We would like to express our sincerest appreciation to them.

To promote common understanding of the key issues and exchange the views among the contributors and with other researchers and policymakers, we organized a series of workshops: in Washington, April 2008, in Tokyo December 2008 and again in December 2009, as a Mini-Symposium at the International Conference of Agricultural Economics in Beijing in August 2009, in Nairobi in April 2010, at the Annual Bank Conference on Development Economics in Stockholm, June 2010, and in Washington, DC in January and March 2012 . We have benefited greatly from the participants in these workshops. In particular, we would like to thank Karen Brooks, Nobumitsu Hayashi, Will Martin, Robert Mendelsohn, Michael Morris, Niggol Seo, and Jinxia Wang. We also thank Paul Kandasamy for editorial insights, Polly Means for improving the graphical presentations, and Mayuko Tanaka and Kazuko Yamamura for their patient preparation of the manuscript.

We received financial supports from the donor-funded Knowledge for Change Program, hosted at the World Bank, under the project title "An African Green Revolution: Finding Ways to Boost Productivity," along with a grant from the World Bank's Research Support Budget. We also received support from the Global Center of Excellent Program of the Ministry of Education and Science of the Japanese government. We are grateful for the generous support of these institutions.

Contents

1 An Overview 1
Keijiro Otsuka, Donald F. Larson, and Peter B.R. Hazell

Part I Climate and the Transferability of Asian Green Revolution to Sub-Saharan Africa

2 Lessons from the Asian Green Revolution in Rice 17
Jonna P. Estudillo and Keijiro Otsuka

3 The Possibility of a Rice Green Revolution in Large-Scale Irrigation Schemes in Sub-Saharan Africa 43
Yuko Nakano, Ibrahim Bamba, Aliou Diagne, Keijiro Otsuka, and Kei Kajisa

4 The Declining Impacts of Climate on Crop Yields During the Green Revolution in India, 1972–2002 71
Takuji Tsusaka and Keijiro Otsuka

5 The Impact of Technological Changes on Crop Yields in Sub-Saharan Africa, 1967–2004 95
Takuji Tsusaka and Keijiro Otsuka

Part II Prospects for Upland Rice and Maize Green Revolution in Sub-Saharan Africa

6 Causes and Consequences of NERICA Adoption in Uganda 123
Yoko Kijima and Keijiro Otsuka

7 Impact of NERICA Adoption on Rice Yield: Evidence from West Africa 143
Aliou Diagne, Soul-Kifouly Gnonna Midingoyi, and Florent M. Kinkingninhoun-Medagbe

8 **Maize Revolutions in Sub-Saharan Africa** 165
Melinda Smale, Derek Byerlee, and Thom Jayne

9 **Maize, Soil Fertility, and the Green Revolution in East Africa** 197
Tomoya Matsumoto and Takashi Yamano

Part III The Role of Fertilizer Markets and Fertilizer Application

10 **Chemical Fertilizer, Organic Fertilizer, and Cereal Yields in India** 225
Kei Kajisa and N. Venkatesa Palanichamy

11 **The Demand for Fertilizer When Markets Are Incomplete: Evidence from Ethiopia** 243
Daniel Zerfu Gurara and Donald F. Larson

12 **Technology Adoption in Agriculture: Evidence from Experimental Intervention in Maize Production in Uganda** 261
Tomoya Matsumoto, Takashi Yamano, and Dick Sserunkuuma

Part IV Conclusion

13 **Towards a Green Revolution in Sub-Saharan Africa** 281
Keijiro Otsuka and Donald F. Larson

Erratum E1

Contributors

Ibrahim Bamba Programme Management Department, International Fund for Agricultural Development, Rome, Italy

Derek Byerlee Washington, DC, USA

Aliou Diagne Policy and Impact Assessment, Africa Rice, Cotonou, Benin

Jonna P. Estudillo National Graduate Institute for Policy Studies, Tokyo, Japan

Daniel Zerfu Gurara African Development Bank, Tunis-Belvedère, Tunisia

Peter B.R. Hazell Imperial College, London, UK

Centre for Development, Environment and Policy, School of Oriental and African Studies, University of London, Ashford, Kent, UK

Thom Jayne Department of Agricultural Food, and Resource Economics, Michigan State University, East Lansing, MI, USA

Kei Kajisa School of International Politics, Economics, and Communication, Aoyama Gakuin University, Tokyo, Japan

Yoko Kijima Graduate School of Systems and Engineering, University of Tsukuba, Tsukuba, Ibaraki, Japan

Florent M. Kinkingninhoun-Medagbe Africa Rice Center (AfricaRice), Cotonou, Benin

Donald F. Larson World Bank, Washington, DC, USA

Tomoya Matsumoto National Graduate Institute for Policy Studies, Tokyo, Japan

Soul-Kifouly Gnonna Midingoyi Institut National pour la Recherche Agricole du Benin (INRAB), Porto-Novo, Benin

Yuko Nakano Faculty of Humanities and Social Sciences, University of Tsukuba, Tsukuba-shi, Ibaraki, Japan

Keijiro Otsuka National Graduate Institute for Policy Studies, Tokyo, Japan

N. Venkatesa Palanichamy Department of Agricultural and Rural Management, Centre for Agricultural and Rural Development Studies, Tamil Nadu Agricultural University, Coimbatore, Tamil Nadu, India

Melinda Smale Department of Agricultural, Food and Resource Economics, Michigan State University, East Lancing, MI, USA

Dick Sserunkuuma Department of Agricultural Economics & Agribusiness, Makerere University, Kampala, Uganda

Takuji Tsusaka International Rice Research Institute, Makati City, Philippines

Takashi Yamano International Rice Research Institute, New Delhi, India

List of Figures

Fig. 2.1	The conceptual interrelationships between technology, extension, input markets, and irrigation investments	20
Fig. 2.2	Cereal crop yields in Asia and Sub-Saharan Africa, 1961–2008	21
Fig. 2.3	Adoption of modern rice varieties and irrigation ratio in the Philippines, Vietnam, Bangladesh, and India	25
Fig. 2.4	Fertilizer use and rice yield in the Philippines, Vietnam, Bangladesh, and India	27
Fig. 3.1	Location of irrigation schemes	46
Fig. 4.1	Proportions of harvested areas by cereal crop (%), 2003–2007 average	74
Fig. 4.2	Average cereal yields in India, Southeast Asia, and SSA, 3-year moving averages	74
Fig. 4.3	Cereal yields in India by crop, 3-year moving averages	75
Fig. 4.4	Changes in harvested areas by cereal crop in India	76
Fig. 4.5	Proportion of irrigated area by crop in India	76
Fig. 4.6	Proportion of area sown to modern varieties by crop	77
Fig. 5.1	Proportions of harvested areas by cereal crop (%), 2003–2007 average	98
Fig. 5.2	Average cereal yields in SSA, Asia, North America, South Asia, and North Africa, 3-year moving averages	99
Fig. 5.3	Cereal yields in SSA by crop, 3-year moving averages	100
Fig. 5.4	Global wheat map: amount of wheat output per hectare of total agricultural land (Source: Compiled by the University of Minnesota Institute on the Environment with data from: Monfreda et al. 2008	102
Fig. 5.5	Changes in harvested areas by cereal crop in SSA, 3-year moving averages	102

Fig. 6.1 Cumulative distribution of per capita income 139
Fig. 9.A.1 Log of maize yield (kg/ha) by seed type 218

Fig. 12.1 Survey villages in Uganda 266
Fig. 12.2 Predicted maize yield with the use of purchased inputs in 2nd cropping season of 2009 in Uganda. The maize yields in the graph are the predicted values given the use of 25 kg/ha of the hybrid seed and the average purchase quantitiesof chemical fertilizersreported at the sales experiment in August and September 2009 by household type and by sales arrangement (in terms of discount level and credit avariability). As a referecnce, the average yield of local variety without fertilizers is also given................. 275
Fig. 12.A.1 Time line of surveys and field experiments in Uganda 277

Fig. 13.1 Composition of diet in South and Southeast Asia in 1965 compared to Africa 2005...................................... 284
Fig. 13.2 The amount of grain needed to purchase fertilizers based on international price data, 1960–2009 286
Fig. 13.3 Actual and predicted regional urbanization in Africa and Asia, 1961–2020.. 289

List of Tables

Table 2.1	Number of improved rice varieties released in selected Asian countries by time period	28
Table 2.2	Producer price of rice and fertilizer price in selected Asian and African countries	30
Table 2.3	Decomposition of the changes in paddy yield in selected Asian countries	35
Table 2.4	Adoption of generations of modern varieties of rice (% area) in the central Luzon provinces, the Philippines, 1966–2004	37
Table 3.1	Basic statistics on rice production in survey countries	45
Table 3.2	Characteristics of irrigation schemes and survey data	47
Table 3.3	Paddy yields, input use per hectare, and input prices at the study sites and in Laguna in the Philippines	52
Table 3.4	Costs and returns in study schemes (in US$)	56
Table 3.5	Definition of the variables at each study site	60
Table 3.6	Determinants of input use in Chokwe	62
Table 3.7	Determinants of input use in DRS	64
Table 3.8	Determinants of paddy yield in Chokwe (structural form estimation)	66
Table 3.9	Determinants of paddy yield in Doho (reduced-form estimation)	68
Table 4.1	Yield, climate, and irrigation coverage in India, by crop, 1998–2002 5-year averages	79
Table 4.2	Yield, climate, and irrigation coverage in India, by crop, 1972–1976 5-year averages	80
Table 4.3	First-step probit regression, by crop, 1972–2002	85
Table 4.4	Second-step outcome regression result for wheat, rice, and maize, 1972–2002	86

Table 4.5 Second-step outcome regression result for sorghum and millet, 1972–2002 88

Table 5.1 Crop yields in SSA and South Asia and their growth 100
Table 5.2 Country fixed-effect regression results for yield of wheat, 1967–2004 107
Table 5.3 Country fixed-effect regression results for yield of rice, 1967–2004 109
Table 5.4 Country fixed-effect regression results for yield of maize, 1967–2004 111
Table 5.5 Country fixed-effect regression results for yield of sorghum, 1967–2004 114
Table 5.6 Country fixed-effect regression results for yield of millet, 1967–2004 116

Table 6.1 Rainfall, access to nearby town, and land use in sample areas of NERICA production in Uganda 127
Table 6.2 NERICA rice yield and seed use by district in Uganda 128
Table 6.3 Household and community characteristics by type of adopters of NERICA in Uganda 128
Table 6.4 NERICA yields and sources of seeds by type of adopters in Uganda 131
Table 6.5 Determinants of NERICA adoption in Uganda (multinomial logit model, relative risk ratio) 135
Table 6.6 Yield function (ton/ha) estimation by instrumental variables (2SLS) regression 137

Table 7.1 Performance of NERICA progenies and their parents under low- and high-input conditions and major stresses in upland rice ecosystems 149
Table 7.2 Average yield and difference in mean yield between NERICA Adopters and non-adopters 152
Table 7.3 Socio-demographics of NERICA adopters and non-adopters in the four countries 154
Table 7.4 ATE/LATE estimates of the impact of NERICA adoption on average rice yield 155

Table 8.1 Maize area, production, yield and consumption in regions of Sub-Saharan Africa, 1961–2008 168
Table 8.2 Variability in maize and rice yields and prices around trend in major maize producing countries compared to rice in Asia, 1991–2007 170
Table 8.3 Staple food budget shares, urban centers in Kenya, Mozambique, and Zambia 172
Table 8.4 Adoption of improved maize varieties (% of maize area) in Eastern and Southern Africa, 1990, 1996 and 2006 176

Table 8.5	Adoption of improved maize varieties (% of maize area) in Western and Central Africa, 2005, 1998, and 1981	177
Table 9.1	Summary statistics of maize production in Kenya and Uganda	205
Table 9.2	Input and output level in maize production by soil carbon content in Kenya and Uganda	207
Table 9.3	Determinants of the newly purchased HYV seed adoption in Kenya and Uganda	208
Table 9.4	Determinants of log of maize yield (kg/ha) in Kenya	210
Table 9.5	Determinants of log of maize yield (kg/ha) in Uganda	212
Table 9.6	Relative prices and marginal returns of nitrogen application in Kenya and Uganda	215
Table 9.A.1	Determinants of sample attrition in Kenya and Uganda (probit)	219
Table 9.A.2	Input and output prices on maize production in Kenya and Uganda by region	220
Table 10.1	Summary statistics of dairy sector development in Tamil Nadu from 1993–2003	229
Table 10.2	Estimation results of paddy and upland cereal yield functions in Tamil Nadu	234
Table 10.3	Estimation results of FYM demand functions for paddy and upland cereals in Tamil Nadu	235
Table 10.4	Results of manure and rice price transmission analyses in Tamil Nadu	236
Table 10.5	The determinants of village-level farm-yard manure (FYM) price	237
Table 10.A.1	Description, measure, and summary statistics of the variables for yield functions in Tamil Nadu	239
Table 10.A.2	Description, measure, and summary statistics of the variables for farm-yard manure (FYM) demand functions in Tamil Nadu	239
Table 10.A.3	Description, measure, and summary statistics of the variables for manure and rice price transmission functions in Tamil Nadu	239
Table 10.A.4	Description, measure, and summary statistics of the variables for the determinants of village-level farm-yard manure (FYM) price	240
Table 11.1	Sample statistics on output and fertilizer use in Ethiopia	250
Table 11.2	Procurement channels for organic and chemical fertilizer in Ethiopia in 2004 (% of total)	252
Table 11.3	Estimation results of fertilizer application function in Ethiopia	253

Table 11.A.1 Estimation results of fertilizer application function in Ethiopia from the base model and alternative model 257

Table 12.1 Comparison of input use in maize production between Kenya and Uganda 264
Table 12.2 Number of households by participation group in experiments in Uganda 269
Table 12.3 Summary statistics of key variables in maize production in Uganda 270
Table 12.4 Purchase quantity of modern inputs at sales experiment held in August-September 2009 in Uganda 272
Table 12.5 Determinants of maize yield in 2nd cropping season of year 2009 in Uganda 274

Table 13.1 Regional composition of diet in Africa, 2005 285
Table 13.2 Average yields for rice and maize by region in Sub-Saharan Africa and Asia, 1965 and 2005 286
Table 13.3 Installed irrigation as share of cropland by region in Sub-Saharan Africa and Asia, 1965 and 2005 287
Table 13.4 Agricultural land relative to rural and urban populations by region in Sub-Saharan Africa and Asia, 1965 and 2005 288
Table 13.5 Average farm size in India and Indonesia during Asia's Green Revolution 290
Table 13.6 The dominant role of smallholder farming in Asia and Africa 291

Chapter 1
An Overview

Keijiro Otsuka, Donald F. Larson, and Peter B.R. Hazell

Abstract On average, agriculture accounts for 70% of full-time employment in Africa, 33% of national income, and 40% of total export earnings, and its importance is even greater in the poorest countries. Yet its performance in recent decades has been one of the worst in the world. Africa has some of the lowest levels of land and labor productivity and these have barely changed in 30 years. And, while it might appear obvious that accelerating agricultural growth should figure prominently in any strategy to reverse Africa's decline, agriculture virtually fell off the development agenda for Africa until recently. Now that agriculture is back in the spotlight, important differences about the nature of the effective agricultural development strategy have emerged around four key issues; whether to prioritize small or large farms; whether to focus on food staples or high value products; whether to promote technologies and farming practices that require fertilizers and modern seeds; and the degree to which governments should intervene in markets. By reviewing the existing studies in these areas, this chapter sets the stage for detailed empirical studies conducted in subsequent chapters.

K. Otsuka
National Graduate Institute for Policy Studies,
7-22-1 Roppongi, Minato-ku, Tokyo 106-8677, Japan
e-mail: otsuka@grips.ac.jp

D.F. Larson (✉)
Development Research Group, World Bank, 1818 H Street NW,
Washington, DC 20433, USA
e-mail: dlarson@worldbank.org

P.B.R. Hazell
Imperial College, London, UK

Centre for Development, Environment and Policy,
School of Oriental and African Studies (SOAS), University of London,
Fallowfield, Westwell, Ashford, Kent, TN25 4LQ, UK
e-mail: p.hazell@cgiar.org

K. Otsuka and D.F. Larson (eds.), *An African Green Revolution: Finding Ways to Boost Productivity on Small Farms*, DOI 10.1007/978-94-007-5760-8_1,

Keywords Small farms • Large farms • Food staples • High value crops • High-input farming • Low-input farming • State-led development • Market-led development

1.1 Introduction

The situation in Africa is dire. Even before the recent world food crisis, a staggering one in three people and a third of all children were undernourished and more than one half of all Africans (about 300 million people) lived on less than one dollar per day, and the continent was becoming increasingly dependent on relief aid from abroad. As a result of the recent food price increases, the FAO estimates an additional 100 million Africans were driven further into poverty (FAO 2009). Yet again, the lives of millions of Africans were dependent upon emergency handouts from abroad.

Although there are many contributing factors, the poor performance of the agricultural sector lies at the heart of the problem. On average, agriculture accounts for 70% of full-time employment in Africa, 33% of national income, and 40% of total export earnings, and its importance is even greater in the poorest countries. Yet its performance in recent decades has been one of the worst in the world. There are many indicators of agricultural performance and Africa ranks poorly by most of them. Africa has some of the lowest levels of land and labor productivity and these have barely changed in 30 years; the continent has declining per capita output levels, especially of staple foods; it has some of the lowest chemical fertilizer use rates, with serious nutrient mining and declining soil fertility; and Africa is badly losing world market shares for its traditional export crops (Hazell and Wood 2008; World Bank 2008).

Paradoxically, there is enormous potential for agricultural growth in Africa. The continent is blessed with abundant natural resources (e.g., it has 12 times the land area of India and about two-thirds as many people to feed). Even if Africa were only to double its average cereal yield to about 2 t/ha, this would lead to an extra 100 million t/year of cereals, shifting Africa from a food deficit to a major food surplus region. With a rapidly growing labor force (despite HIV/AIDS), there is growing scope for adopting higher-yielding but more labor-intensive technologies and farming systems. Markets are expanding, with rapid population growth and urbanization at home and new export opportunities as a result of trade liberalization and globalization. Furthermore, agricultural markets are becoming competitive and they are becoming better integrated regionally (Yamano et al. 2011). Moreover, with few exceptions, the distribution of land is still equitable by international standards and small farms that are efficient but poor dominate the continent.

Given this context it might appear obvious that accelerating agricultural growth should figure prominently in any strategy to reverse Africa's decline, much as it helped kick start economic growth and poverty reduction across Asia in the 1970s. Yet agriculture virtually fell off the development agenda for Africa during the past

25 years. Official development assistance for African agriculture was allowed to decline from about $2 billion to less than $1 billion per year from the mid-1980s to the late 1990s, and remained largely stagnant thereafter, while public investment in agriculture by African governments averaged only 5% of total public spending, about half the level maintained in Asian countries over the same period (Fan and Rao 2003).

Encouragingly, a renewed interest in agriculture has emerged. The African Heads of State signed the Maputo agreement in 2003, in which they pledged to double their support for agriculture to 10% of their budgets. Similarly, the G8 countries have pledged additional support for African agriculture, first at Glen Eagles in 2006 and again at L'Aquila 2009 during the recent world food crisis. There are tentative signs that the amounts of additional funding pledged for agriculture are actually beginning to materialize, while at the same time Africa led initiatives have been developed and launched (NEPAD 2008; AGRA 2010) that seek to bring a green revolution to Africa. But even as the battle for a recommitment to agriculture has been won, there remain important differences about the nature of the agricultural development strategy that should be adopted, which could undermine the effectiveness of future agricultural development efforts.

1.2 Policy Questions

Debate continues around four key issues; whether to prioritize small or large farms; whether to focus on food staples or high value products; whether to promote technologies and farming practices that require fertilizers and modern seeds; and the degree to which governments should intervene it markets. Taken together, the questions address whether Asia's Green Revolution can serve as a model for Africa.

1.2.1 Small Farms or Large?

Small farms lay at the heart of the GR strategy in Asia, and many advocates of an African green revolution presume a similar role for Africa's small farms. This is based in part on the observation that about 80% of Africa's farms are smaller than 2 hectares, and they account for large shares of agricultural production and rural and national employment (Nagayets 2005). Small farms have also been shown to be more efficient producers in poor, labor-surplus economies, including many African countries (Heltberg 1998; Eastwood et al. 2009; Binswanger-Mkhize and McCalla 2009). Indeed, the inverse relationship between farm size and productivity, which indicates the higher efficiency of small farms, is widely found in SSA (Holden et al. 2009). Targeting small farms should therefore be a "win-win" proposition for growth and poverty alleviation.

Skeptics counter that small farms have become too small to be competitive in today's market chains. Farmers are increasingly being asked to compete in markets that are more demanding in terms of quality and food safety, more concentrated and integrated, and much more open to international competition. Supermarkets, for example, are playing an increasingly dominant role in controlling access to urban retail markets (Reardon et al. 2003), and African markets for major cereals are increasingly dominated by low priced imports from abroad. As small farms struggle to sell their products and diversify into higher-value products, they are increasingly being out matched by larger farms and food imports. Skeptics also argue that Africa's small farmers have anyway diversified their livelihoods away from agriculture to the point where farming now accounts for only small shares of their total income, and it is better to invest in helping them exit agriculture and encourage land consolidation and the emergence of more large scale, mechanized farms (Maxwell et al. 2001; Ellis and Harris 2004; Collier 2009).

Counter arguments are based on the fact that Africa's domestic markets lag the rest of the world in terms of their integration and spread of supermarkets, and that most small farmers still grow and sell traditional foods (many of which are not traded internationally) in local markets. Moreover, other than in and around coastal cities, the price of cereals often remains high because of high transport and marketing costs. It is also argued that for many African countries, the problem is not that all their small farms are inherently unviable in today's marketplace, but that they face an increasingly tilted playing field that, if left unchecked, could lead to their premature demise. If greater action were taken to provide essential public goods like rural roads and agricultural R&D, and to help organize small farmers for marketing purposes, then many more small farms would again become competitive. As for rural income diversification into non-farm activities, this is not necessarily a positive phenomenon, unlike in Asia where non-farm sectors have been rapidly growing (Otsuka et al. 2009). It would be positive if it reflected a successful structural transformation in which agricultural workers are gradually being pulled into more lucrative non-farm jobs (Larson and Mundlak 1997; Mundlak et al. 2004). But too often in Africa, diversification into the non-farm economy is driven by growing land scarcity, declining rural wages, and poor agricultural growth, with many workers moving into low skill, low paid jobs in the service sector (Bryceson and Jamal 1997; Start 2001; Headey et al. 2010). Lipton characterizes this as "the migration of despair", and argues it "depresses wage rates, denudes rural areas of innovators, and hence, while it may briefly relieve extreme need, seldom cuts chronic poverty" (Lipton 2005 p.7).

Resolution of the small farm debate depends on recognizing that country economic characteristics have an important bearing on the outcomes and opportunities (Hazell et al. 2007). For example, the case for small farms is more compelling in densely populated countries (e.g. Ethiopia and Rwanda) than in land surplus countries (e.g. Zambia and Mozambique), and in lesser rather than more developed countries. It is also important to keep in mind that land is only one of several factors employed in agricultural production and it is possible to scale-up farms by making other types of investments (Larson and León 2006). Even so, the case for prioritizing small farms would be stronger if there were additional evidence about:

- Not just their current efficiency as producers, but also their willingness to seize new technological and market opportunities and spearhead a green revolution
- The nature of the constraints holding small farms back, and how easily and at what cost these might be overcome through appropriate interventions.
- The possibility that the best farms to target for a green revolution are neither the smallest nor the largest but are the middle sized farms. It is conceivable these offer the best combination of cost efficiency and market and technological prowess, and would still lead to pro-poor outcomes.[1]
- Finally, questions remain about what should be done about the smallest subsistence oriented farms that are too poor and inefficient to become commercially viable? Including these farms in an agricultural development strategy might prove an effective and sustainable alternative to safety nets (Hazell et al. 2007), but might also add considerably to the cost and difficulty of implementing an agricultural development strategy.

1.2.2 High Value Agriculture or Food Staples?

With historically low world prices for food staples despite the recent price spike, and rapid expansion in international agricultural trade, some see the best opportunities for African farmers in high-value commodities such as fruits, flowers, vegetables, and livestock. In many successfully growing Asian and Latin American countries, domestic demand for these products is growing rapidly, providing ready market outlets for increased domestic production. In contrast, growth in domestic demand is much weaker in Africa, primarily because of low and stagnant per capita incomes, and the best high-value market opportunities lie with exports.

The high value sector is particularly attractive to some donor agencies because it fits with their market led philosophy in which the private sector provides the leadership and much of the required investment, and the public sector is asked mostly to keep out of the way.

The rapid and private sector driven growth in horticultural exports from South Africa, Kenya and Côte d'Ivoire has raised hopes for similar opportunities for other African countries. Moreover, at least in Kenya, small farmers have participated in the growth of the sector; it is estimated that over half of Kenyan horticultural exports are produced by smallholders (Minot and Ngigi 2010). Non-traditional high value agricultural exports now account for about 10% of the total value of agricultural output in Africa (Diao and Hazell 2004), though they are concentrated in relatively few countries that have good infrastructure as well as historical and institutional links to world markets.

[1] Small farms today are only about half as big as they were at the time of Asia's green revolution, so focusing on middle sized farms today would be equivalent to the small farm focus of the Green Revolution.

Critics question the potential for less fortunate countries to achieve similar success, question the pro-poor nature of high value export agriculture, and doubt the future profitability of these exports given intense competition from the many Asian and Latin American countries. They argue that the opportunities for high value diversification for most African farmers will have to wait until domestic demand grows more rapidly, and see much greater market potential in domestic and regional markets for food staples (Diao and Hazell 2004). Africa wide, demand for food staples is expected to grow by about 4% per annum over the next 20–25 years (Rosegrant et al. 2005), and with rapid urbanization, an increasing share of that demand will need to be met through market transactions rather than subsistence production. These trends provide reasonable scope for more rapid growth in food staples production without depressing prices. But that is an Africa-wide perspective, and many African countries are small or cannot grow the more popular cereals consumed (e.g. wheat and rice) and can only offer their farmers limited market prospects.

There are several promising ways of growing Africa's markets for food staples. One is through increasing the competitiveness of Africa's farmers against imports so that they can capture larger shares of their growing domestic and regional markets, especially for maize and rice. Cost reducing technologies will need to play an important role, but also needed are investments that slash domestic transport and marketing costs and a willingness to keep trade barriers among African countries low. Steps that encourage investments in storage facilities and agro-processing are also promising avenues for expanding food staple markets.

Although the per hectare returns from food staples are too low to lift many small farms out of poverty, they can still serve as an important early step. Because high transport and marketing costs make local food purchases pricey and uncertain, small farms are often reluctant to rely on food markets and only engage in cash crop production once they have produced enough food for themselves (Hazell and Poulton 2007). Increases in the productivity of food staples enable them to achieve this on less land, freeing up land, labor and other resources for diversification into higher value alternatives.

Simulations with economy-wide models of four African countries[2] show that food staples offer more realistic pathways for achieving growth and poverty reduction within the time frame of the MDGs than do high value export production (Diao et al. 2007). This strategy is not only more feasible for achieving a sustained 5% agricultural growth rate (even after allowing for market absorption), but also outperforms a strategy built around increasing production of high value exports. This is because productivity enhancements for staple crops (e.g., through technological change) benefit a broad range of small farmers and agricultural areas, and lower food prices, whereas growth in high-value export crops only reaches farmers in the better-connected areas and has little impact on the food costs of the poor.

Although market opportunities vary by type of country, and are more favorable in countries with good access to world markets and/or increasing per capita incomes, still food staples would seem to provide a key small farm opportunity, even an imperative for food security in many countries. For these reasons, food staples have been prioritized by the Alliance for a Green Revolution in Africa (AGRA), and represent a

[2] Ethiopia, Ghana, Uganda and Zambia.

first step in achieving the broader CAADP agenda. In short, for most countries the choice may not be between high value products and food staples, but rather the pursuit of food staples first plus high value crops wherever market opportunities permit.

1.2.3 Low-External Input or High-Input Farming

Another important debate centers on the type of agricultural intensification appropriate to Africa, with views ranging from the need to bring intensive Asian-style green revolution technologies to Africa, through improving existing low-external-input farming methods, to a more ideologically driven organic farming agenda.

Although it has been actively promoted for nearly four decades in Africa through research and small farm development projects, high-external input (HEI) agriculture has made only modest inroads. There is overwhelming evidence that fertilizers and improved crop varieties can dramatically increase yields in many parts of Africa, as demonstrated, for example, by the hybrid maize revolution in Eastern and Southern Africa (Smale and Jayne 2010), and the activities of the Sasakawa SG2000 program. Yet fertilizer use across Africa remains remarkably low, averaging a meager 13 kg/ha compared to 98 kg/ha in South Asia and 190 kg/ha in East Asia and the Pacific (World Bank 2008). Key reasons for this include the high costs of fertilizer given the small volumes purchased, poor infrastructure and long distances to markets, inadequate small farm access to credit and complementary inputs, low crop yield response and high risk in many rainfed areas (Morris et al. 2007).

Most African farmers already use low-external input (LEI) farming practices by default and this is largely responsible for the very low yields they achieve. Considerable work has been done to improve and intensify these options. Available documented evidence suggests there is considerable potential to increase yields through improved LEI practices (Pretty et al. 2006; Tripp 2006), but their uptake has been disappointing. A difficulty is that they are typically labor and land intensive, both seasonally and in total (Tripp 2006). Improved LEI systems rely on the generation of organic matter (OM) and nutrients within the farming system, and these require land and labor to produce. Both can be costly. In heavily populated areas, land is the scarce resource and smallholder farmers may not have sufficient land to use land-intensive methods of OM generation (e.g. through fallows or grazing livestock). In fact, when corrected for the total land area needed to support each hectare of crop, improved LEI yields can be lower than current practices. On the other hand, in sparsely populated areas, labor may be too expensive for composting and other labor intensive forms of OM generation. The challenge is to develop LEI technologies that boost both labor and land productivity, and that is much easier to achieve with the use of inorganic fertilizer (Lipton 2005).

Tripp (2006) describes a wide range of LEI approaches, and two recent and fully documented successes are conservation farming in Zambia and *zaï* planting pits in Burkina Faso (Haggblade et al. 2010). Once these kinds of improvements have been made, the improved soil conditions achieved could well enhance the potential for a higher crop yield response to fertilizer, making it more profitable to adopt higher but still modest amounts of purchased fertilizer. LEI are also knowledge intensive and often require collective action amongst neighboring farmers (Tripp 2006). Many of the successful examples have resulted when grass roots organizations have taken the initiative and brought in well trained experts who train and organize communities. Unfortunately this approach is not easily replicated or sustained at scale.

Organic farming is promoted by some green lobbies as an extreme form of LEI. Although good yield gains are claimed (Badgley et al. 2007), the approach suffers all the disadvantages of other LEI methods, but without the option of exploiting the higher yield response to fertilizers under improved soil management conditions. Except in cases where significant market premium can be obtained for organic products, currently largely limited to export markets, organic farming does not seem to hold much promise for most African farmers.

In conclusion, it seems unlikely that crop yields can be increased on any scale in Africa without greater use of inorganic fertilizer. On the other hand, Africa's complex and risky rainfed farming systems, limited irrigation potential, and often degraded and fragile soils make it unsuitable for the kinds of input intensive monocultures found in Asia's irrigated Green Revolution areas. African farmers need to strike the right balance between managing soil organic matter, fertility and moisture content and the use of inorganic fertilizers. This will require a wide range of locally adapted practices, or a rainbow revolution (IAC 2004). Given also the high cost of fertilizer for most African farmers, the transition to more intensive farming may initially be restricted to bread basket areas with good agricultural potential and ready access to markets. For the longer term transition, greater priority will need to be given to developing rural infrastructure and transport systems, strengthening fertilizer distribution and credit systems for small farmers, and breeding improved crop varieties that can give higher yields under African conditions with low to moderate fertilize use and uncertain rainfall.

A difficulty in pursuing this agenda is lack of knowledge about the best approaches to promote in different types of areas. While an eclectic approach to integrated soil fertility management is clearly desirable, it can be costly to research and may require considerable capacity building at farm and community levels. This can be expensive and slow to implement, forestalling any quick gains with a green revolution. More empirical evidence is needed about the conditions under which more complex approaches are needed, of the likely payoffs to their development and dissemination, and the constraints to adoption. Similarly, more empirical information is needed about when to use higher levels of fertilizer, and how to overcome the constraints that prevent higher use.

1.2.4 State or Market Led Development?

A fourth issue of contention centers on the role of the state in Africa's agricultural development. Copying the Asian green revolution model, Africa has a past history of heavy government involvement in agriculture which unfortunately led to high costs and little gain, and which finally collapsed under the weight of its high fiscal costs in the 1980s (Djurfeldt et al. 2005). Much of this involvement was removed through the structural adjustment programs that followed; state monopolies that dominated agricultural markets and input distribution systems were downsized or disbanded, along with associated input subsidies and price controls (Akiyama et al. 2003). It is now fashionable among some donors and experts to think that the private sector and producer organizations can perform most market chain functions in agriculture and that the government's role should be limited to creating an enabling environment, such as setting and regulating grades and standards, ensuring food safety, and registering and enforcing contracts.

Some economists argue that few interventions are needed, when farmers are homogeneous and well organized (Pineiro and Trigo 1983). In contrast, when the farming community is dispersed and politically disengaged, the authors argue that a gap emerges that the state might fill. Still, it is a difficult challenge for any government to identify the needs of farmers in a heterogeneous setting without an effective institutional setting. Eicher (1995) argues that political, technological, economic and institutional preconditions must be met for a technology to succeed. Pointing to smallholder successes in boosting maize yields in Zimbabwe during 1980–1986, he argues that technical innovations can linger until appropriate policies fall in place. Smale (1995) makes a similar case, arguing that maize innovations in Malawi failed to catch on due to institutional constraints, despite support from international researchers and donors. Mollinga and Bolding (2004) argue that governments willing to make needed investments in irrigation subsequently faced obstacles in their subsequent management.

Part of what drives the debate about the role of government in Africa is disagreement about the Asian experience. For example, Djurfeldt et al. (2005) argue that Asia's Green Revolutions were state led and contrast the conflicted policy environment in Africa during the later part of the twentieth century with the consistent support provided by the state in Asia. Still, many of the examples of successful state interventions are drawn from countries where rice grown under homogeneous conditions played a dominant role for farmers and consumers. Consequently, it may be the case state institutions may work best when state interventions are less needed. Moreover, there is disagreement as to whether the state led or followed innovation. In Chap. 2, the authors argue that Asia's Green Revolution in rice was technology-led and subsequently supported by policies. This is a view consistent with Hayami and Ruttan's (1985) argument that technological innovations induced institutional innovations by enhancing the rates of return to innovational reforms including policy reforms.

There is plenty of evidence to show that the private sector can be very effective in driving high value market chains, especially for export (e.g. Minot and Ngigi 2010). These involve relatively few farmers, are easily integrated through contract farming arrangements to ensure that farmers have access to all the required inputs and meet credence requirements, and offer relatively high returns. But there is much less evidence to show that the private sector can successfully drive market chains for staple foods during the early stages of agricultural development. As large numbers of farmers struggle with low productivity and high subsistence needs, low input use, low incomes, poor infrastructure, and high risks, the amount of profit to be made in market chains for food staples remains low and unattractive for much private investment. Left to market forces alone, the chains for food staples tend to be supplied by commercial farms in areas with good market access and by imports, while the vast majority of small farmers and more remote areas get left out.

In theory, allowing markets to evolve in this way might seem desirable, enabling the most efficient types of agriculture, commodities, regions, and farm sizes to prevail. But this fails to reckon with the many institutional and market failures commonly found in Africa, and these failures can lead to discriminatory and inefficient outcomes. For example, if market failures penalize small farms over large ones in accessing markets and inputs, then unfettered markets may favor large-farm outcomes that are less efficient as well as less equitable than those that could result from small farm led growth. In this case, targeted policy interventions that correct the underlying market failures might be win-win solutions for efficiency and equity.

Some argue that a wide range of failures in input and output markets exist in Africa, and many of these are linked and spill over from one market to another (Dorward et al. 2005, 2009). A green revolution requires a process of sustainable intensification in which farmers combine a package of inputs – land, labour, technical skills and information, purchased inputs, and fixed and working capital – to produce outputs for sale. It also requires that markets exist for the outputs to be sold, and that farmers receive a level of remuneration that makes the whole investment worthwhile. If farmers are to invest in sustainable intensification they need to be assured of reasonably reliable access to a complete set of production and marketing services, on reasonable terms. If one element of the package is missing, then investments in all the others may be lost or significantly reduced (Hazell et al. 2007).

Analysts differ in the extent to which they believe these complementarities pose a problem for the development of private service suppliers. Conventional liberalisation policy does not recognize this as a problem, assuming the "invisible hand" of the market will effectively bring about all the needed transactions as if they had been centrally coordinated. Other analysts (especially, Dorward et al. 2005, 2009) observe that potential service suppliers face uncertain demand for their services unless farmers are assured of access to other complementary services. Such assurance is lacking in poor rural areas that have not yet achieved a widespread transition out of low input/low output farming unless some external agent undertakes to provide the important missing services or coordinates provision of the missing services by other actors. Such coordination mechanisms must be credible to farmers and to all service providers. Without such mechanisms, private investors will not invest significant capital in developing agricultural service businesses and will only provide agricultural services on an opportunistic basis.

There have been few attempts to test empirically the extent to which markets fail to provide the coordinated package of production and marketing services needed to support a green revolution in Africa. Nor is there agreement on what constitutes a minimal set of supporting services. For farmers to take advantage of modern varieties, markets for fertilizer and surplus output are required, but it is hard to generalize further. Formal credit mechanisms are rare in rural areas in developing countries, even where modern farming methods are widely used. And in some instances, even seed can be replicated on farm if sufficient care is taken. At the same time, general government investments in security, education, transportation and communication infrastructure, and institutions that promote property rights and enforce contracts all foster the endogenous growth of markets.

Still, waiting for markets to evolve endogenously comes at a cost and governments may hope to speed the process. Yet there has also been little work on how the public sector can step in to solve the problem where it exists. Large government systems intended to provide credit, insurance or other services have proved subject to political capture and too expensive to maintain (Cole 2009; Skees et al. 1999). Clearly Africa should not go back to the failed public interventions of the past. But there is also a much richer institutional landscape today within which new kinds of institutional and policy innovations can be developed and tested. Local governments, NGOs, community-based organizations, and some types of private firms have grown to play important roles, larger than was the case in much of Asia at a comparable stage of development. In this context, African countries have an opportunity to take a more flexible and innovative approach to solving market failures that builds strategic partnerships between key actors at local, meso, and macro levels. Such mechanisms are being developed and tested, for example, with public-private partnerships for agricultural research, market mediated approaches to insurance and agricultural credit, smart fertilizer subsidies, social enterprise approaches for supporting agricultural service providers, and legal recognition of property rights at community levels. But much greater efforts are needed, including systematic evaluation of lessons from the many creative and experimental efforts already underway.

1.3 The Remaining Chapters

The remaining chapters explain Africa's productivity gap and explore ways to close the gap. This is done by examining recent experience in Africa and also by drawing on lessons from Asia. The remaining chapters are organized into three parts.

Part I is concerned with the transferability of Asia's Green Revolution to SSA and focuses on whether an African Green Revolution should focus on staple crops and small farms. The section's four chapters look closely at the roles of technology, irrigation, climate endowments, and agricultural policies in the process of the Asian Green Revolution in rice, maize and other cereal crops, and ask whether and to what extent geography explains why a Green Revolution has eluded Africa. More specifically, Chap. 2 reviews the evolutionary process of improving rice productivity in Asia with a view to drawing lessons for SSA. Chapter 3 reports that Asian-style lowland rice farming system results, at least potentially, in sustainable and high productivity in Mozambique and Uganda as well as selected countries in West Africa. While irrigated paddy areas accounts for mere 15% or so of paddy fields in SSA, it accounts for well more than 50% in Asia. Thus, direct transferability of Asian technology to irrigated area in SSA does not necessarily imply the high technology transferability to SSA where rainfed paddy field dominates. It must be also pointed out that although large irrigation schemes were often mismanaged in the past, which was taken to imply the failure of attempt at the Green Revolution, those irrigation schemes are relatively well managed at present. Comparisons of the impacts of climate on crop yields are made between India and SSA in Chaps. 4 and 5, and the potential for successful transfer of technologies from Asia is examined. In particular, these studies attempt to assess whether and to what extent the adoption of new technologies mitigates the impact of climate on crop yields. Also examined in Part I are strategic crops in which Green Revolutions are likely to be realized in SSA.

Part II is concerned with the prospects for upland NERICA and maize Green Revolutions, both of which were once considered highly promising, while paying attention to the issue of low-input vs. high-input agriculture. The story of NERICA, a variety of upland rice developed for Africa, is reviewed and experiences from East and West Africa are explored in Chaps. 6 and 7. A case study of Uganda in Chap. 6 examines the importance of preserving seed quality, once modern varieties are introduced. It also looks at the provision of new milling capacity by the private sector following the introduction of NERICA and how milling capacity affects the initial stage of NERICA dissemination. Chapter 7 looks at whether, in practice, NERICA provides higher yields once adopted. It also looks at the attractiveness of NERICA in drought-prone areas because of its short maturity, which suites well the region's short rainy season. Chapter 8 attempts to draw lessons from the adoption of high-yielding maize varieties in SSA, whereas Chap. 9 reports the results of case studies on the possibilities of maize Green Revolution in Kenya and Uganda. The chapter looks to see whether chemical fertilizer application rates are optimal, given fertilizer's high price, and the role that soil fertility plays in sustaining yield gains.

The poor quality of Africa's soils and the limited reach of fertilizer markets are often cited as a significant obstacle to the adoption of the type of high-yielding varieties that boosted global food supplies. Three studies in Part III explore how markets shape farmer incentives and fertilizer demand. Using farm households' 3-year rotating panel data from 1993 to 2003, Chap. 10 examines the market characteristics of farmyard manure and its impact on cereal yields in India. Chapter 11 explores whether missing markets play a role in limiting chemical fertilizer applications among Ethiopians. Chapter 12 reports the results of experiment, in which packages of free hybrid maize seeds and chemical fertilizers were distributed to randomly selected farmers free of charge. These studies are expected to provide implications for the role of governments in achieving substantial productivity growth in African agriculture.

Finally in Chap. 13, we summarize the major findings of case studies in such a way to draw strategy to realize a Green Revolution in SSA.

References

Akiyama T, Baffes J, Larson DF, Varangis P (2003) Commodity market reform in Africa: some recent experience. Econ Syst 27(1):83–115

Alliance for Green Revolution in Africa (AGRA) (2010) Strategy for an African Green Revolution. http://www.agra-alliance.org/section/about/agrastrategy

Badgley C, Moghtader J, Quintero E, Zakem E, Chappell MJ, Aviles-Vazquez K, Samulon A, Perfecto I (2007) Organic agriculture and the global food supply. Renew Agri Food Syst 22(2):86–108

Binswanger-Mkhize HP, McCalla A (2009) The changing context and prospects for agricultural and rural development in Africa. In: Pingali PL, Evenson RE (eds) Handbook of agricultural economics, vol 4. Elsevier, Amsterdam

Bryceson DF, Jamal V (1997) Farewell to farms: deagrarianisation and employment in Africa. Ashgate, Aldershot

Cole S (2009) Fixing market failures or fixing elections? Agricultural credit in India. Am Econ J: Appl Econ 1(1):219–250

Collier P (2009) Africa's organic peasantry: beyond romanticism. Harv Int Rev 31(2):62–65

Diao X, Hazell P (2004) Exploring market opportunities for African smallholders, vol 6, 2020 Africa Conference Brief. International Food Policy Research Institute, Washington, DC

Diao X, Hazell P, Resnick D, Thurlow J (2007) The role of agriculture in development: implications for Sub-Saharan Africa, vol 153, Research Report. International Food Policy Research Institute, Washington, DC

Djurfeldt G, Holmen H, Jirstrom LR (eds) (2005) The African food crisis; lessons from the Asian Green Revolution. CAB International, Wallingford, Oxford

Dorward A, Kydd JG, Morrison JA, Poulton CD (2005) Institutions, markets and economic coordination: linking development policy to theory and praxis. Dev Change 36(1):1–25

Dorward A, Kydd JG, Poulton C, Bezemer D (2009) Coordination risk and cost impacts on economic development in poor rural areas. J Dev Stud 45(7):1093–1112

Eastwood R, Lipton M, Newell A (2009) Farm size. In: Pingali PL, Evenson RE (eds) Handbook of agricultural economics, vol 4. Elsevier, Amsterdam

Eicher CK (1995) Zimbabwe's maize-based green revolution: preconditions for replication. World Dev 23(5):805–818

Ellis F, Harris N (2004) New thinking about urban and rural development. Keynote paper prepared for the U.K. Department for International Development Sustainable Development Retreat

Fan S, Rao N (2003) Public spending in developing countries: trends, determination and impact. International Food Policy Research Institute, Washington, DC

Food and Agriculture Organization of the United Nations (FAO) (2009) The state of food insecurity in the world 2009. http://www.fao.org/publications/sofi/en/

Haggblade S, Tembo G, Kaboré D, Reij C, Ajayi OC, Franzel S, Mafongoya P, Place F (2010) Sustainable soil fertility management systems. In: Haggblade S, Hazell PBR (eds) Successes in African agriculture. Johns Hopkins University Press, Baltimore

Hayami Y, Ruttan VW (1985) Agricultural development: an international perspective. Johns Hopkins University Press, Baltimore

Hazell P, Poulton C (2007) All-Africa review of experiences with commercial agriculture: case study on food staples. Background paper for the Competitive Commercial Agriculture in Sub–Saharan Africa (CCAA) study, World Bank, Washington, DC. http://siteresources.worldbank.org/INTAFRICA/Resources/257994-1215457178567/Ch10_Food_Staples.pdf

Hazell P, Wood S (2008) Drivers of change in global agriculture. Philos Trans R Soc B 363(1491):495–515

Hazell P, Poulton C, Wiggins S, Dorward A (2007) The future of small farms for poverty reduction and growth, vol 42, 2020 Discussion Paper. International Food Policy Research Institute, Washington, DC

Headey D, Bezemer D, Hazell P (2010) Agricultural employment trends in Asia and Africa: too fast or too slow. World Bank Res Obs 25(1):57–89

Heltberg R (1998) Rural market imperfections and the farm size–productivity relationship: evidence from Pakistan. World Dev 26(10):1807–1826

Holden S, Otsuka K, Place F (2009) The emergence of land markets in Africa: assessing the impacts on poverty, equity, and efficiency. Resources for the Future, Baltimore

Inter-Academy Council (IAC) (2004) Realizing the promise and potential of African agriculture: science and technology strategies for improving agricultural productivity and food security in Africa. IAC Secretariat, Royal Netherlands Academy of Arts and Sciences, Amsterdam
Larson DF, León M (2006) How endowments, accumulations, and choice determine the geography of agricultural productivity in Ecuador. World Bank Econ Rev 20(3):449–471
Larson DF, Mundlak Y (1997) On the intersectoral migration of agricultural labor. Econ Dev Cult Change 45(2):295–319
Lipton M (2005) The family farm in a globalizing world: the role of crop science in alleviating Poverty. 2020 Discussion Paper 40. International Food Policy Research Institute, Washington, DC
Maxwell S, Urey I, Ashley C (2001) Emerging issues in rural development. An issues paper. Overseas Development Institute, London
Minot N, Ngigi M (2010) Are horticultural exports a replicable success story? evidence from Kenya and Côte d'Ivoire. In: Haggblade S, Hazell P (eds) Successes in African agriculture. Johns Hopkins University Press, Baltimore
Mollinga PP, Bolding A (2004) The politics of irrigation reform: contested policy formulation and implementation in Asia, Africa and Latin America. Ashgate, Aldershot
Morris ML, Kelly VA, Kopicki RJ, Byerlee D (2007) Promoting increased fertilizer use in Africa: lessons learned and good practice guidelines. World Bank, Washington, DC
Mundlak Y, Larson DF, Butze R (2004) Agricultural dynamics in Thailand, Indonesia and the Philippines. Aust J Agric Res Econ 48(1):95–126
Nagayets O (2005) Small farms: current status and key trends in the future of small farm. Paper in proceedings of a research workshop, Wye, UK, International Food Policy Research Institute, Washington, DC, pp 26–29 June 2005
New Partnership for Africa's Development (NEPAD), Comprehensive Africa Agricultural Development Programme (CAADP) (2008) Strategic plan of the NEPAD secretariat agriculture unit 2008–2013. Johannesburg. http://www.nepad-caadp.net/pdf/NEPAD%20Strategic%20 Plan%20070208%20_DRAFT_.pdf
Otsuka K, Estudillo JP, Sawada Y (2009) Rural poverty and income dynamics in Asia and Africa. Routledge, London
Pineiro M, Trigo E (1983) Technical change and social conflict in agriculture: Latin American perspectives. Westview Press, Boulder
Pretty JN, Noble AD, Bossio D, Dixon J, Hine RE, de Vries FWTP, Morrison JIL (2006) Resource conserving agriculture increases yields in developing countries. Environ Sci Technol 40(4):1114–1119
Reardon T, Timmer CP, Barrett C, Berdegué J (2003) The rise of supermarkets in Africa, Asia, and Latin America. Am J Agric Econ 85(5):1140–1146
Rosegrant MW, Cline SA, Weibo L, Sulser TB, Valmonte-Santos R (2005) Looking ahead: long-term prospects for Africa's agricultural development and security, 2020 Discussion Paper 41. International Food Policy Research Institute, Washington, DC
Skees J, Hazell PBR, Mario M (1999) New approaches to crop yield insurance in developing countries, vol 55, EPTD discussion papers. International Food Policy Research Institute, Washington, DC
Smale M (1995) Maize is life: Malawi's delayed green revolution. World Dev 23(5):819–831
Smale M, Jayne TS (2010) Seeds of success in retrospect: hybrid maize in Eastern and Southern Africa. In: Haggblade S, Hazell P (eds) Successes in African agriculture. Johns Hopkins University Press, Baltimore
Start D (2001) The rise and fall of the rural non-farm economy: poverty impacts and policy options. Dev Policy Rev 19(4):491–505
Tripp R (2006) Self-sufficient agriculture: labour and knowledge in small-scale farming. Earthscan, London
World Bank (2008) World development report 2008: agriculture for development. World Bank, Washington, DC
Yamano T, Otsuka K, Place F (2011) Emerging development of agriculture in East Africa: markets, soil, and innovations. Springer, Amsterdam

Part I
Climate and the Transferability of Asian Green Revolution to Sub-Saharan Africa

Chapter 2
Lessons from the Asian Green Revolution in Rice

Jonna P. Estudillo and Keijiro Otsuka

Abstract The Asian Green Revolution in rice entailed a long-term evolutionary process spanning more than four decades since the mid-1960s. The purpose of this chapter is to identify important lessons from the Asian Green Revolution in rice and examine whether the modern rice technology in Asia could be appropriately transferred to contemporary sub-Saharan Africa (SSA). While there are many lessons to learn, this study focuses on high-yielding well-adapted lowland rice varieties, appropriate fertilizer application, and favorable institutional and policy environment that played pivotal roles in launching and sustaining the Asian Green Revolution in rice. The Green Revolution in SSA could include more than one commodity as none of which dominates; we argue that such Green Revolution should include rice for a number of reasons.

Keywords Green Revolution • Asia • Sub-Saharan Africa • Modern variety • Lowland rice • Rice yield • Cereal yield

2.1 Introduction

There were fears in the 1950s and the early 1960s that the tropical Asian rice-based economy would be experiencing massive famine and starvation because the region had already reached its cultivation frontier – the limit of new land available for rice production – without any favorable prospect for increasing rice production from the existing paddy fields. In fact, rice production growth in South and Southeast Asia for the decade prior to1965 was mainly brought about by the expansion of cultivated area (Barker and Herdt 1985). The major source of rice production growth, however,

J.P. Estudillo • K. Otsuka (✉)
National Graduate Institute for Policy Studies,
7-22-1 Roppongi, Minato-ku, Tokyo 106-8677, Japan
e-mail: jonna@grips.ac.jp; otsuka@grips.ac.jp

K. Otsuka and D.F. Larson (eds.), *An African Green Revolution: Finding Ways to Boost Productivity on Small Farms*, DOI 10.1007/978-94-007-5760-8_2,

shifted from area expansion to yield increase because of the Green Revolution (GR), which started in Asia in 1966, when the International Rice Research Institute (IRRI) released IR8, the first modern variety (MV) of rice. MVs refer to the short-statured, stiff-strawed, fertilizer-responsive, and non-photoperiod-sensitive rice varieties (Chandler 1982). Later MVs incorporated better traits such as short growth duration, multiple disease and insect resistance, superior grain quality, and tolerance for problem soils characterized by nutritional deficiencies and toxicities (Khush 1995; Evenson and Gollin 2003).

Farmers in tropical Asia quickly adopted the earlier released MVs, particularly those located in irrigated and favorably rainfed environment (David and Otsuka 1994; Byerlee 1996), where MVs tend to be more productive. Farmers were also quick in replacing older MVs with newer ones because newer MVs with better characteristics were privately more profitable (Estudillo et al. 1999). There was an increased demand for fertilizer partly because the yields of MVs were responsive to high application of fertilizer and partly because of favorable fertilizer prices brought forth by fertilizer subsidy programs (David 1976; Hossain and Singh 2000). Public investments in irrigation also accelerated because MVs increased the profitability of irrigated rice culture (Hayami and Kikuchi 1978; Kikuchi et al. 2002). In addition, credit programs were instituted along with the establishment of national research and extension systems in many countries (Barker and Herdt 1985; Herdt 2010). In short, the very essence of the GR in Asia is the development and diffusion of a series of MVs in irrigated and favorably rainfed areas and subsequent acceleration in public-sector investments in complementary infrastructures and institutions (Hayami and Otsuka 1994; Hazell 2009).

The Asian GR in rice entailed a long-term evolutionary process spanning more than four decades since the mid-1960s. The purpose of this chapter is to identify important lessons from the Asian GR in rice and examine whether the modern rice technology in Asia could be appropriately transferred to contemporary sub-Saharan Africa (SSA). While there are many lessons to learn, this study focuses on high-yielding well-adapted lowland rice varieties, appropriate fertilizer application, and favorable institutional and policy environment that played pivotal roles in launching and sustaining the Asian GR in rice.

The GR in SSA could include more than one commodity as none of which dominates; we argue that such GR should include rice for a number of reasons. First, Asia has already accumulated a huge stock of scientific knowledge and useful production methods in rice propagation.[1]

Second, there is an increasing demand and production of rice in SSA, but low self-sufficiency of only 60%. Third, lowland rice varieties suitable to lowland irrigated and favorably rainfed ecosystems that were earlier developed in Asia or elsewhere based on cross-breeding between Asian MVs and local varieties have exhibited high yield potential in lowland areas in SSA even with minimal fertilizer application. For example, the yield of IRRI-type varieties in a few well-irrigated areas of Senegal, Kenya,

[1] See Chandler (1982) and Khush (1987, 1995) for a review of the history of rice breeding program.

and Tanzania is as high as 5–6 t/ha, which exceeds the average irrigated rice yield in tropical Asia (Otsuka and Kijima 2010). Fourth, and finally, the New Rice for Africa (NERICA) shows high yield potential in the upland areas, where NERICA thrives well in the presence of high moisture stress (Chap. 6 and Chap. 7, in this volume). For example, in the uplands of Uganda, the average NERICA yield is 2.5 t/ha (Kijima et al. 2008), which is much higher than the average upland rice yield of 1 t/ha in SSA, and even higher than the yield of rice in upland areas in the Philippines, which is about 2 t/ha (Estudillo and Otsuka 2006). Thus, we hypothesize that the evolutionary process of rice GR could be launched in SSA, essentially because the potentially profitable new rice technologies are already available.

This chapter has five remaining sections. Section 2.2 presents the conceptual framework and our hypotheses. Section 2.3 compares the yield trends in major cereal crops in Asia and SSA. Section 2.4 briefly describes Asian GR in rice in terms of the evolution and spread of MVs, changes in irrigation, fertilizer use and trends in rice yield as well the subsequent changes in institutions and policies in Asia. Section 2.5 explores the sources of yield growth and describes the evolutionary processes underlying the GR in Central Luzon in the Philippines. Section 2.6 describes the current state of rice yield in SSA and the potential of transferring the Asian GR. Finally, Sect. 2.7 concludes this chapter.

2.2 Conceptual Framework and Hypotheses

The whole evolutionary process of GR in Asia could be best described in a simple schematic diagram shown in Fig. 2.1. Our main argument is that the GR in Asia is essentially technology-led and policy-supported, rather than policy-driven as is oftentimes assumed (Djurfeldt et al. 2005). The development of MVs induced subsequent public-sector investments in extension and national research programs, irrigation, and credit programs by increasing the rates of returns to such investments (Fig. 2.1). Responding to profitable opportunities created by the new technologies, both factor and product markets gradually developed in Asia. Such developments, in turn, led to enhanced profitability in research and extension investments that further induced the development of newer and better technologies.

We have two major hypotheses regarding the evolutionary processes of the Asian GR.

Hypothesis 1
The Asian GR was essentially technology-led, i.e., the development and adoption of early generation MVs and high fertilizer application in the presence of a well-developed irrigation infrastructure led to an increase in yield.

Hypothesis 2
The Asian GR was sustained by innovations in MVs and subsequent improvements in supportive policies and institutions, i.e., increased investments in irrigation, adaptive research, extension, and subsidy programs for purchased inputs.

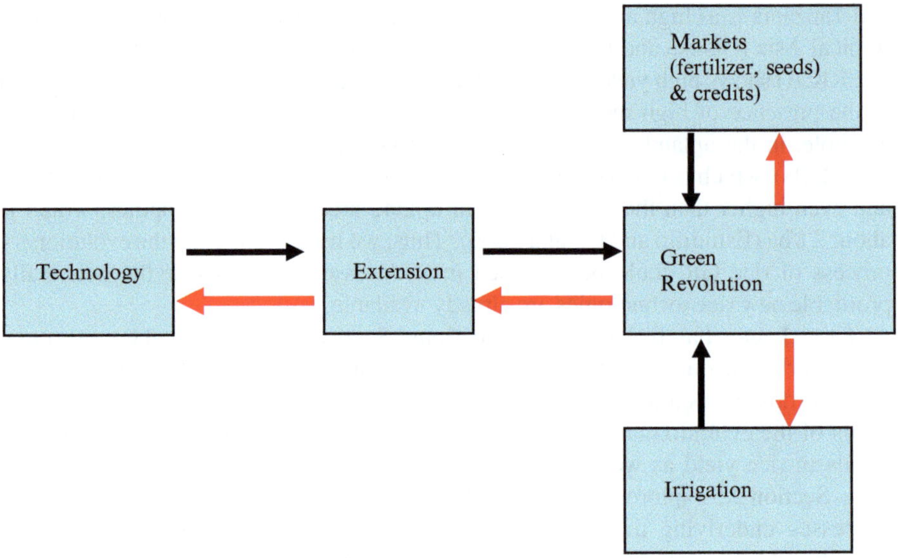

Fig. 2.1 The conceptual interrelationships between technology, extension, input markets, and irrigation investments

Based on the Asian experience, we would like to argue that the induced institutional innovations in the sense of Hayami and Ruttan (1985), such as the establishment of effective irrigation management systems, national agricultural research and extension systems, and marketing institutions, could take place in SSA, once new technological breakthrough is made. Since useful rice technologies are already available from Asia, it could be possible to initiate the technology-led evolutionary processes of GR in SSA.

2.3 Cereal Yields in Asia and SSA

In order to launch the GR in SSA, it is necessary to identify strategic crops where new technology has been effective in increasing crop yield at the farm level. Figure 2.2 shows the yield of rice, maize, wheat, sorghum and millet in Asia and SSA. The difference in rice yield between Asia and SSA was about 80% around 1960 which could be partly attributed to a relatively well-developed irrigation infrastructure system in Asia and partly to the prevalence of low-yielding upland rice in SSA whose yield has been stagnant at 1 t/ha.[2] The small yield gap before the Asian GR indicates that it is the advent of MVs that essentially propelled yield growth in Asia.

[2] According to Balasubramanian et al. (2007), upland rice accounts for roughly 40% of rice area in SSA.

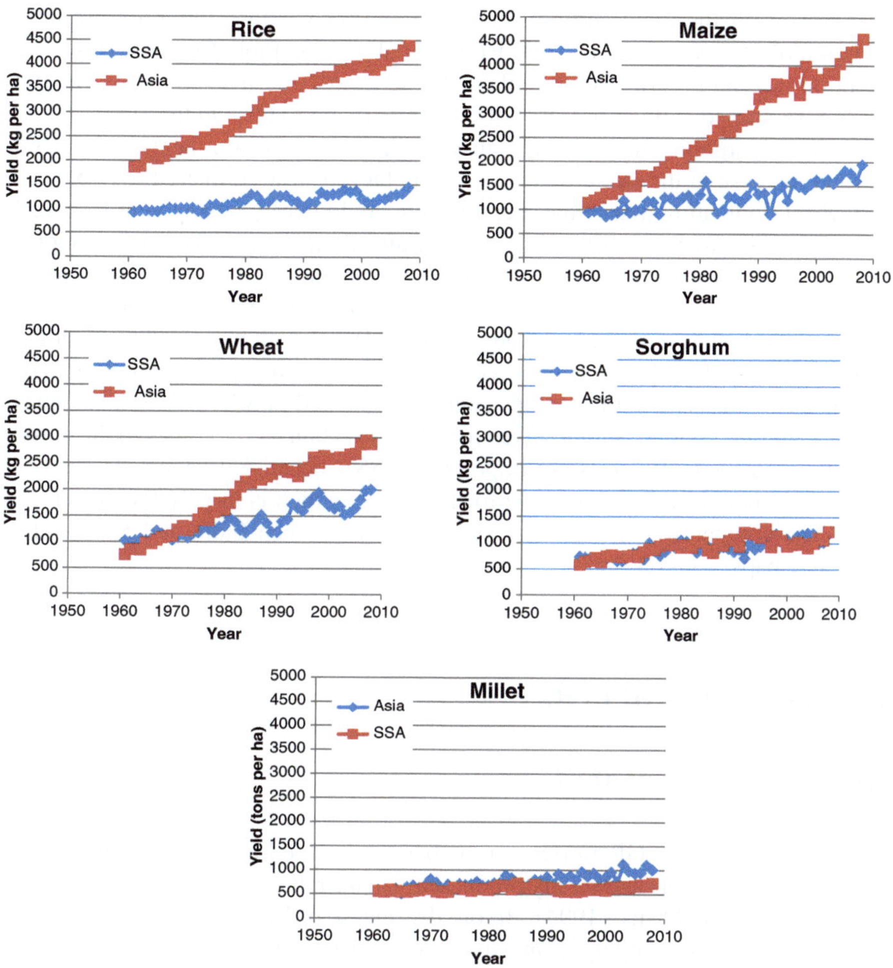

Fig. 2.2 Cereal crop yields in Asia and Sub-Saharan Africa, 1961–2008 (Source: FAO Stat online)

The difference in rice yield grew larger beginning in the mid-1970s, when the later generation of MVs that incorporated multiple pest- and disease-resistant traits became available. The difference in yield between the two continents was similarly visible in the case of maize at about the same period although maize yield was growing faster compared to that of rice in SSA. Indeed, Byerlee (1996) argues that maize is a single success story for the diffusion of MVs in SSA, where rice and wheat are less important food staples.

Yield of wheat in Asia continued to grow from about 1 t/ha in 1960 to 3 t in 2008. SSA similarly started at 1 t and achieved a modest progress with marked geographical concentration in Southern Africa and Ethiopia, reaching an average yield of about 2.0 t/ha, while Middle Africa lagged behind (with an average yield of 1.1 t).

One reason for the yield gap between Asia and SSA is the slow progress in the adoption of MVs of wheat in SSA. Heisey et al. (2003, Table 6) reported that MV adoption in Asia in terms of area planted rose from 19% in 1970 to 86% in 1997 while SSA started from a low level of 5% while rising to 66% in the 2000s.

Yields of sorghum and millet are fairly similar in the two continents and progress in yield increase in the two crops has been by far too slow compared to rice, maize and wheat, indicating that in these crops there could hardly be any vent for a technology transfer to SSA. According to Deb and Bantilan (2003), the adoption of MVs in sorghum and millet takes place largely in India, where farmers continuously change older MVs to newer ones. In SSA, Nigeria, Niger, and Sudan are the major producers of both crops where the adoption of MVs and yield growth were slow.

Overall, it is clear that Asia possesses a good stock of matured technology in rice and maize so that GR in these crops could be launched in SSA provided an appropriate adaptive research is undertaken to tailor the Asian technologies to the current conditions in SSA. Wheat is not important in SSA because it cannot be grown in tropical climate. In this chapter, we focus on rice, which is largely the GR crop in Asia. The comparative analyses of yield growth between India and SSA will be made in Chaps. 4 and 5.

2.4 The Asian GR in Rice

2.4.1 Evolution of MVs

The process of Asian GR in rice can be best understood as a long history of rice varietal improvements and steady productivity gains rather than a one-shot phenomenon (Hayami and Otsuka 1994). Before the release of IR8 in 1966, the increase in rice production came largely from expanding land area devoted to rice culture or by improving the resource base through expansion of irrigation. During this period, there was hardly any significant yield increase because MVs were not available and there was hardly any adoption of improved crop management practices. The earliest released MVs, the so-called "first-generation MVs" (e.g., IR5 and IR8 developed by IRRI and C4 developed by the College of Agriculture of the University of the Philippines) are semi-dwarfed varieties, photoperiod insensitive, and had medium growth duration (130 days) much less than the growth duration of the more common traditional varieties (TVs) (160–170 days). MV1 doubled the yield potential of tropical rice but its yield fluctuates greatly because it is susceptible to attacks of numerous diseases and pests (IRRI 1985; Pingali et al. 1990).

The "second-generation of MVs" (MV2) consisting of IR36 to IR62, which were developed from the mid-1970 to the mid-1980s, incorporated wide-spectrum of pests and disease resistance traits and early maturity period of 110–115 days (Khush 1987, 1995). Resistant MVs contributed significantly to the acceleration of yield growth by reducing yield variability thereby increasing the expected yield particularly during the dry season (Otsuka et al. 1994). The "third generation of MVs" (MV3) incorporated improved grain and nutritional quality with such traits as multiple pests and diseases resistance and shorter-growing period (Hossain et al. 2003). Typified of this innovation is IR64, which gained broad acceptance for many years since its release in 1985 because of its good cooking quality – it contains intermediate level of amylase that makes it moist and remains soft even when cool.

Rice breeders believe there are two new prospects for further improving MVs: (1) development of hybrids and (2) the use of new of biotechnologies that produce genetically modified (GM) rices.[3] Good quality hybrid rice seed could be purchased from private seed growers and is relatively costly whereas the yield advantage of hybrids over conventional MVs, which can be propagated by the farmers, tend to be modest (about 15% only) (Byerlee 1996). Thus, hybrid rice adoption is observed to be high only in favorable areas in China where the seed industry is well developed. Many scientists believe that biotechnology would be helpful in developing rices that are suitable to unfavorable production environments, notably flood-prone areas, for which the conventional breeding method has produced only a small number of rices (Khush 1995).

We can conveniently divide the GR into two phases: – (1) the replacement of TVs by MV1, which provided the potential for dramatically increasing yield ("revolutionary" phase) and (2) the replacement of MV1 by the newer and better MVs ("evolutionary" phase). Resistance against multiple pests and diseases contributed perhaps half of all the gains in total yield growth (Byerlee 1996) while the rise in cropping intensity (i.e., the number of rice crops that can be grown per year) because of shorter growth duration and photoperiod insensitivity was the major contributor to a substantial increase in rice production per year (Barker and Herdt 1985). Improved pest resistance also contributed to environmental sustainability through reduced usage of pesticides, which include some of the most environmentally harmful chemicals.

It is during the evolutionary phase when knowledge-intensive crop management practices were introduced and have started to substitute for input use and improved input efficiency (Byerlee 1994). These knowledge-intensive cultural practices can be effectively included in the package of inputs and production practices complementary to fertilizer in realizing the yield potential of MVs. In Asia, the most common crop management practices are selection of good quality seeds, leveling and bunding of fields, straight-row planting, direct-seeding, use of an appropriate mix of a variety of chemical fertilizers and manure, appropriate amount and timing of

[3] Byerlee and Fisher (2002) explore the policy and institutional option for biotechnology in developing countries given the presence of market failures in developing countries in accessing the new tools and technologies. The authors argue for a public-private partnership and market segmentation with active participation of the national agricultural systems to access proprietary tools and technologies.

pesticide and weedicide application, and water control management. While manure and chemical fertilizer are generally complementary in increasing yield of cereals (i.e., sorghum and maize) in India (Chap. 10 in this volume) and Ethiopia (Chap. 11 in this volume), the use of manure is generally not widely spread in rice-growing areas in Asia perhaps because of the low price of chemical fertilizer brought forth by subsidies. Overall, the evolution of MVs and adoption of improved crop management practices give support to Hypothesis 1.

2.4.2 Irrigation and the Spread of MVs

There was a general perception that the Asian GR in rice is a revolution in favorable areas only (Lipton and Longhurst 1989). Earlier released MVs were particularly developed to perform well under irrigated and favorably rainfed environment with intensive use of fertilizers, while rice yield remained stagnant in marginal environments, where water, climate and soil constraints cannot be overcome through varietal improvements (Evenson and Gollin 2003). There is a slow progress in developing improved germplasm for unfavorable environments (i.e., drought-prone rainfed lowland, upland, flood prone, and tidal wetlands) because these environments are extremely heterogeneous while it is necessary to develop rice varieties that are adaptable for each small area with specific growing condition (Khush 1995). The irrigated rice area consisting of 79 million ha remains the major granary producing about 75% of the world's rice output.

Adoption of MVs rose steadily over time in Asia since the mid-1960s reaching over 90% of rice area in the Philippines, Indonesia, and Vietnam in the early 2000s. MV adoption was generally faster in irrigated and favorably rainfed areas because the presence of irrigation is by far the most important factor affecting MV adoption (David and Otsuka 1994). Ruttan (1977) found in the early period of GR that neither farm size nor tenure affected the adoption of MVs. Indeed it is small farmers who actively adopted MVs and introduced intensive rice production management systems.

Figure 2.3 compares the rate of MV adoption and irrigation ratio (i.e., proportion of irrigated rice area) in the Philippines, Vietnam, Bangladesh and India. It is clear that the former far exceeded the latter, implying that MVs were widely adopted, not only in irrigated areas, but also in rainfed areas. In the Philippines, this could be attributed to the release of drought-resistant and early maturing MVs that allow the farmers in favorably rainfed areas to catch the late monsoon rains to plant a second crop of rice (Estudillo and Otsuka 2006). Spread of MVs suitable to unfavorable production environments in this country started in the mid-1990s.

Hossain (1993, 2009) reported that in Bangladesh beginning in the late-1980s, rice was increasingly grown in low-lying areas during the dry season using modern variety *boro* rice with pump irrigation. The rapid adoption of the high-yielding *boro* rice was facilitated by the government's market liberalization policies for minor irrigation equipment, most importantly, the shallow tube well. The rapid expansion

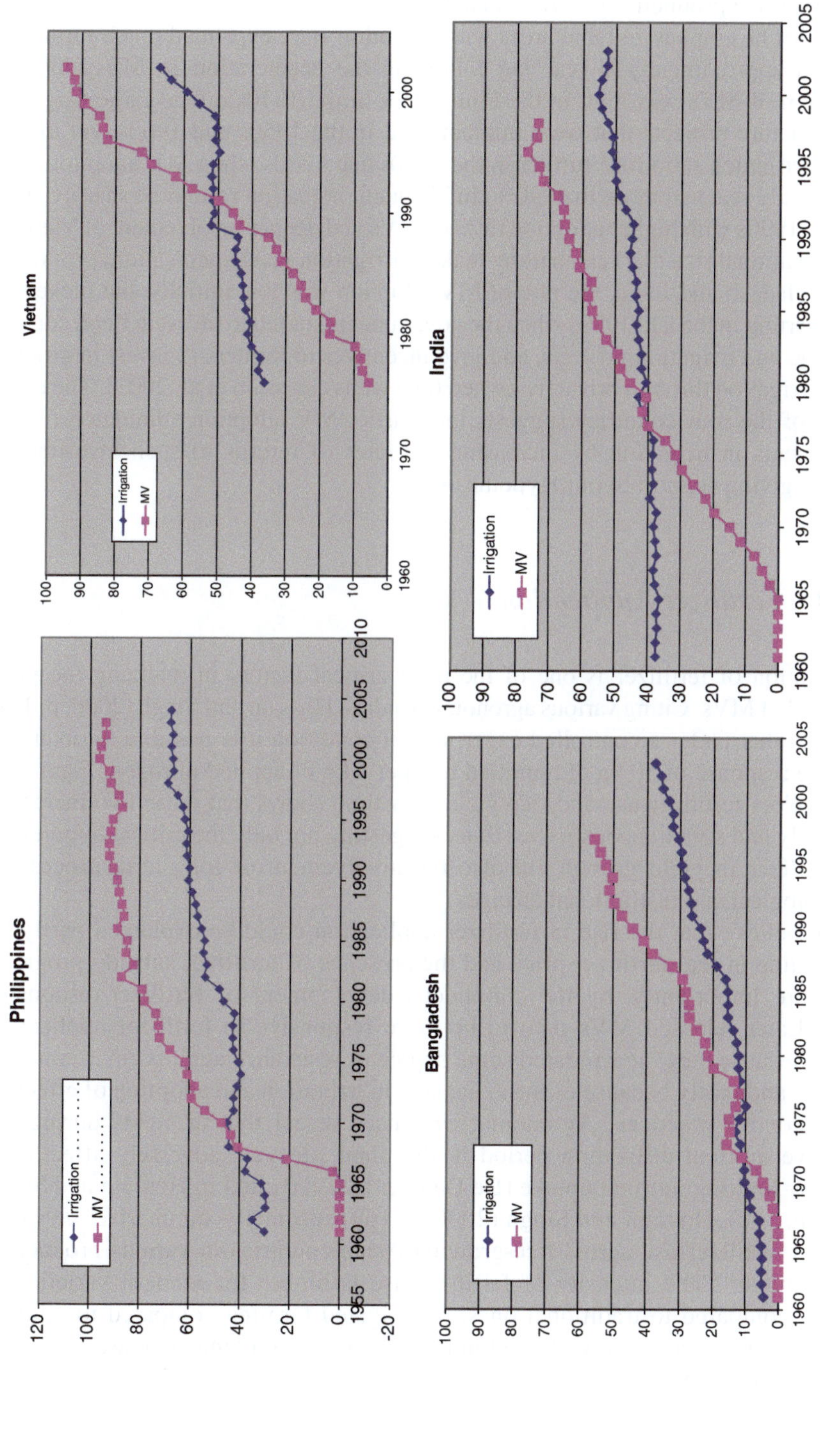

Fig. 2.3 Adoption of modern rice varieties and irrigation ratio in the Philippines, Vietnam, Bangladesh, and India (Data source: World Rice Statistics online)

of irrigated *boro* rice in the dry season contributed importantly to the accelerated growth in rice productivity in Bangladesh.

It must be emphasized that areas with irrigation were expanded much further on with an approximately10-year lag following the acceleration in MV adoption. Adoption of MVs was high in the Philippines in the 1970s due to large irrigation infrastructure projects that were implemented in the 1950s and 1960s, yet importantly, irrigation ratio rose further in the 1980s and 1990s when MV adoption rates had already reached more than 90%. In Vietnam, irrigation ratio rose sharply since the late 1990s with higher adoption rates of MVs and deeper involvement of Vietnam in international rice markets thereby making irrigation investments more profitable. In Bangladesh and India, the rate of MV adoption was low initially, but picked up later, starting in the late 1970s when the governments started to invest in flood control, drainage and irrigation projects, and give incentives to farmers to invest in groundwater irrigation through privately owned tube wells (Hossain et al. 2003). The experience of the four countries suggests that earlier MV adoption stimulated further investments in irrigation by increasing the rates of returns to such investments, thereby giving support to our Hypothesis 2.

2.4.3 Fertilizer Application

Application of fertilizer is one of the most critical factors in realizing the yield potential of MVs. Citing various agronomic studies, Hossain and Singh (2000, p. 159) reported that under a controlled experimental condition it is possible to obtain an average response of 50 kg of unmilled rice per 1 kg of applied nitrogen. Figure 2.4 which plots fertilizer use and rice yield over time shows that these two increased gradually and simultaneously over time, suggesting not only the critical importance of fertilizer in yield growth but also the involvement of long-term process of technological and institutional changes.

We believe that the rise in fertilizer application could be explained partly by the decline in real fertilizer price and the presence of fertilizer subsidy program, but more importantly, by the continuous development of fertilizer-responsive MVs. Later released MVs tend to be more responsive to fertilizer application partly because they incorporated some degree of resistance against environmental stresses and partly because of the expansion in irrigation and adoption of effective water control practices. Agronomic evidences reveal that droughts during the effective nutrient utilization period in the plant lifecycle adversely affect crop yield by limiting nutrient uptake (De Datta et al. 1990 cited in Hossain and Singh 2000, p. 160). Hossain and Singh (2000, p. 168) succinctly summarize the variations in fertilizer use across rice-growing Asian countries in various production environment: "The intensity of fertilizer-use is higher for modern varieties of cereals compared to traditional ones, on irrigated farming compared to rainfed, and on well-drained land with medium elevation than on lands that are subject to droughts and floods".

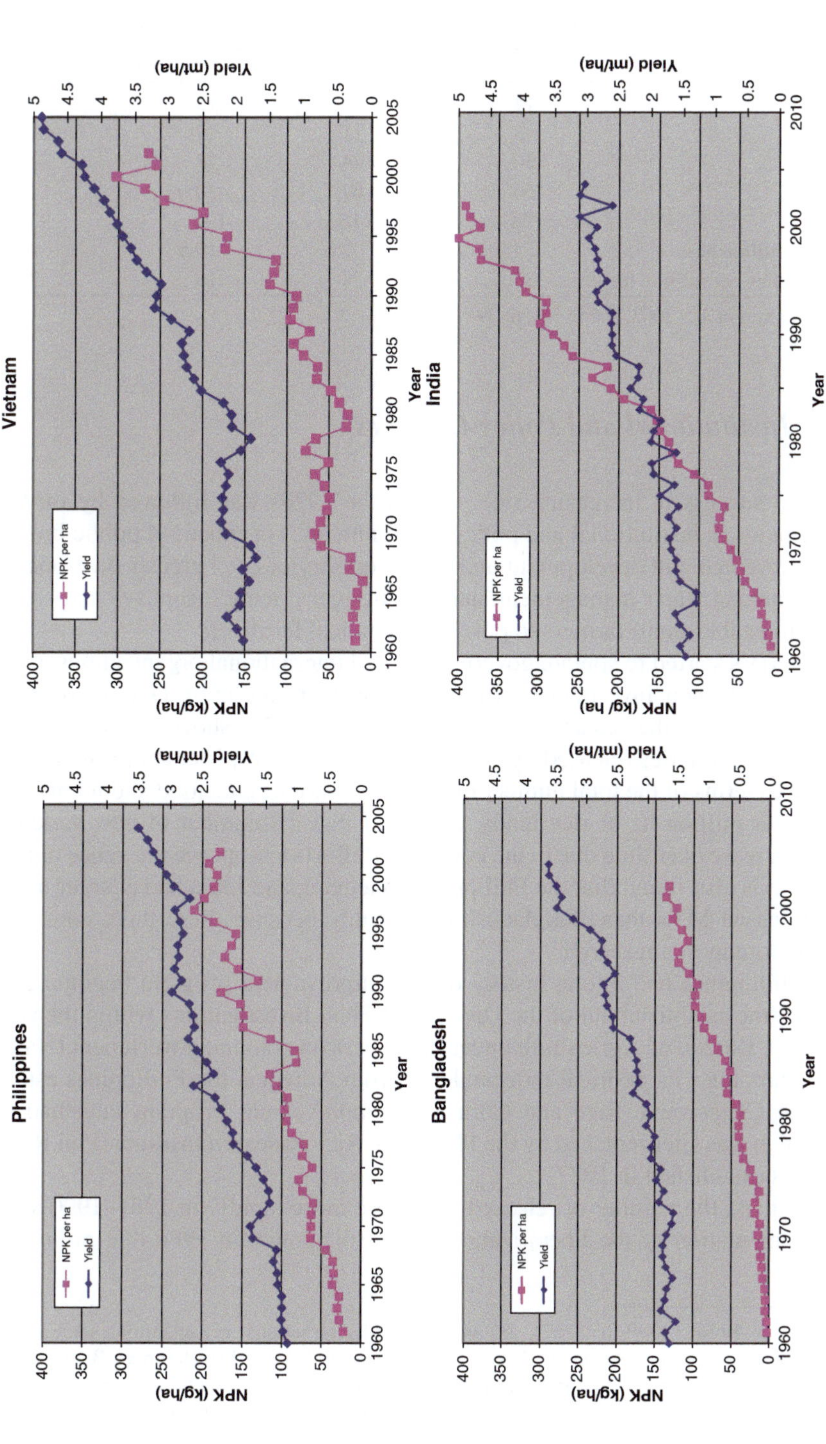

Fig. 2.4 Fertilizer use and rice yield in the Philippines, Vietnam, Bangladesh, and India (Source: World Rice Statistics online)

Table 2.1 Number of improved rice varieties released in selected Asian countries by time period

Time period	Philippines	Vietnam	Bangladesh	India
Pre-1970	14	93	13	208
1971–1980	38	11	29	211
1981–1990	20	44	28	347
1991–1999	56	67	30	170
Total	128	215	100	936
Rice land (million ha)	3.9	7.6	10.5	44.8
No. of varieties per million ha	33	28	10	21

Source: Hossain et al. (2003, Table 5.3, p. 79)

2.4.4 Institutional and Policy Changes

The initial success in increasing rice yield in the 1970s was followed by further improvements in institutional and policy environments: (1) enhanced public investments in research and development and extension services, (2) greater investments in irrigation and water management, and (3) stronger pricing incentives and credit systems to enable small farmers to purchase chemical fertilizer.[4]

When MVs started to spread, governments and international organizations gradually increased their investments in crop improvement research. One measure of such investment is the release of new varieties. Table 2.1 shows the number of improved rice varieties released by period in the Philippines, Vietnam, Bangladesh and India in terms of the total number of improved varieties released and number of varieties per million ha of rice lands. It is clear that the number of new varieties tends to increase over time due to the continuous efforts to improve the genes of rice varieties. It is also found that the Philippines, Vietnam, and India had a larger number of released MVs than Bangladesh presumably because these three countries invested more in rice research.

The Philippines had a long history of rice improvement program beginning in 1901 with the establishment of the Division of Plant Investigations within the auspices of the Bureau of Agriculture under the American colonial government (Halos 2005). Years after its political independence from America, the Philippines established the Cooperative Rice and Corn Seed Improvement Program launched in 1953, which was later replaced by the Philippine Rice Research Institute (Phil Rice) which was established in 1987.

In Vietnam, the number of released MVs rose more rapidly in 1981–1990 after the implementation of the liberalization (Doi Moi) policy in 1986 that promoted

[4] According to Herdt (2010, p. 3267), aid agencies and international organizations such as the United States Agency for International Aid, World Bank, and Asian Development Bank recognized irrigation as one of the most important agricultural development assistance targets. Investments of these institutions on irrigation were at its peak level in the 1970s following the introduction of IR8, the first MV, in 1966.

Vietnam to become a major rice exporter, thereby making rice research a socially profitable venture. Rice production growth in Vietnam is attributed to the continuous improvements of modern varieties. Hybrid and improved varieties imported from China have contributed to the GR in north Vietnam and those developed by IRRI in south Vietnam while the national agricultural research systems have successfully developed location-specific varieties (Ut and Kajisa 2006).

India had the largest number of released improved varieties mainly because it has the largest rice area in the world (46 million ha). This country had a long history of agricultural development programs since its independence in 1947. Prime Minister Jawaharlal Nehru allocated 31% of the country's budget to build massive irrigation projects, power plants, state agricultural universities, national agricultural research systems, and fertilizer plants (Hazell 2009). In Bangladesh, while the dry-season irrigated *boro* rice contributes about 60% of the country's rice production, there have been efforts from the national agricultural systems to develop rices that are locally adaptable to varied production environment susceptible to drought, floods, and salinity (Hossain 2009).

Investments in irrigation started even before the advent of MVs as a land-augmenting strategy to mitigate the high population pressure on land. Investments in irrigation together with farmers' long experience on rice culture were major factors behind the success of the development and diffusion of MVs in tropical Asia (Hayami and Otsuka 1994). Yet major investments in irrigation took place after the acceleration in MV adoption indicating that MVs had increased the rates of returns to such investments (Fig. 2.2). In the early phase of GR, the expansion in irrigation system was in the form of large gravity irrigation systems financed mainly by international organizations (Herdt 2010), which was later replaced by privately owned portable shallow tube wells particularly common in South Asia (Hossain 2009).

According to Hayami and Otsuka (1994), there was a substantial decline in fertilizer-rice price ratio in tropical Asia before the mid-1960s yet the increase in rice yield was very slow. The authors attributed this to the lack of major advancements in developing fertilizer-responsive rice varieties at a time when major public agricultural research centers for rice production were not yet established. This observation points to the importance of the development and diffusion of fertilizer-responsive rice varieties in increasing yield.

To make rice production profitable to the farmers, many Asian governments resorted to fertilizer subsidy and rice price support programs. According to Table 2.2, fertilizer-rice price ratios are different among the four countries because of the presence of fertilizer subsidy and rice market interventions in these countries. The price of fertilizer paid by the farmers had a declining trend since 1980 in the Philippines, Bangladesh and India.

The Philippines started a national fertilizer subsidy program in 1973, when MV adoption had already reached 60%, as part of a national program dubbed as Masagana 99 under the Marcos administration, which is a total package of production incentives to farmers including not only low cost production credit but also a fertilizer subsidy (Esguerra 1981). Fertilizer prices paid by the farmers were below the market price because the government controlled and regulated domestic fertilizer

Table 2.2 Producer price of rice and fertilizer price in selected Asian and African countries

Year	Philippines	Vietnam	Bangladesh	India	Kenya	Tanzania	Nigeria	Madagascar
Price of urea per ton of nitrogen paid by farmers (US$ per ton)								
1961	347		77	337				
1965	179		96	292				
1970	188		132	317	221			
1975	208		242	489	652			
1980	250		398	528	842			
1985	287		373	384	663	865		
1990	248	670	288	292	516			1236
1995	237	522	273	223	719			1108
2000	360	358	235	223	445	601	628	596
2002	358		219	216	497	544		
Farm harvest price of rough rice (US$ per ton)								
1961	114		70	80				
1965	77		86	137				
1970	61		108	97	71			
1975	130		165	91	142			
1980	146		182	134	203			
1985	149		157	119	189	na[b]		
1990	187	323	177	125	189			140
1995	239	176	176	151	107			169
2000	192	134[a]	119	133	300	na	279	190
2002	171		114	124	143	na		
Fertilizer-paddy price ratio								
1961	3.04		1.10	4.21				
1965	2.33		1.12	2.14				
1970	3.09		1.23	3.27	3.11			
1975	1.61		1.47	5.37	4.58			
1980	1.71		2.19	3.95	4.16			
1985	1.93		2.38	3.23	3.51	na		
1990	1.33	2.07	1.63	2.33	2.73			8.82
1995	1.00	2.96	1.56	1.48	6.70			6.55
2000	1.88	2.67	1.98	1.68	1.48	na	2.25	3.13
2002	2.09		1.92	1.74	3.47	na		

Source: FAOStat online for urea price and World Rice Statistics online for farm harvest price and exchange rates. Urea price and farm harvest price in the Philippines from 1980 to 1995 were taken from the Central Luzon Loop Surveys

[a] Refers to 1999

[b] na means not available

prices (Halos 2005). Fertilizer subsidy program continued on to the Aquino administration and in more recent years to the Arroyo administration through a cash voucher of PHP500 (about US$10) per bag of fertilizer purchased. The government in Bangladesh initiated a policy in 1979 to liberalize modern agricultural inputs, allowing privatization in the import and marketing of irrigation equipment and chemical fertilizer (Hossain 2009). In India, key inputs – fertilizer, power and water – were subsidized since the 1960s as part of public investments to launch and sustain the GR (Hazell 2009).

The Philippines has the highest farm harvest price of rough rice among the four countries and has the highest ratio of domestic to border price of rough rice. A comparison of the farm harvest price of rough rice with the border price of Thai rice 5% brokens shows a ratio of 1.41 in 1985, 1.33 in 1990, 1.52 in 1995, 1.94 in 2000, and 1.82 in 2002.[5] The farmers have been receiving a price premium for their rice because of government heavy intervention in the grain market through its marketing parastatal, the National Food Authority, which is involved in the domestic rice market through its buffer stock operations and in setting quantitative restrictions on rice imports on which it has a monopoly control.

The government of Bangladesh embarked into a freer food market regime between 1991 and 1993 by abolishing major food ration channels, reducing domestic rice procurement, and liberalizing foreign trade in food grains. The ratio of farm harvest price of rough rice to the border price of milled rice converted to rough rice equivalent was close to unity in Bangladesh in 1995. According to Ahmed et al. (2000) such regime was put into place because of the surge in the adoption of MVs in the second half of the 1980s coupled with sustained investment in rural infrastructure, which contributed to a steady decline in domestic real rice prices.

Overall, it is clear that the Asian GR is a continuous evolutionary process involving the development and spread of superior MVs accompanied by high application of chemical fertilizer, expansion of irrigated area, and adoption of better cultural practices complemented by public investments in agricultural research and development and price stabilization mechanisms that made modern rice technologies profitable to farmers. In our view, governments are not leaders but supporters of the long-term process of the Asian GR.

2.4.5 Sustainability of the GR

There are a few important issues on the long-term sustainability of the Asian GR: (1) slow down in rice production growth, (2) loss of biodiversity, (3) degradation of soil and water quality, and (4) withdrawal of land and labor from rice production to other uses. Barker and Dawe (2002) enumerated the causal factors on stagnation of rice production growth. The production-related factors are: (1) MV adoption ceiling, (2) full exploitation of yield gains based on the conventional breeding technique (the so-called "yield plateau" (Pingali et al. 1997)), and (3) the decline in soil fertility and greater pest infestations due to many years of continuous mono

[5] We converted the border price of milled rice to rough rice equivalent by adjusting the border price of milled rice for marketing and processing costs of 25% and milling recovery rate of 65% (Estudillo et al. 1999).

cropping of rice. The market-related factors are (1) lower rice prices and (2) rising wages thereby making rice farming a less profitable venture.

Many naysayers argue that GR could lead to a loss of biodiversity because only a few MVs dominate in the farmers' fields thereby narrowing the genetic base of readily available rice varieties. Norman Borlaug (2002) has this to say "The high yields of the Green Revolution also had a dramatic conservation effect: saving millions of acres of wild lands all over the Third World from being cleared for more low-yield crops…We can save the farmers' old varieties through gene banks and small-scale gene farm, without locking up half of the planet's arable land as a low-yield gene museum".

Whereas excessive use of chemical fertilizers and pesticides could lead to a serious degradation of soil and water quality, there have been serious efforts since the 1970s to develop rices that are resistant to various forms of infestations and make integrated pest management locally adaptable. In two villages in Central Luzon in the Philippines the proportion of adoption of integrated pest management rose from nil in 1985 to 25% in 2004.

There are fears that the withdrawal of land and labor from rice production to the growing nonfarm sector may lead to rice supply shortages. Yet there have been improvements in international rice trade facilitated by the GATT Uruguay rounds of talks and the return of Myanmar, Cambodia, and Vietnam in the rice market. Countries located in Asia's major river deltas could have the greatest comparative advantage in rice production because they have plenty of water resources and cheap labor.

2.5 Decomposition of Growth in Rice Yield in Asia

2.5.1 Yield Growth Decomposition

It is by now well-known that yield growth, rather than area expansion, is the most important contributor to rice production growth in Asia. In this section, we explore the sources of yield growth in the Philippines, Vietnam, Bangladesh, and India in separate periods of time (i.e., 1970–1980, 1980–1990, 2000–2005) corresponding to different phases of GR in these countries. The first phase can be described as the early phase when MV adoption rates are still low and released MVs are susceptible to attacks of pests and diseases. The second phase is the period when MV adoption rates have achieved higher levels and released MVs are resistant against pests and diseases. The third phase is the period when MV adoption rates have gone up higher and released MVs have incorporated stronger resistance against pests and diseases and better grain quality. While recognizing the importance of increased fertilizer application in increasing yield, our decomposition technique simply assesses the relative importance of changes in MV adoption and yield growth of MVs and TVs.

We decompose the sources of yield growth using the simple decomposition technique. Since the average yield is a weighted average of yield of MVs and TVs, we have

$$Y = s_{MV} Y_{MV} + s_{TV} Y_{TV}, \tag{2.1}$$

where:

Y = average yield
s_{MV} = ratio of MV area
s_{TV} = ratio of TV area
Y_{MV} = MV yield
Y_{TV} = TV yield.

Since $s_{TV} = 1 - s_{MV}$, we can write Eq. 2.1 as,

$$Y = s_{MV} Y_{MV} + (1 - s_{MV}) Y_{TV} \tag{2.2}$$

Taking the difference of each variable and dividing by Y, we can derive

$$\frac{\Delta Y}{Y} = s_{MV} \frac{\Delta Y_{MV}}{Y} + (1 - s_{MV}) \frac{\Delta Y_{TV}}{Y} + \Delta S_{MV} \left(\frac{Y_{MV} - Y_{TV}}{Y} \right). \tag{2.3}$$

The first term on the right hand side, $s_{MV} \frac{\Delta Y_{MV}}{Y}$, is the contribution of change in MV yield, the second term $(1 - s_{MV}) \frac{\Delta Y_{TV}}{Y}$ is the contribution of the change in TV yield, and the last term $\Delta S_{MV} \left(\frac{Y_{MV} - Y_{TV}}{Y} \right)$ is the contribution of the change in MV ratio. To the extent that continuous improvement of MVs yield is achieved assisted by increased fertilizer application, the first term tends to be large. On the other hand, the last term tends to be large, when the yield gap between MVs and TVs is large, which is particularly the case when MVs are grown under irrigated condition and TVs are grown under rainfed condition. We expect that the contribution of the change in TV yield is small.

2.5.2 *Sources of Yield Growth*

MVs have been continuously improved by IRRI, which is located in the Philippines. Thus, we expect to observe that the change in MV yield is a major contributor to the overall yield growth in the Philippines even in the early years of GR. In contrast, improvement of locally adaptable MVs was carried out mainly by the national research programs in Vietnam and India, which have been strengthened over time. Thus, we expect to observe the increasing contribution of the change in MV yield in these countries. In Bangladesh, TVs grown under rainfed condition were often replaced by MVs, which were adopted during the dry season by using irrigation pumps and tube-wells. Thus, the contribution of the change in MV ratio is likely to be large.

Table 2.3 shows the decomposition of changes in rice yield in the Philippines, Vietnam, and Bangladesh and India.[6] In the Philippines in 1970–1980, the main contributors are the change in MV yield and change in MV ratio. The first MV2, which is IR36, was released in the Philippines in 1976, when it gained wide acceptance immediately because of the widespread tungro infestation in 1971–1972 (Chandler 1982; Barker et al. 1985). MV yield rose in 1970–1980 partly because of the initiation of the Masagana 99. As a result, fertilizer use in the Philippines rose sharply in 1975–1980, thereby increasing MV yield, which responds favorably to high application of fertilizer. There was a slow and gradual increase in irrigation ratio when the National Irrigation Administration (NIA), which was established in 1964, embarked into a program in expanding irrigated rice area through investments in gravity irrigation system (Halos 2005). The NIA program and the introduction of pests- and disease-resistant MVs contributed to rice self-sufficiency in the country in the period 1978–1983. From 1980 to 1990, the contribution of change in MV yield to overall yield growth had been sustained. Meanwhile the contribution of the change in MV ratio declined continuously over time because MV adoption reached the ceiling of well over 90%. It is also interesting to note that the change in TV yield somehow contributed positively to yield increase beginning in 1980–1990, albeit small, and this is attributed to the release of improved TVs particularly those coming from the Phil Rice.

GR in Vietnam shares similar features with the Asian proto-type GR that involves the development and rapid spread of MVs followed by increase in irrigated area and increased chemical fertilizer application. Early spurt of GR started in the late 1960s, with the introduction of IR8, but was interrupted by the Vietnam War that ended in 1975. Rice production grew faster than the harvested area in the 1980s and 1990s. The GR in Vietnam has been sustained by the continuous improvement of modern varieties by the regional research institutes (Ut and Kajisa 2006). In the south, The Cuu Long Delta Rice Research Institute, Can Tho University, Southern Agricultural Sciences, and IRRI are involved in the development and dissemination of improved MVs that thrive well in saline-affected rainfed rice fields with semi-deep water, which are prevalent in Mekong River Data. In the north, the Vietnam Agricultural Science Institutes, Plant Protection Institute and agricultural research institutes in China are the major players in the development and dissemination of newer and improved MVs. Such adaptive research contributed to the sustainable growth of rice yields in this country, which is reflected in the increased contribution of changes in MV yield over time in this country (see Table 2.3).

[6] Data on the adoption of MVs in India are available from 1961 to 1998 only while separate data on the yield of MVs and TVs are not available. Thus, we simply extrapolated the yield of MVs and TVs by regressing yield using MV ratio, time, and interaction between MV ratio and time as explanatory variables.

Table 2.3 Decomposition of the changes in paddy yield in selected Asian countries[a]

Source of change in yield	Philippines	Vietnam	Bangladesh	India
	1970–1980		1975–1980	1970–1980
Total change in yield	0.2730		0.0756	0.1391
	(100)		(100)	(100)
Due to change in MV yield	0.1629		−0.0208	0.0755
	(60)		(−27)	(54)
Due to change in TV yield	0.0019		0.0324	0.0141
	(1)		(42)	(10)
Due to change in MV ratio	0.1082		0.0640	0.0495
	(39)		(85)	(36)
	1980–1990	1980–1990	1980–1990	1980–1990
Total change in yield	0.2262	0.3993	0.226	0.2159
	(100)	(100)	(100)	(100)
Due to change in MV yield	0.1586	0.1267	−0.0082	0.1467
	(70)	(32)	(−3)	(67)
Due to change in TV yield	0.0209	0.2098	0.065	0.0007
	(9)	(52)	(29)	(1)
Due to change in MV ratio	0.0467	0.0628	0.1692	0.0685
	(21)	(16)	(74)	(32)
	1990–2000	1990–2000	1990–2000	1990–1998
Total change in yield	0.0996	0.2995	0.2505	0.1861
	(100)	(100)	(100)	(100)
Due to change in MV yield	0.0647	0.1572	0.0708	0.1334
	(65)	(52)	(28)	(72)
Due to change in TV yield	0.0084	−0.0291	0.0352	0.0039
	(8)	(−9)	(14)	(2)
Due to change in MV ratio	0.0265	0.1714	0.1445	0.0488
	(27)	(57)	(58)	(26)
	2000–2005	2000–2002	2000–2005	
Total change in yield	0.1637	0.0522	0.1232	
	(100)	(100)	(100)	
Due to change in MV yield	0.1647	0.0381	0.0533	
	(100)	(73)	(43)	
Due to change in TV yield	0.0060	0.0000	0.0339	
	(4)	(0)	(28)	
Due to change in MV ratio	−0.0073	0.0141	0.036	
	(−4)	(27)	(29)	

Source: Authors' calculations
[a]Numbers in parentheses are percentages

In Bangladesh, yield growth in 1975–1980 could be largely attributed to the change in MV ratio. Adoption of MVs was rather slow in Bangladesh till the late-1980s when irrigation investments accelerated (Fig. 2.3). There was hardly any increase in TV yield (1 t in 1975 and 1980) and MV yield even declined from 2.35 t in 1975 to 2.21 t in 1980. There was huge variability in the yield of earlier MVs that are susceptible to a wide spectrum of pests and diseases. The contribution of the change in MV ratio was by far the largest in 1980–1990, when MV adoption was gaining momentum and, as maybe expected, its contribution declined over time when both the MV and TV yields rose. Like the Philippines, Bangladesh has a stock of improved TVs that contributed to yield increase. By 2000–2005, however, the increase in MV yield was clearly the dominant source of yield growth.

Punjab and Haryana in northwestern India were the earliest to adopt MVs and became the breadbaskets for the entire country. Because of the rapid spread of MVs in these two states the change in MV ratio contributed to 36% of the change in yield between 1970 and 1980 (Table 2.3). Yet more importantly, the contribution of the change in MV yield to the total change in yield continued to rise from 54% between 1970 and 1980 to 67% between 1980 and 1990 and further to 72% between 1990 and 1998. Similar to the case of the Philippines, more improved MVs were released and adopted by Indian farmers coupled with increased usage of inputs such fertilizers and tube wells as the Indian government has been heavily involved in the production, dissemination, and adoption of these inputs since the mid-1960s. In brief, there is no doubt that the main contributor to yield increase in the four Asian countries had been the continuous MV improvements.

2.5.3 The GR in Central Luzon, the Philippines

Here we discuss the gradual evolutionary processes of GR in Central Luzon in the Philippines, which was one of the earliest to experience the GR in Southeast Asia. As early as 1966 the irrigation ratio among our sample farmers was as high as 60% and it increased further to 71% with the opening of Pantabangan Dam in 1975 and since the early 1990s there was a spread of small-scale deep well water pumps. Rice cropping intensity in Central Luzon increased from 1.1 in 1966/1967 to 1.6 in 1979/1980, and then decreased to 1.5 in 1998/1999 coinciding with the trends in irrigation ratio.

When the first survey was conducted in Central Luzon in 1966, the entire area was planted with TVs (Table 2.4). There was a fast rate of adoption of MV1 immediately after its first release that by 1970, 66% of sample farmers were adopting MV1. The speed of MV diffusion in Central Luzon in the early period of release of MVs was almost the same as in the irrigated ecosystem in the Philippines as a whole (Estudillo and Otsuka 2006). The adoption of MV2 was quicker than MV1 because MV2 are more resistant to attacks of major insects and diseases. A few years after the release of IR36 in 1976, 92% of sample farmers in Central Luzon were already adopting MV2 in 1979, when TVs were completely replaced by MVs. The adoption of MV3 was also fast and their adoption reached 90% by 1990. It is noticeable that the adoption rates of MVs in Central Luzon far exceeded the irrigation ratio suggesting that MVs were grown well in rainfed production environments.

Table 2.4 Adoption of generations of modern varieties of rice (% area) in the central Luzon provinces, the Philippines, 1966–2004

Year	% area planted				NPK (kg/ha/season)	Average rice yield (mt/ha/season)
	TV	MV1	MV2	MV3		
1966	100	0	0	0	9	2.3
1967	94	6	0	0	17	1.9
1970	34	66	0	0	29	2.5
1971	8	92	0	0	59	2.4
1974	27	73	0	0	39	2.2
1979	0	8	92	0	62	3.6
1980	0	9	91	0	78	4.3
1982	0	3	97	0	63	4.1
1986	0	1	38	61	67	3.6
1987	0	2	6	92	88	4.3
1990	0	2	8	90	70	4.6
1991	0	3	15	82	103	4.5
1994	0	0	6	94	93	3.9
1995	0	0	0	100	125	4.6
1998	0	0	0	100	150	4.8
1999	0	0	0	100	143	3.4
2003	0	0	0	100	136	4.2
2004	0	0	0	100	165	4.7

Data source: Central Luzon Loop Surveys, International Rice Research Institute

The most important factor affecting farmers' adoption of MVs is the availability of irrigation, which became much more important in the adoption of MV2 and MV3. A possible reason could be that MV2 and MV3 are more resistant to attacks of various pests and diseases, which tend to be frequent in irrigated environments because of continuous mono-cropping of rice, the decline in genetic diversity of the more common MVs, and the ability of pests to evolve genetically over successive crops of rice (Khush 1995). Socio-economic factors such as age and schooling of head, farm size and tenancy were, in general, not significant determinants of the adoption of MVs (David and Otsuka 1994).

There was an increasing trend in the application of chemical fertilizer starting in the mid-1970s and rice yield rose remarkably in this period. TV has an average yield of about 2 t/ha per season, MV1 has about 2.5 t, MV2 has about 4 t, and MV3 has more than 4.5 t. According to Otsuka et al. (1994), MV2 had a statistically significant yield advantage over MV1 and the adoption of MV2 contributed to yield growth under irrigated condition and during the dry season.

Overall, the experience in Central Luzon demonstrates that the diffusion of new MVs has been remarkably fast and there has been no reversal back to TVs and older MVs indicating that the use of newer and betters MVs has been more profitable. Indeed, the achievements in varietal improvements and farmers' acceptance of newer MVs played a catalytic role in sustaining the GR in Central Luzon.

2.6 Rice in SSA

There are many episodes of success stories in rice production in SSA, even though these successes are inadequate to fully launch a GR in rice. They indicate, however, that there is a high potential of rice production and sustainability even without heavy government support. We would also like to argue that markets are now working much better than in the past and that the increasing population pressure on limited land resources makes intensive Asia-type farming systems more profitable. It must be also pointed out that there has been little effort in the past to develop lowland rice sectors in SSA in contrast to cassava, maize, cotton, and horticultural crops (Haggblade 2004). In fact, even the very basic production practices, such as leveling, bunding, and flooding are not adopted by many farmers in SSA, which indicates inadequate extension efforts as well as the lack of efforts to transfer production methods from Asia to SSA. Indeed, there are very few extension workers in SSA who are knowledgeable about rice production. In our observation, rice yield could double by simply introducing bunding, leveling, and flooding without using additional external inputs.

There has been an on-going debate on the strategies to launch a GR in SSA. Djurfeldt et al. (2005) argue that a broader process that is "state-driven, market-mediated, and small-farmer based" is necessary while Dorward et al. (2009) argues to identify constraints imposed by market and state and explore their complementary roles in technology transfer. On the contrary, Borlaug (2002) believes that the African GR could be simple and straightforward consisting only of two fundamental ingredients (1) modern technology and (2) remunerative and stable prices to farmers. By modern technology, Borlaug (2002) refers to high-yielding seeds well-adapted to local conditions, irrigation, chemical fertilizers, and integrated pest management. In order to follow Borlaug's paradigm, Otsuka and Yamano (2005) recommend increasing investments in agricultural research and enhancing the capacity of national systems for adaptive research. Indeed, according to Pardey et al. (2007), real research spending in Africa as a whole has stagnated since the 1970s with the spending per scientist declining continuously in the past 30 years and most dramatically during the 1980s.

Otsuka and Kijima (2010) compare the states of agricultural development in Asia and SSA. First, in the early 1960s, the yield difference in rice between Asia and SSA was in the order of about 20–30%, which could be largely attributed to a well-developed irrigation system in Asia. Second, yield difference in rice appeared since the mid-1960s with the advent of IR8, which began the GR in Asia. Irrigation investments and government-sponsored credit programs, and the establishment of the national research and extension services were accelerated following high adoption rates of MVs. And thirdly, irrigated rice yields in SSA are highly comparable to irrigated yields in Asia despite the absence of major technological breakthroughs (Chap. 3 in this volume). This likely reflects the increasing intensification of farming systems, including the adoption of IRRI-type MVs in selected areas, because of the continued population pressure that makes land-saving and yield-enhancing technologies profitable.

Markets for inputs and outputs have started to develop in SSA. For example, prices of bean seeds in Kenya and milk prices in Kenya and Uganda are determined largely by transportation costs, which indicates that market prices are competitive (Staal and Baltenweck 2008). Kijima et al. (2008) reported that in Uganda, where NERICA were adopted, access to rice millers was greatly improved owing to the rapid increase in the number of rice millers. Tsuboi (2008) reported that the total number of private rice mills in Uganda rose from 183 in 2000 to 591 in 2007. Seeds have been increasingly available from seed suppliers and purchase from neighboring farmers (Chap. 6 in this volume), indicating the development of both formal and informal seed markets. African maize farmers have become more aware of the importance of seed selection in increasing yield that they are willing to purchase good quality maize seeds even at a premium price (Chaps. 9 and 12 in this volume). These are good examples to show that markets could respond favorably to the diffusion of new profitable technology in SSA.

There are numerous success stories on the use of Asian rice technology that indicates the possibility of inducing the evolutionary processes of a GR in SSA. First, MVs that were developed in Asia and grown in areas with simple irrigation canal in Côte d'Ivoire have an average yield of 3.6 t/ha while those varieties that were grown in areas without canal have an average yield of 2.5 t/ha with minimum or even zero application of chemical fertilizer (Sakurai 2006). Second, Kajisa and Payongayong (2011) demonstrate that paddy yields can be potentially high (3.8 t/ha) in irrigated areas of Mozambique, where irrigation facilities are poorly maintained. Third, the NERICA could potentially increase the yield potential of upland fields from 1 t to 2–3 t/ha as in the case in Uganda (Chap. 6 in this volume). NERICA is an upland variety that thrives well in fields that tend to experience water stress and believe to be the most appropriate variety suitable for nearly 40% of the total rice area in SSA. Yet a resurvey of the same farmers in 2006 (the first survey was undertaken in 2004), Kijima and Otsuka (Chap. 6 in this volume) reported that dropout rate was generally high even as high as 75% in areas where the Vice President distributed free NERICA seeds. This is quite different from the experience in Asia where farmers never reverted back to TVs but were instead continuously adopting newly released MVs. The dropout rate in the case of NERICA is high because of lower yield performance particularly of NERICA seeds that are self-propagated by the farmers. Fourth, and finally, a study in Doho irrigation scheme in Eastern Uganda reveals that paddy yields are as high as 3 t/ha even without application of chemical fertilizer and despite continuous double cropping of rice for the last few decades (Nakano and Otsuka 2011).

It appears that the GR in rice in SSA is not an impossible dream at all given the comparable yield performance of SSA vis-à-vis that of Asia. Indeed, what is needed for a take-off in SSA is first to develop rice varieties that are better adaptable to local African condition and then to strengthen extensions system, enhance the development of seed and fertilizer markets, increase investments in irrigation and water management, and give price incentives to farmers as is illustrated in Fig. 2.1.

2.7 Conclusions

This study has attempted to demonstrate that the fundamental lesson from the Asian GR is the continuous development of MVs and the subsequent acceleration in public sector investments in supporting measures. The Asian GR is essentially technology-led and policy-supported, rather than policy-driven as is oftentimes assumed. The continuous development of MVs induced subsequent acceleration in public sector investments in irrigation, credit and fertilizer subsidy programs, and national research and extension services as the rates of returns to these investments rose. Farmers were induced to adopt the GR technologies because the favorable policy environment made the new technologies privately profitable. Factor and product markets were also induced to develop further to internalize the gains from profitable opportunities associated with the new technologies. We believe that if GR is to start in SSA, it is likely to include rice because rice has a matured transferable technology that is readily available from Asia.

In order to realize a GR in SSA, we believe that a focused, concrete, and carefully designed strategy is needed, rather than the commonly accepted comprehensive approach, which allocates scarce budget thinly to a large number of purposes. As Ruttan (1984) properly emphasizes, the economists ought to design effective strategies to develop agriculture with a particular focus on institutional reforms and with a clear recognition of the inducement effects of technological and institutional changes. In the case of rice in SSA, it is critically important to recognize that productive technologies are either already available or transferable from Asia. What is badly needed now are adaptive research to transfer Asian rice technology and extension systems to disseminate available new technologies to farmers. Subsequently, such efforts should be supported by investing in low-cost irrigation and marketing infrastructure, while recognizing that expected rates of return to investments in extension, irrigation, and marketing will be enhanced by the *inducement effects* of new promising technologies.

References

Ahmed R, Steven H, Chowdhury T (2000) Out of the shadow of famine: evolving food markets and food policy in Bangladesh. Johns Hopkins University Press, Baltimore

Barker R, Dawe D (2002) The transformation of the Asian rice economy and directions for future research: the need for increased productivity. In: Sombilla M, Hossain M, Hardy B (eds) Developments in the Asian rice economy. International Rice Research Institute, Los Banos

Barker R, Herdt R with Rose B (1985) The rice economy of Asia. Resources for the Future, Washington, DC

Borlaug N (2002) We can feed the world: here's how. Wall Street J, 13 May

Byerlee D (1994) Technology transfer systems for improved crop management: lessons for the future. In: Anderson JR (ed) Agricultural technology: policy issues for the international community. CAB International, Wallingford, Oxford

Byerlee D (1996) Modern varieties, productivity, and sustainability: recent experience and emerging challenges. World Dev 24(4):697–718

Byerlee D, Fisher K (2002) Accessing modern science: policy and institutional options for agricultural biotechnology in developing countries. World Dev 30(6):931–948

Chandler R Jr (1982) An adventure in applied science: a history of the international rice research institute. International Rice Research Institute, Los Banos

David CC (1976) Fertilizer demand in the Asian rice economy. Food Res Inst Stud 5(1):109–124

David CC, Otsuka K (eds) (1994) Modern rice technology and income distribution in Asia. Lynne Rienner, Boulder

De Datta SK, Buresh RJ, Mamaril CP (1990) Increasing nutrient use efficiency in rice with changing needs. Fertil Res 26(1):157–167

Deb UK, Bantilan MCS (2003) Impacts of genetic improvement in sorghum. In: Evenson RE, Gollin D (eds) Crop variety improvement and its effect on productivity. CAB International, Wallingford, Oxford

Djurfeldt G, Holmen H, Jirstrom LR (eds) (2005) The African food crisis; lessons from the Asian Green Revolution. CAB International, Wallingford, Oxford

Dorward A, Kydd JG, Poulton C, Bezemer D (2009) Coordination risk and cost impacts on economic development in poor rural areas. J Dev Stud 45(7):1093–1112

Esguerra E (1981) An assessment of the Masagana 99 credit subsidy as an equity measure. Philipp Rev Econ 17(3 and 4):168–191

Estudillo JP, Otsuka K (2006) Lessons from three decades of green revolution in the Philippines. Dev Econ 44(2):123–148

Estudillo JP, Fujimura M, Hossain M (1999) New rice technology and comparative advantage in rice production in the Philippines, 1966–94. J Dev Stud 35(5):162–185

Evenson RE, Gollin D (2003) Crop variety improvement and its effect on productivity: the impact of international agricultural research. CAB International, Wallingford, Oxford

Haggblade S (2004) Building on the successes in African agriculture: 2020 focus 12. International Food Policy Research Institute, Washington, DC

Halos SC (2005) A century of rice innovations, vol 2005-12, Discussion paper series. Philippine Institute of Development Studies, Manila

Hayami Y, Kikuchi M (1978) Investment inducement to public infrastructure: irrigation in the Philippines. Rev Econ Stat 60(1):70–77

Hayami Y, Otsuka K (1994) Beyond the green revolution: agricultural development strategy into the new century. In: Anderson JR (ed) Agricultural technology: policy issues for the international community. CAB International, Wallingford, Oxford

Hayami Y, Ruttan VW (1985) Agricultural development: an international perspective. Johns Hopkins University Press, Baltimore

Hazell P (2009) Transforming agriculture: the green revolution in Asia. In: Spielman DJ, Pandaya-Lorch R (eds) Millions fed: proven successes in agricultural development. International Food Policy Research Institute, Washington, DC

Heisey PW, Lantican MA, Dubin JJ (2003) Wheat. In: Evenson RE, Gollin D (eds) Crop variety improvement and its effect on productivity. CAB International, Wallingford, Oxford

Herdt RW (2010) Development aid and agriculture. In: Evenson RE, Pingali PL (eds) Handbook of agricultural economics, vol 4. Elsevier, Amsterdam

Hossain M (1993) Production environments, MV adoption, and income distribution in Bangladesh. In: David CC, Otsuka K (eds) Modern rice technology and income distribution in Asia. Lynne Rienner, Boulder

Hossain M (2009) Pumping up production: shallow tube wells and rice in Bangladesh. In: Spielman DJ, Pandaya-Lorch R (eds) Millions fed: proven successes in agricultural development. International Food Policy Research Institute, Washington, DC

Hossain M, Singh VP (2000) Fertilizer use in Asian agriculture: implications for sustaining food security and the environment. Nutr Cycl Agroecosyst 57(2):155–169

Hossain M, Gollin D, Cabanilla V, Cabrera E, Johnson N, Khush GS, McLaren G (2003) International research and genetic improvement in rice: evidence from Asia and Latin America. In: Evenson RE, Gollin D (eds) Crop variety improvement and its effect on productivity. CAB International, Wallingford, Oxford

International Rice Research Institute (IRRI) (1985) International rice research: 25 years of partnership. IRRI, Los Banos

Kajisa K, Payongayong E (2011) Potential and constraints to the rice green revolution in Mozambique: a case study of the Chokwe irrigation scheme. Food Policy 35(5):614–625

Khush G (1987) Rice breeding: past, present and future. J Genet 66(3):195–216

Khush G (1995) Modern varieties – their real contribution to food supply and equity. Geo J 35(3):275–284

Kijima Y, Otsuka K, Sserukuuma D (2008) Assessing the impact of NERICA on income and poverty in Central and Western Uganda. Agric Econ 38(3):327–337

Kikuchi M, Maruyama A, Hayami Y (2002) Investment inducements to public infrastructure: irrigation in the Philippines and Sri Lanka since independence, no 2002-003, Foundation for Advanced Studies in International Development discussion paper series on international development strategies. FASID, Tokyo

Lipton M, Longhurst R (1989) New seeds and poor people. Unwin Hyman, London

Nakano Y, Otsuka K (2011) The determinants of household contribution to the collective irrigation management: the case of Doho Rice Scheme in Uganda. Env Dev Econ 16(5):521–551

Otsuka K, Kijima Y (2010) Technology policies for a green revolution and agricultural transformation in Africa. J Afr Econ 19(S2):60–96

Otsuka K, Yamano T (2005) Green Revolution and regional inequality: implications of Asian experience for Africa. In: Djurfeldt G, Holmen H, Jistrom M, Larsson R (eds) The African food crisis: the lessons from the Asian Green Revolution. CAB International, Wallingford, Oxford

Otsuka K, Gascon F, Asano S (1994) Second-generation MVs and the evolution of the green revolution: the case of Central Luzon, 1966–90. Agric Econ 10(3):283–295

Pardey PG, Roseboom J, Beintema NM (2007) Investments in African agricultural research. World Dev 25(3):409–423

Pingali PL, Moya P, Velasco L (1990) The post-green revolution blues in Asian rice production: the diminished gap between experiment station and farmer yields, vol 90-01, IRRI social sciences division paper. International Rice Research Institute, Los Banos

Pingali PL, Hossain M, Gerpacio R (1997) Asian rice bowls: the returning crisis? CAB International, Wallingford, Oxford

Ruttan VW (1977) The green revolution: seven generalizations. Int Dev Rev 19(1):16–23

Ruttan VW (1984) Social science knowledge and institutional change. Am J Agric Econ 66(5):549–559

Sakurai T (2006) Intensification of rainfed lowland rice production in West Africa: present status and potential green revolution. Dev Econ 44(2):232–251

Staal S, Baltenweck I (2008) Determinants of milk producer prices in Kenya and Uganda, Mimeo. International Livestock Research Institute, Nairobi

Tsuboi T (2008) Towards rice green revolution in Africa: the case of NERICA promotion Uganda. Paper presented in Tokyo international conference on African development, Yokohama, Japan

Ut TT, Kajisa K (2006) The impact of green revolution on rice production in Vietnam. Dev Econ 44(2):167–189

Chapter 3
The Possibility of a Rice Green Revolution in Large-Scale Irrigation Schemes in Sub-Saharan Africa

Yuko Nakano, Ibrahim Bamba, Aliou Diagne, Keijiro Otsuka, and Kei Kajisa

Abstract The importance of rice in Sub-Saharan Africa (SSA) is increasing rapidly, as the consumption of rice is increasing and the imbalance between domestic production and consumption has been growing in SSA. Therefore, national and international attention now centers on how to increase rice production in SSA as an important component of the region's strategies on food security. One possible strategy to achieve this goal is to take an Asian-style approach as Asia has successfully achieved a rice Green Revolution over the last several decades. This study aims to investigate the potential of SSA's large-scale irrigation schemes for a rice Green Revolution in the region as well as the conditions for achieving the potential. For this study, we use household-level data collected in six SSA countries: Uganda, Mozambique, Burkina Faso, Mali, Niger, and Senegal.

Y. Nakano (✉)
Faculty of Humanities and Social Sciences, University of Tsukuba,
1-1-1, Tennoudai, Tsukuba-shi, Ibaraki 305-8577, Japan
e-mail: nakano.yuko.fn@u.tsukuba.ac.jp

I. Bamba
Programme Management Department, International Fund for Agricultural Development,
Via Paolo di Dono 44, 00142 Rome, Italy
e-mail: i.bamba@ifad.org

A. Diagne
Policy and Impact Assessment, Africa Rice, 01 B.P. 2031, Cotonou, Benin
e-mail: A.Diagne@cgiar.org

K. Otsuka
National Graduate Institute for Policy Studies,
7-22-1 Roppongi, Minato-ku, Tokyo 106-8677, Japan
e-mail: otsuka@grips.ac.jp

K. Kajisa
School of Internationanl Politics, Economics, and Communication,
Aoyama Gakuin University, 4-4-25 Shibuya, Shibuya-ku, Tokyo 150-8366, Japan
e-mail: k.kajisa@sipeb.aoyama.ac.jp

K. Otsuka and D.F. Larson (eds.), *An African Green Revolution: Finding Ways to Boost Productivity on Small Farms*, DOI 10.1007/978-94-007-5760-8_3,

Keywords Green Revolution • Sub-Saharan Africa • Food security • Irrigation • High-yielding varieties • Lowland rice • Paddy yield

3.1 Introduction

The importance of rice in Sub-Saharan Africa (SSA) is increasing rapidly (Otsuka and Kijima 2010). The consumption of rice is increasing, and the imbalance between domestic production and consumption has been growing in SSA. The total milled rice production in SSA increased from 2 million tons in 1961 to 17 million tons in 2012 (FAO 2012). At the same time, milled rice imports into SSA increased from 0.5 million tons in 1961 to 10 million tons in 2009 due to inadequate local production to meet the growing demand. SSA accounts for a third of global rice imports at a cost of more than US$4.3 billion per year, which otherwise could be used to finance infrastructure development and other productive purposes. Therefore, national and international attention now centers on how to increase rice production in SSA as an important component of the region's strategies on food security.

One possible strategy to achieve this goal is to take an Asian-style approach as Asia has successfully achieved a rice Green Revolution over the last three decades (Otsuka 2006; Otsuka and Kalirajan 2005). At the same time, however, many studies are skeptical about this strategy (Spencer 1994; World Bank 2008). Among many reasons, the central reason behind the skepticism is the under-development of irrigation in the region (Hayami and Godo 2005; Spencer 1994; World Bank 2008). Although large- and middle-scale irrigation played a significant role in facilitating the diffusion of fertilizer-responsive high-yielding modern varieties (MVs) in Asia, high investment costs, declining rice prices, and the failures of past large-scale government-led gravity irrigation projects are believed to be the main reasons for the reluctance of donors and governments to invest in large-scale irrigation in SSA (Inocencio et al. 2007).

However, with the passage of time, conditions for growing irrigated rice have changed dramatically. The price of rice is expected to increase in the long run (USDA 2008). In addition, the reform process initiated in the past two decades by African countries has tremendously changed the institutional and policy environment for growing rice in large irrigation schemes. For example, the Office du Niger irrigation scheme in Mali is now touted as a "success story" (Aw and Diemer 2005). In fact, a recent assessment of existing irrigation schemes by Inocencio et al. (2007) found that the costs of irrigation projects are not significantly higher in SSA than in other regions and that irrigation investments can provide good returns under the right conditions. Therefore, it is worth examining empirically whether large-scale irrigation schemes can be a cradle for a rice Green Revolution (GR) in SSA, as was the case in Asia. However, the number of micro-level studies in SSA is limited.[1]

[1] Sakurai (2006) examines the possible constraints to lowland rain-fed rice cultivation in Côte d'Ivoire and Kijima et al. (2006, 2008, 2011) investigate the potential of upland NERICA cultivation in Uganda. However, none of them are about rice cultivation in large- or medium-scale irrigation.

Table 3.1 Basic statistics on rice production in survey countries

Country	Uganda	Mozambique	Burkina Faso	Mali	Niger	Senegal
Harvested rice area (000 ha)	93.0	179.0	51.0	451.0	27.8	95.0
Production (t)	140	201.0	95.2	877.0	76.5	264.5
Yield (t/ha)	1.51	1.12	1.87	1.94	2.75	2.78
Rice production ecology (%)						
Irrigated wetland	2	2	46	22	80	50
Rainfed wetland	53	59	50	13	0	40
Dry land	45	39	4	1	0	0
Deepwater and mangrove	0	0	0	64	20	10

Source: Balasubramanian et al. (2007)

This study aims to investigate the potential of SSA's large-scale irrigation schemes for a rice GR in the region as well as the conditions for achieving the potential. We use household-level data collected in six SSA countries: Uganda, Mozambique, Burkina Faso, Mali, Niger, and Senegal. The study sites are large-scale irrigation schemes and are considered as areas with high potential for rice cultivation in terms of availability of water and agro-climatic conditions. However, we observe wide variations in the availability of irrigation water, cultivation practices, and rice productivity within a scheme or between schemes. This provides us with a good opportunity to examine under what conditions the potential of an irrigation scheme can be fully realized.

This chapter is organized as follows. After the explanation of the study countries and study sites in Sect. 3.2, we descriptively analyze the characteristics of rice production at each irrigation scheme in Sect. 3.3. In Sect. 3.4, in order to identify the possible constraints to rice production statistically, we estimate the regression function showing the determinants of rice yield and input use for Uganda and Mozambique, for which data sets are available for regression analyses for our purposes. Section 3.5 concludes by providing the policy implications.

3.2 Study Sites and Data

The six countries for our study come from two regions: East Africa (Uganda and Mozambique) and the Sahel region of West Africa (Burkina Faso, Mali, Niger, and Senegal). Table 3.1 shows the production and ecology of the rice sector in each country. Generally speaking, the rice sector is less developed in East Africa than in the Sahel region in terms of the extensiveness of irrigated area as well as productivity. Irrigated area consists of only 2% in both Uganda and Mozambique, with an average rice yield of 1.51 and 1.12 t/ha, respectively. Meanwhile, the four Sahelian countries show higher proportions of irrigated area and higher average yields, and, among them, Niger's and Senegal's figures look superior and as good as those of Asian countries.

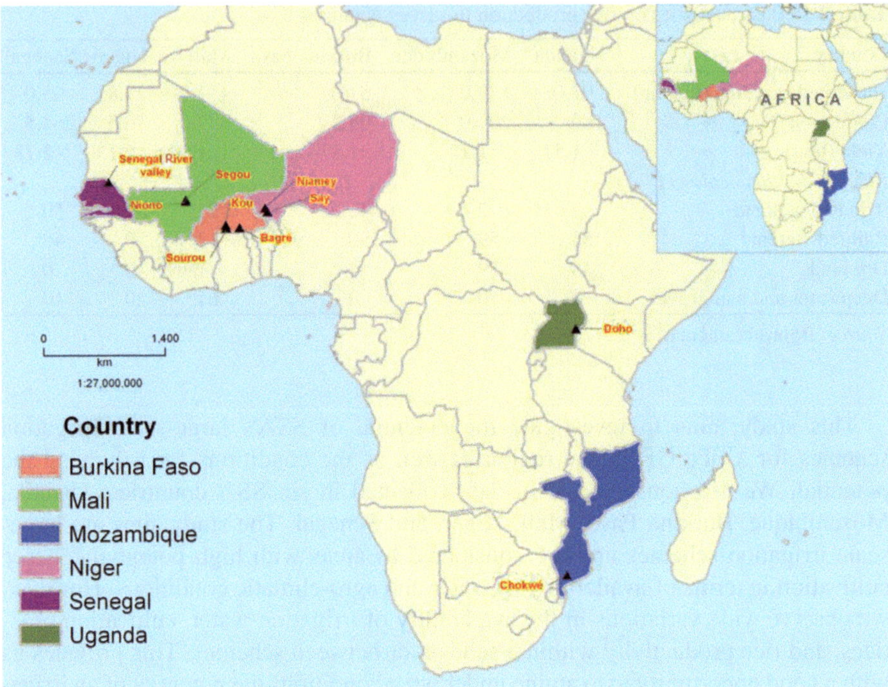

Fig. 3.1 Location of irrigation schemes

Our analyses rely on the following ten irrigation schemes (Fig. 3.1).

- Doho rice scheme in Uganda
- Chokwe scheme in Mozambique
- Kou valley, Sourou, and Bagré schemes in Burkina Faso
- Ninon and N'Débougou in Mali
- Say and Daibéri in Niger
- Senegal River Valley in Senegal

Table 3.2 summarizes the major characteristics of each scheme and the features of the survey data. In addition to the differences in water delivery systems, size, climatic conditions, management body, and proximity to city, one notable difference can be found in supporting institutions. Neither Uganda nor Mozambique provides support for fertilizer purchase, agricultural credit, and output procurement to rice farmers. Meanwhile, all four Sahelian countries received an informal (Burkina Faso and Mail) or formal (Niger and Senegal) subsidy on fertilizer. In addition, in Mali and Senegal, there exist savings and loan programs or farmer organizations that facilitate group purchase of fertilizer. Niger's government provides price support (Kore 2006). Generally speaking, institutions are more supportive in the four Sahelian countries than in the two East African countries.

In all surveys except the one in Niger, data were collected by random or stratified random sampling methods to obtain representative farmers in each irrigation

Table 3.2 Characteristics of irrigation schemes and survey data

Country	Uganda	Mozambique	Burkina Faso			Mali		Niger		Senegal
Scheme	Doho	Chokwe	Kou Valley	Sourou	Bagré	Niono	N'Débou-gou	Say2	Daibéri	Senegal River Valley
Survey year	2007	2007	2005–2006	2005–2006	2005–2006	2005–2006	2005–2006	2005	2005	2006/07
Irrigation system	River and gravity	Dam and gravity	River and gravity	River and diesel pump	Dam and gravity	Dam and gravity	Dam and gravity	River and electrical pumping	River and electrical pumping	River and electrical pumping
Potential irrigated area (ha)	1,000	26,000	n.a.	35,000	8,158	900, 000		70,000		240,000
Current irrigated area (ha)	952	4,000	1,400	3,200	1,885		11,757	186	295	60,000
Annual rainfall (mm)	1,150	650	1,200	800	900	550		400–700	200–400	300–400
Management body	Farmer organi zations	Para state	Farmer organi zations	Para state	Para state	Para state	Para state	Para state	Para state	Para state
Fertilizer subsidy	No	No	No public fertilizer program, small-scale program by NGOs[a]			No public fertilizer program, small-scale program by NGOs[a] Support by the Office du Niger through farmer organizations[b]		30% subsidy rate on fertilizer price		50% subsidy rate on fertilizer and herbicide prices
Credit program	No	A small-scale program targeted to large-scale farmers only.	No			Group purchases of fertilizer through non-public savings and loan programs and farmer organizations		No public credit program		Credit available through the CNCAS (agricul-tural bank) and some rural private micro-finance institutions

(continued)

Table 3.2 (continued)

Country	Uganda	Mozambique	Burkina Faso			Mali		Niger		Senegal
Scheme	Doho	Chokwe	Kou Valley	Sourou	Bagré	Niono	N'Débou-gou	Say2	Daibéri	Senegal River Valley
Other related policies								Price support provided through government purchase of a 75-kg sack of paddy at 10,000 CFA		Input voucher available (delivered by the National Extension Agency, SAED)
The nearest large city	Mbale	Maputo	Bobo Dioulasso	Ouagad-ougou	Ouagad-ougou	Bamako	Bamako	Niamey	Niamey	Saint Louis
Distance to the nearest large city (km)	30	220	30	250	2	330	346	56	105	101
Sampling unit	Rice farming house hold	Rice farming house-hold	Rice farming house-hold	Rice farming house-hold	Rice farming house-hold	Rice farming household	Rice farming household	Rice farming house-hold	Rice farming house-hold	Rice farming househ-old
Sampling method	Stratified random	Random	Random	Random	Random	Random	Random	Purposive	Purposive	Random
Sample size	288 (103)[c]	176	78	40	30	49	50	60	50	100

[a]Subsidy was introduced after the food crises in 2008

[b]The field agents of the office du Niger approve annual fertilizer budgets of farmer organizations in good standing with local financial institutions. It assists also farmers bidding and bulk purchase through farmer organizations

[c]Sample size declines to 103 when we include water depth

scheme.[2] The sampling unit is farming households. We use only rice farmers in the schemes for our analyses. The Doho's survey was conducted by one of the authors in 2007.[3] The International Rice Research Institute conducted the survey in Chokwe in 2007. The surveys of four Sahelian countries were conducted by the Africa Rice Center and its country partners in 2005–2006 or in 2006–2007. Since the surveys were conducted independently with their own research focus, the available variables are not completely comparable. Moreover, as of 2010, the data sets for the four Sahelian countries are not yet cleaned for the estimation of the determinants of yield and input use. Hence, the regression analyses in Sect. 3.4 rely on only the data from Uganda and Mozambique. However, all the surveys still share some key common variables for descriptive analyses in Sect. 3.3.

3.3 Descriptive Analyses

3.3.1 Features of Surveyed Irrigation Schemes

Table 3.3 compares rice production and production environments among the surveyed irrigation schemes. In order to investigate the importance of irrigation water for rice cultivation, we divide farmers in Doho and Chokwe into those who have good access to irrigation water and those who do not.[4] Since we do not have a corresponding variable for the Sahelian studies, we show the change in water access at each scheme since the last crop season. Although some show deterioration and others show improvement, we observed in our survey that water access in all surveyed schemes in the four Sahelian countries was generally good. Some might have claimed deterioration but this does not seem to mean a severe water shortage as they used to have sufficient water and the deterioration was marginal. Hence, in descriptive analyses, we treat all the Sahelian schemes as "good access." As another case of good water access, we show data of an irrigated rice-growing area in Asia, in our case, Laguna Province in the Philippines in 1976, 1982, and 1987 (Hayami and Kikuchi 2000). This enables us to assess the potential of SSA's irrigated rice in comparison with Asia when they were at a similar stage of the Green Revolution. The similarity of the stage is determined based on the type of modern varieties cultivated by farmers at each study site (either MV1, MV2, or MV3).[5]

[2] The sampling in the Doho rice scheme is stratified by irrigation blocks. The other studies use simple random sampling.

[3] See Nakano (2009) for more details.

[4] In Doho, farmers facing main canals are classified into the group of good access and, otherwise, the group of not-good access. In Chokwe, those who claimed "receiving enough water in 2007" were classified into the group of good access.

[5] The first-generation MVs (MV1s) were released from the mid-1960s to the mid-1970s and were more fertilizer-responsive than traditional varieties. Yet, they were susceptible to pests and diseases. The second-generation MVs (MV2s), which were designed to ensure stable yields by incorporating multiple pest and disease resistance, were released from the mid-1970s to the mid-1980s. The third-generation MVs (MV3), which incorporated better grain quality and stronger host-plant resistance, were released from the mid-1980s to the late 1990s.

They are reported in the second row of the table. A comparison reveals that Doho's current stage corresponds to the period between 1976 and 1982 in Laguna,[6] Chokwe does so in 1976, and the Sahelian schemes do so somewhere between 1982 and 1987 in Laguna.

One of the most important findings from Table 3.3 is the importance of irrigation water to the productivity of rice. As long as water access is good, the paddy yield at both Doho and Chokwe (3.2 and 2.2 t/ha, respectively) is not much lower than the yield at the corresponding stage in Laguna. It is worth noting also that, although Doho and Chokwe have cultivated rice for a long time (since the late 1970s in Doho and since the 1950s in Chokwe), they achieved this level of yield in the survey year. This is consistent with the finding in agronomy which claims that irrigation water maintains soil fertility and rice can be cultivated sustainably without suffering a yield decline. The importance of irrigation is also found in the Sahelian schemes. Water access is generally good in all the Sahelian schemes and they achieve very attractive yields. Among them, the Senegal River Valley shows amazingly high yield (5.3 t/ha). Note also that the irrigated area of this scheme is huge (60,000 ha). These facts imply that the availability of sufficient irrigation water is a key to achieving yield similar to or even higher than Asia and it is not impossible to achieve this on a large scale like the case of the Senegal River Valley.

Related to this, we would like to stress also that the varieties from Asia or the ones based on Asian parental varieties perform well in SSA under irrigated conditions. The most popular variety in Chokwe is ITA312, which was developed by the International Institute of Tropical Agriculture (IITA) and it has its parental variety in Asia. The next popular variety, C4, is a variety developed in the Philippines. In Niger, IR1529 from IRRI is used. The modern varieties cultivated in Senegal use *Oryza sativa* germplasm imported from Asia, are widely accepted, and achieve superior yield.[7]

Another important finding is that low fertilizer use is one of the constraints to increasing yield in Doho and Chokwe. Farmers in Doho and Chokwe apply much less chemical fertilizer than farmers in the Philippines or those in the Sahelian countries. One of the reasons for the low input use at both study sites may be the high price of chemical fertilizer. The real prices of nitrogen in terms of kilograms of paddy are 4.3 in Doho, 7.9 in Chokwe, but 3.7–3.5 in Laguna or 1.5 in Senegal. Moreover, the low fertilizer use and resulting low yield are associated with the

[6] The major rice varieties cultivated in DRS were modern varieties introduced by a Chinese aid agency in the 1970s and crossed with local varieties in the nearby experiment station. Although we cannot be decisive, we may be able to categorize them into MV1 or MV2.

[7] In 1994, three improved varieties, Sahel 108, Sahel 201, and Sahel 202, were released by Africa Rice and its national partners after screening more than 1,000 lines of *Oryza sativa* germplasm accessions imported from Asia (Africa Rice Center 2006). The Asian parents of the short-duration improved variety Sahel 108 are IR305, Babawee, and IR36, which came from IRRI. The medium-duration varieties Sahel 201 and 202 were developed using lines that originated, respectively, from Sri Lanka and IITA. The Sahel varieties rapidly gained producers' acceptance in Senegal and Mauritania as they replaced earlier introduced varieties. Currently, these three varieties occupy about 70% of irrigated rice area in the Senegal River Valley in both Senegal and Mauritania (Africa Rice Center 2006).

availability of irrigation water. In areas where the water supply is not reliable, farmers hesitate to use fertilizer as the marginal product of fertilizer depends on the availability of sufficient water (Estudillo and Otsuka 2006). In Chokwe, farmers apply 23 kg of nitrogen per hectare when they receive sufficient irrigation water, whereas they apply only 13 kg when they receive insufficient irrigation water. Therefore, the availability of irrigation water is likely to have not only a direct impact on rice yield but also an indirect impact through the increase in fertilizer application.

On the other hand, by African standards, fertilizer use in the Sahelian schemes is remarkably high. In fact, in Niger, the average fertilizer application rates are well above the recommended rate, which is 400 kg/ha. This can be partly attributed to a relatively low fertilizer price and institutional support in these countries. According to national statistics, the ratio of urea price to paddy price is 2.5 in Burkina Faso, 1.3 in Mali, and 1.6 in Niger, which are close to the fertilizer price ratios in Table 3.3.[8] Note that the countries with strong institutional supports for fertilizer (see Mali, Niger, and Senegal in Table 3.2) show very low fertilizer price ratios. This ratio is not disadvantageous at all compared with the ratio of about 3.5 in Laguna in 1976–1982 and 2 in major rice producers in Asia such as India and Pakistan in 2001 (Minten et al. 2006).

Related to this, we would like to stress that farming practices appear to be homogeneous in the Sahelian schemes. Regarding chemical fertilizer application, the standard deviation relative to the mean (i.e., the coefficient of variation) is much smaller than that in Doho and Chokwe. This may be one of the benefits of a well-managed irrigation scheme. This homogeneity implies that a serious constraint to rice production in the Sahel relative to Uganda and Mozambique might not exist, and thus many farmers use a large amount of chemical fertilizer to achieve yield comparable with yield in Asia.

Table 3.3 also shows labor use and the real daily wage in terms of kilograms of paddy.[9] A notable feature is found in Chokwe. The wage rate is higher and the labor input, excluding bird scaring and the proportion of hired labor (75 or 77 days), is lower than those in Asia, especially in the 1970s (105 days). In Asia, the introduction of labor-using modern varieties increased labor demand, and that increase was met by an abundant supply of landless wage laborers (David and Otsuka 1994; Hayami and Kikuchi 2000). Generally speaking, few landless households exist in

---

[8] Chemical fertilizer reported in Table 3.3 consists of urea and other kinds of complete fertilizer packages.

[9] Note that these are the real wages in terms of paddy. If we compare wages in US$ at official exchange rates of the survey years, they become 2.94 (Doho, Uganda), 1.73 (Chokwe, Mozambique), 1.90 (Burkina Faso), 1.90 (Mali), 1.89 (Niger). The wages in Sahelian countries become higher than Chokwe, Mozambique partly due their higher paddy prices (thus, resulting in lower real wages) and partly due to overvaluation of CFA franc. Doho's wage is still much higher than the other countries presumably due to the fact that wage labor is used mainly in labor intensive works such as transplanting and harvesting and the peak labor season is overlapped in short period among the farmers as irrigation rotation is not well coordinated.

Table 3.3 Paddy yields, input use per hectare, and input prices at the study sites and in Laguna in the Philippines

Country	Uganda		Mozambique		Burkina Faso			Mali		Niger		Senegal	Philippines		
Scheme	Doho		Chokwe		Kou Valley	Sourou	Bagré	Niono	N'Débougou	Say2	Daibéri	Senegal River Valley	Laguna		
Survey year	2007		2007		2005–2006			2005–2006		2005-06		2006-07	1976	1982	1987
Water access	Facing the main channel	Not facing the main channel	Receive enough water	Do not receive enough water	Deteriorated access to water	Mixed results	Improved access to water	Improved access to water	Improved access to water	Mixed results	Deteriorated access to water		n.a.		
Rice variety	MV1 and MV2[a]		MV1 (ITA312, C4)		MV2 (FKR14)	MV3 (FKR28)		MV3 (Gambiaka Kogoni 91-1)		MV1(IRI529) MV3(WITA 8)		MV3(Sahel 108, 201, 202)	MV1	MV2	MV3
Paddy yield (t/ha)	3.2 (1.3)	2.7 (1.7)	2.2 (1.2)	1.3 (1.0)	3.2 (0.7)	3.6 (1.7)	3.2 (1.2)	3.2 (1.6)	3.1 (1.3)	4.3 (1.5)	3.6 (1.2)	5.3	2.8	3.6	4.3
Input use															
Nitrogen (kg/ha)	1.6 (3.5)	2.0 (5.6)	23 (27.6)	13 (24.5)								114	58	80	94
Chemical fertilizer (kg/ha)					309 (209.6)	318 (72.7)	299 (88.7)	248 (109.0)	241 (107.3)	446 (201.0)	460 (183.7)	313			
Labor (person day/ha)[c]	179 (76.5)	173 (103)	75 (72.3)	77 (78.0)									105	80	69

Proportion of hired labor (%)	55	52	36	35									71	70	83
Proportion of HHs using hired labor (%)					84.6	90.0	86.7	100	46	93.3	82.0				
Animal use (%)	0	5	47	52	97.4	56.5	43.3	97.8	91.8	91.4	90.0		78	73	86
Tractor use (%)	0	0	51	76	0	43.5	6.7	2.2	0	0	0		100	98	98
Input price															
Price of Nitrogen in terms of kg of paddy	4.3		7.9		2.3	2.4	3.5	1.2	1.6	1.7	1.8	1.5	3.7	3.5	2.1
Labor wage in terms of kg of paddy	10		12		7.8	8.4	10.9	4.8	5.2	7.9	7.9		9.8	15	13.2
Household characteristics															
Total farm size (ha)	2.11	2.27	1.5	2.5	1.0	1.3	1.9	5.4	1.4	3.9	4.0	1.56			
	(1.7)	(1.8)	(2.2)	(4.4)	(0.2)	(1.1)	(0.2)	(3.5)	(1.2)	(2.4)	(2.0)				
Age of hh head	46.2	47.1	51.6	51.9	54.3	40.2	43.4	52.4	52.2	47.8	49.1				
	(11.6)	(14.1)	(14.6)	(12.8)	(14.3)	(8.3)	(7.7)	(13.7)	(12.1)	(11.0)	(14.6)				
Average schooling year of hh member	6.0	5.2	4.27	4.59									5.0		6.5
	(2.7)	(3.0)	(2.02)	(1.8)											

(continued)

Table 3.3 (continued)

Country	Uganda		Mozambique		Burkina Faso			Mali		Niger		Senegal	Philippines		
Scheme	Doho		Chokwe		Kou Valley	Sourou	Bagré	Niono	N'Débougou	Say2	Daibéri	Senegal River Valley	Laguna		
Survey year	2007		2007		2005–2006			2005–2006		2005-06		2006-07	1976	1982	1987
Schooling year of hh head					4.0	3.3	3.0	4.9	4.0	2.1	1.6				
					(4.5)	(3.4)	(3.0)	(3.5)	(4.1)	(3.2)	(3.6)				
HH size			8.23	7.04	18.8	11.6	11.9	21.3	19.7	10.3	10.6		5.9		5.3
			(3.92)	(3.45)	(7.0)	(6.8)	(4.7)	(14.3)	(15.4)	(4.6)	(6.4)				
Sample size	111	177	151	25	78	40	30	49	50	60	50				

Standard deviations in parentheses
The major rice varieties cultivated in Doho were modern varieties introduced by a Chinese aid agency in the 1970s and crossed with local varieties at the nearby experiment station. Although we cannot be decisive, we may be able to categorize them into MV1 or MV2
Village-level questions about relative access to irrigation water compared to the previous year (mixed results refer to a situation where an equal number of villages point to improved and deteriorated access to irrigation water)
Excluding labor for bird scaring
Local price of paddy per kg in survey year
Uganda, Doho: 505 Ush
Mozambique, Chokwe: 3.9 MT
Burkina Faso, Kou Valley: 128 Fcfa, Sourou: 119 Fcfa, Bagré: 92 Fcfa
Mail, Ninon: 208 Fcfa, N'Débougou: 192 Fcfa
Niger, Say: 126 Fcfa, Daibéri: 126 Fcfa

Africa. Although exchange labor between farming households is a common practice in Chokwe, coordination of the timing of the exchange is difficult during peak labor periods such as transplanting and harvesting/threshing periods because of the synchronization of such peak periods among farmers. For these reasons, the wage rate becomes high, especially during the peak labor periods, and this may hinder farmers in Chokwe from applying a sufficient amount of labor for cultivating MVs.

Under such circumstances, household size relative to farm size could affect production performance in the Sahelian countries. For example, among them, Burkina Faso's and Niger's relative wage rate is higher than that of Mali. This may stem from Mali's larger household size and smaller farm size than the others. We expect that the labor constraint may be more severe in Burkina Faso and Niger than in Mali. Meanwhile, other household characteristics such as the age of the household head and average years of schooling of adult household members are similar between schemes. Therefore, these factors do not seem to be important in explaining the difference in performance across countries.

3.3.2 Competitiveness of Rice Production in Surveyed Irrigation Schemes

We now turn to the costs and returns of the study schemes presented in Table 3.4 in order to show the profitability and competitiveness of irrigated rice of SSA against *imported* rice from Asia. Similar to Table 3.3, Doho and Chokwe are divided into two groups depending on water access. For the comparison among schemes, all figures are converted to US$ using the official exchange rate in the survey year. Since the necessary data for the imputation of labor and owned capital costs was not made for the Sahelian irrigation schemes, we show only net return or income, while we show profit as well for Doho and Chokwe. Senegal is not included as the data are not ready for this analysis yet.

In Doho and Chokwe, where water access is not good, the gross value of output is low due to low yield, whereas the total cost does not change regardless of water access conditions. Therefore, income and profit become lower when water access is not good. Particularly, profit in Chokwe in the case of unfavorable water access becomes negative, indicating that these farmers cannot be competitive with imported rice in local markets. To examine this point more clearly, we show the production cost per ton of milled rice in comparison with the international f.o.b. price in the survey year in the lower part of the same table. Note that the unit cost would be higher if we included the cost of irrigation (hence, generous for Doho and Chokwe to judge the competitiveness) and that the price for imported rice in local markets would be higher due to transportation costs (hence, generous for Asia). Nevertheless, those figures give some idea of the competitiveness of the irrigated rice of Doho and Chokwe.

According to the figures, although some divergence exists in the international price (US$290–335), generally speaking, domestic irrigated rice seems to be able to

Table 3.4 Costs and returns in study schemes (in US$)

	Uganda		Mozambique		Burkina Faso			Mali		Niger	
	Doho		Chokwe		Kou Valley	Sourou	Bagré	Niono	N'Débougou	Say2	Daibéri
	2007		2007		2005–2006			2005–2006		2005–2006	
	Facing the main channel	Not facing the main channel	Receive enough water	Do not receive enough water	Deteriorated access to water	Mixed results**	Improved access to water	Improved access to water	Improved access to water	Mixed results**	Deteriorated access to water
Costs and returns (per ha)											
Gross output value (paddy) (A)	952	786	307	203	790	841	552	1,262	1,298	1,030	858
Seed	33	29	9	12	16	15	42	28	22	0	0
Fertilizer	4	5	23	14	150	228	216	140	152	227	149
Pesticide	6	4	3	2	2	29	11	7	2	0	0
Hired labor	271	269	80	71	32	153	63	51	90	293	292
Family labor, imputed	261	274	85	78							
Capital (tractor, thresher, animal) paid out	0	2	49	74	26	48	55	36	49	72	55
Capital (tractor, thresher, animal) imputed	0	0	23	24							
Total paid-out cost (B)	281	280	163	174	226	473	386	261	315	593	497
Total cost (C)	574	583	272	276							
Net return (A)-(B)	671	506	143	29	564	368	166	1,000	983	438	361
Profit (A)-(C)	377	203	35	−73							
Unit production cost of milled rice (US$/t)	299	358	302	407							

	Uganda		Mozambique		Burkina Faso			Mali		Niger	
	Doho		Chokwe		Kou Valley	Sourou	Bagré	Niono	N'Débougou	Say2	Daibéri
	2007		2007		2005–2006			2005–2006		2005–2006	
	Facing the main channel	Not facing the main channel	Receive enough water	Do not receive enough water	Deteriorated access to water	Mixed results**	Improved access to water	Improved access to water	Improved access to water	Mixed results**	Deteriorated access to water
Int'l rice price (US$/t f.o.b.) in survey year											
Thai 2nd grade			335					291			
Thai A1 super			275					219			
Pakistan 25%			290					235			
Vietnam 5%			313					255			
Sample size	111	177	144	32	78	40	30	49	50	60	50

Chokwe: The milling cost of 1,765 MT per ton of paddy and 65% recovery rate are assumed
Exchange rates are: Chokwe: $1 = MT27 in 2007 Doho: Exchange rate: $1 = Ush 1,716 in 2007 West Africa: $1 = Fcfa 526 in 2005–2006 average

offer a lower price in local markets if water access is good and thus productive (US$299 in Doho and US$302 in Chokwe). This implies that, under proper management, large-scale irrigation can provide good returns as emphasized by Inocencio et al. (2007). Although we cannot perform a similar exercise for Sahelian countries, noting relatively high net returns in all schemes except Bagré, their competitiveness should also be high.[10]

In summary, descriptive analyses indicate that irrigated rice of large-scale irrigation schemes has potential to achieve high yield and thus to be competitive if farmers have good access to irrigation water and use adequate crop management practices. Note that such efficient rice farming is achieved by small farmers, as in the case of Asia (see Table 3.3 for farm size). In the following section, using data from Doho and Chokwe, we conduct more detailed statistical analyses, to explore what kind of constraints hinder adequate management for high yield and how they are related to water access.

3.4 Regression Analyses

3.4.1 Methodology

In order to examine the conditions to achieve high yield at our study sites, we estimate the yield function and input use functions. In a structural form, yield per ha can be expressed as a function of inputs per ha, given technology and the management ability of farmers:

$$y_i = \beta_0 + x_i\beta_1 + H_i\beta_2 + u_i$$

where y is yield per ha, X is a vector of inputs, and H is a vector of household and farming characteristics. Our econometric concern, however, is that inputs are endogenous variables and OLS is not an appropriate approach. To circumvent this problem, we apply the instrumental variable (IV) method, regressing input use on the exogenous variables that farmers cannot change at least in the short run in accordance with the current season's production decision:

$$x_i = \gamma_0 + H_i\gamma_1 + Z_i\gamma_2 + v_i,$$

where x is the use of a particular input in X and Z is a vector of the exogenous variables that serve as identifying instrumental variables for X in the yield function. In this approach, the first-stage regressions can be regarded as the estimation of the

[10] Bagre's low income (US$166) stems from the much lower paddy price in local markets (92 Fcfa) than the other schemes (128 and 119 Fcfa). Meanwhile, excessively high income in Mali (US$1,000 and $983) is due to the high paddy price (208 and 192 Fcfa).

reduced-form input use functions. We use the results of input use functions to identify the constraints to input use. If the factor markets function perfectly, the level of inputs should be determined solely by input prices relative to the output price, technology, and farmers' farming ability (as the determinants of marginal returns), but not by factor endowments and wealth. Thus, if we find that any endowments and wealth have significant coefficients, we can conjecture that there are imperfections in the factor market. Combining such results with the results of the yield function, we assess how such constraints affect yield.

However, in Doho, we could not find appropriate identifying instrumental variables to explain the variation in possible endogenous variables. Therefore, we turned to the estimation of the reduced-form yield function. Hence, our yield function and input use function for Doho are expressed as

$$y_i = \delta_0 + H_i\delta_1 + Z_i\delta_2 + w_i,$$

$$x_i = \gamma_0 + H_i\gamma_1 + Z_i\gamma_2 + v_i,$$

Although we cannot estimate the direct and indirect impact of irrigation water on paddy yield separately in this approach, we can still estimate the aggregate impact of irrigation water on yield, which is the major interest of our analysis.

3.4.2 *Variable Construction*

For Chokwe, the input use vector (*X*) consists of (1) chemical fertilizer, (2) labor, (3) proportion of hired labor, (4) machinery use, and (5) the method of crop establishment. For Doho, we use only the first three inputs, as the use of machinery is uncommon and the common method of crop establishment is transplanting.

For both sites, the vector *H* consists of (1) plot size, (2) availability of irrigation water, (3) human capital, and (4) season dummy (if the survey covers multiple seasons). Since the size of the cultivated area is primarily determined by the availability of water at the initial stage of farming, we can practically treat it as an exogenous variable. Irrigation water, which is managed by the state, and farm location are assumed to be exogenously given to the farmers. The average schooling years and age of the household head are included to capture the ability of farm management and experience, which would affect yield at a given level of inputs. Since these are pre-determined, we treat them as exogenous variables.

As identifying instrumental variables, we include (1) land endowment, (2) other asset endowment, (3) membership in a cooperative, (4) access to market and extension service, and (5) gender of the household head. The list of variables and detailed definitions for each survey are presented in Table 3.5. The factor prices and output price are not included because our data sets were collected in one area in a particular year where prices are practically the same for all the households.

Table 3.5 Definition of the variables at each study site

Study site	Chokwe	Doho
Dependent variable (y)	Paddy yield (t/ha)	Paddy yield (t/ha)
Input use (X)		
Fertilizer	Total amount of N+P+K (kg/ha)	Total cost for fertilizer (100 thousand Ush/ha)
Labor	Total labor input (days/ha)	Total labor input (days/ha)
Hired labor	Proportion of hired labor	Proportion of hired labor (%)
Machinery	Tractor (=1 if use tractor)	
	Threshing machine (=1 if use threshing machine)	
Seeding	Crop establish method (=1 if transplanted; =0 if direct seeding)	
Environment and HH characteristics (H)		
Availability of irrigation water	Insufficient irrigation (=1 if hh receives insufficient water)	Directly measured water depth in the plot (cm)
	Downstream parcel (=1 if the plot is located downstream)	Distance from the main channel to the intake of the strip (km)
		Distance from the intake of the strip to each plot (km)
Plot size	Size of the cultivated area in the sample plot (ha)	Size of the cultivated area in the sample plot (ha)
Human capital endowment and farming ability	HH size	Number of adult household members
	Female-headed household (=1 if female-headed)	Female-headed household (=1 if female-headed)
	Average schooling years of adult household members	Average schooling years of adult household members
	Average schooling years of adult household members squared	Average schooling years of adult household members squared
	Age of head	Age of head
	Age of head squared	Age of head squared
Season dummy		2nd season 2007 (=1 if 2nd season 2007)
Instrumental variables (Z)		
Land endowment	Unirrigated owned area (ha)	Unirrigated cultivated area (ha)
	Unirrigated owned area (ha) squared	Unirrigated cultivated area (ha) squared
	Irrigated owned area (ha)	Other cultivated area in DRS (ha)
	Irrigated owned area (ha) squared	Other cultivated area in DRS (ha)
Other assets and access to cash	Value of non-agricultural assets	
	Proportion of salary earners	Proportion of salary earners
	Proportion of salary earners squared	Proportion of salary earners squared
Membership of co-op	Member of water user group (=1 if member)	
	Member of agricultural association (=1 if member)	
	Member of cooperative (=1 if member)	

3.4.3 Input Use Function

We begin with the interpretation of results in Chokwe in Table 3.6. Besides the OLS results, when the dependent variable is either censored or binary, we show the Tobit or Probit results for checking robustness of the estimation results. The results of the NPK function in Chokwe indicate that farmers do not apply chemical fertilizer unless they receive sufficient irrigation water, due to the strong complementary relationship between them. The positive and significant coefficient of the value of the non-agricultural asset in Chokwe seems to imply that farmers with good credit access can purchase sufficient amounts of chemical fertilizer. The availability of cash on hand, which is measured by the proportion of salary earners, significantly increases fertilizer application until the proportion becomes 20% in Chokwe. Based on these results, we argue that improvement in access to irrigation water and in credit/cash would increase fertilizer application.

Labor input is related positively to household size and negatively to the size of the cultivated area in Chokwe. These determinants would not be significant if farmers were able to hire labor as much as they wished. Although the proportion of hired labor increases with the size of the cultivated area (positive and significant coefficient), it would not reach the level that farmers wished to apply. Another reason for the labor constraint could be the credit constraint for payment to hired labor as implied by the positive and significant coefficient of the non-agricultural asset value in hired labor regression, because a piece-rate cash payment is the most common labor contract in Chokwe.[11]

In Table 3.6, we also show the regression results for the use of a tractor, thresher, and transplanting in Chokwe. Similar to the other results, the coefficient of the non-agricultural assets is positive and significant for the use of either tractors or threshers, suggesting the importance of credit access for renting these machines. The probability of tractor use increases with the average schooling years partly because tractors (all 4-wheels in Chokwe) must be managed and operated skillfully and partly because the opportunity cost of educated labor is high, which induces substitution of tractors for labor. A puzzling result is the U-shape relationship between the use of threshers and the proportion of salary earners, which is opposite to the case of NPK. Farmers are less likely to practice the transplanting method as the size of the cultivated area becomes larger because transplanting is a more labor-intensive method of crop establishment than direct seeding.

Table 3.7 shows the regression results of input use functions in Doho. The negative and significant coefficient of the distance from the main channel to the intake of the strip for the cost of current inputs indicates that farmers apply more current inputs when they have better access to irrigation water, which is consistent with the results of Chokwe. Therefore, irrigation water has not only a direct impact on rice

[11] During field interviews, we encountered several farmers who claimed that they could not hire labor since they did not have cash on hand.

Table 3.6 Determinants of input use in Chokwe

	NPK (kg/ha)		Labor (days/ha)	Prop. hired lab (%)	Use of tractor (dummy)		Use of thresher (dummy)		Transplanting (dummy)	
	(1)	(2)	(3)	(4)	(5)	(6)	(7)	(8)	(9)	(10)
	OLS[a]	Tobit	OLS[a]	OLS[a]	OLS[a]	Probit	OLS[a]	Probit	OLS[a]	Probit
Size of cultivated area in the sample plot (ha)	0.748	0.488	−10.301	0.049	0.039	0.264	0.043	0.522	−0.050	−0.303
	(0.60)	(0.26)	(2.80)***	(2.74)***	(1.58)	(1.97)**	(3.76)***	(2.07)**	(2.46)**	(2.76)***
Insufficient irrigation	−14.140	−29.610	21.356	0.005	0.249	0.903	−0.008	−0.274	−0.006	−0.105
	(2.52)**	(2.93)***	(1.28)	(0.06)	(2.22)**	(2.39)**	(0.15)	(0.18)	(0.07)	(0.25)
Downstream parcels	−1.092	−3.265	5.694	0.028	−0.116	−0.396	−0.093		0.047	0.412
	(0.18)	(0.33)	(0.32)	(0.33)	(0.97)	(1.03)	(1.71)*		(0.49)	(0.84)
HH size	−1.095	−1.542	5.448	−0.010	−0.002	−0.018	−0.007	0.054	0.008	0.054
	(2.06)**	(1.82)*	(3.46)***	(1.33)	(0.20)	(0.55)	(1.45)	(0.50)	(0.97)	(1.41)
Ave sch years	1.360	3.946	−0.531	0.030	0.166	0.548	−0.010	−1.033	−0.056	−0.229
	(0.44)	(0.78)	(0.06)	(0.67)	(2.69)***	(2.75)***	(0.35)	(1.88)*	(1.12)	(0.86)
Ave sch years sq	−0.065	−0.288	0.202	0.001	−0.014	−0.047	0.000	0.067	0.006	0.027
	(0.22)	(0.61)	(0.23)	(0.29)	(2.42)**	(2.49)**	(0.09)	(1.57)	(1.32)	(0.98)
Age of head	0.330	0.694	0.702	0.000	0.008	0.026	0.003	0.162	−0.003	−0.053
	(0.82)	(1.01)	(0.59)	(0.00)	(0.99)	(0.97)	(0.76)	(0.72)	(0.48)	(0.85)
Age of head sq	−0.002	−0.005	−0.015	0.000	−0.000	−0.000	−0.000	−0.002	0.000	0.000
	(0.39)	(0.71)	(1.15)	(0.02)	(0.67)	(0.63)	(0.70)	(0.95)	(0.16)	(0.70)
Female-headed HH	−4.116	−8.785	−1.446	−0.151	−0.040	−0.161	−0.033	−0.693	0.156	0.780
	(1.01)	(1.32)	(0.12)	(2.58)**	(0.49)	(0.63)	(0.87)	(0.86)	(2.37)**	(2.29)**
Unirrig owned area	7.975	26.641	45.845	−0.162	0.010	−0.227	0.020	1.229	0.211	1.289
	(0.98)	(1.69)*	(1.89)*	(1.37)	(0.06)	(0.24)	(0.26)	(0.43)	(1.59)	(2.07)**
Unirrig owned area sq	−0.936	−3.543	−3.497	0.037	0.022	0.132	0.000	0.210	−0.026	−0.166
	(0.80)	(1.30)	(1.01)	(2.19)**	(0.92)	(0.69)	(0.02)	(0.67)	(1.38)	(1.93)

Irrig owned area	2.798	4.160	−1.035	0.010	−0.002	−0.025	0.015	0.353	−0.041	−0.138
	(3.03)***	(2.99)***	(0.38)	(0.71)	(0.13)	(0.41)	(1.74)*	(1.93)	(2.71)***	(2.06)**
Irrig owned area sq	−0.037	−0.054	0.031	−0.000	−0.000	−0.000	−0.000	−0.007	0.001	0.002
	(2.92)***	(2.85)***	(0.84)	(2.29)**	(0.46)	(0.06)	(2.94)***	(2.58)***	(3.13)***	(0.97)
Value of non-ag assets	0.243	0.302	0.120	0.005	0.004	0.014	0.002	0.032	−0.001	−0.003
	(3.09)***	(2.51)**	(0.51)	(3.97)***	(2.31)**	(2.51)**	(3.17)***	(2.07)**	(0.52)	(0.62)
Prop of salary earners	55.349	109.958	−22.513	−0.908	−0.276	−1.462	−1.119	−35.771	0.663	3.266
	(1.63)*	(1.83)*	(0.22)	(1.85)*	(0.41)	(0.57)	(3.59)***	(2.00)**	(1.21)	(1.14)
Prop of salary earners sq	−145.635	−288.810	−65.124	2.186	0.531	4.187	2.577	61.922	−1.753	−7.843
	(1.81)*	(1.90)*	(0.27)	(1.88)*	(0.33)	(0.58)	(3.48)***	(1.65)	(1.35)	(1.06)
Member of WUG	0.836	−1.499	10.837	0.009	−0.133	−0.386	−0.031	−2.297	0.064	0.266
	(0.22)	(0.25)	(0.97)	(0.17)	(1.77)*	(1.67)*	(0.88)	(2.41)**	(1.04)	(0.94)
Member of ag assoc	5.358	9.253	−12.212	−0.027	0.112	0.342	0.029	2.116	−0.002	−0.062
	(1.34)	(1.47)	(1.03)	(0.47)	(1.40)	(1.42)	(0.79)	(2.16)**	(0.04)	(0.22)
Member of co-op[b]	−8.749	−10.675	−68.939	0.087	0.065		0.578		0.693	
	(0.37)	(0.31)	(0.97)	(0.25)	(0.14)		(2.63)***		(1.79)	
Constant	−28.088	−106.541	−31.980	0.262	−0.597	−3.304	0.023	−18.219	0.677	0.624
	(1.23)	(2.50)**	(0.47)	(0.79)	(1.31)	(1.77)	(0.11)	(0.67)	(1.83)*	(0.28)
First-stage F test for IV	2.99		1.78	4.35	2.02		3.60		1.81	
	[0.00]***		[0.03]***	[0.00]***	[0.01]**		[0.00]***		[0.03]**	
Observations	176	176	176	176	176	176	176	176	176	176

*Significant at 10%; **Significant at 5%; ***Significant at 1%

[a]Results of the first-stage estimation of the instrumental variable analysis for the estimation of paddy yield function in Table 3.8

[b]Probit analysis cannot include an explanatory variable that predicts the dependent variable perfectly. For this reason, this variable is dropped from the probit estimation

Table 3.7 Determinants of input use in DRS

	Cost for fertilizer (100,000 Ush/ha)	Labor (days/ha)		Prop. hired lab (%)
	(1)	(2)	(3)	(4)
	OLS	Tobit	OLS	OLS
Plot size (ha)	–0.334	–0.34	–84.357	0.094
	(2.63)***	(2.67)***	(2.56)**	(0.94)
Distance from main channel to the intake of the strip (km)	–0.048	–0.05	–7.652	–0.011
	(2.46)**	(2.55)**	(1.51)	(0.72)
Distance from the intake of the strip to each plot (km)	0.073	0.075	–41.326	0.089
	(0.68)	(0.70)	(1.49)	(1.06)
Number of adult household members	–0.02	–0.02	7.849	–0.017
	(1.82)*	(1.85)*	(2.80)***	(1.93)*
Ave sch years	0.016	0.016	–17.087	0.029
	(0.65)	(0.65)	(2.74)***	(1.54)
Ave sch years sq	–0.001	–0.001	1.428	–0.001
	(0.63)	(0.60)	(2.88)***	(0.47)
Age of head	–0.003	–0.004	–4.312	–0.027
	(0.25)	(0.34)	(1.55)	(3.21)***
Age of head sq	0.000	0.000	0.036	0.000
	(0.03)	(0.12)	(1.30)	(3.11)***
Female-headed HH	0.061	0.066	61.067	0.01
	(0.70)	(0.76)	(2.70)***	(0.15)
Unirrig owned area	0.172	0.183	26.412	0.03
	(3.92)***	(4.12)***	(2.32)**	(0.82)
Unirrig owned area sq	–0.023	–0.025	–3.596	–0.001
	(3.56)***	(3.80)***	(2.16)**	(0.16)
Irrig owned area	–0.031	–0.04	–36.419	0.229
	(0.41)	(0.52)	(1.85)*	(3.82)***
Irrig owned area sq	0.023	0.025	5.385	–0.045
	(1.19)	(1.30)	(1.06)	(2.92)***
Prop of salary earners	–1.122	–1.174	–759.96	0.688
	(1.37)	(1.43)	(3.58)***	(1.06)
Prop of salary earners sq	2.16	2.298	1,754.21	–1.342
	(1.16)	(1.23)	(3.65)***	(0.92)
2nd season	0.034	0.029	–30.864	–0.023
	(0.88)	(0.73)	(3.03)***	(0.73)
Constant	0.776	0.793	351.69	0.96
	(3.08)***	(3.14)***	(5.38)***	(4.86)***
Observations	288	288	288	279
R-squared	0.13		0.19	0.19

Absolute value of t statistics in parentheses
*Significant at 10%; **Significant at 5%; ***Significant at 1%

yield but also would have an indirect positive impact through an increase in current input application. The size of the unirrigated cultivated area in Doho has an inverted U-shape relationship to fertilizer application, with the peak at 3 ha. Considering that only 15% of sample households cultivate more than 3 ha, it is almost a positive relationship, which may imply that farmers with larger upland cultivated area may have better access to credit or cash and hence can purchase more fertilizer.

Similar to Chokwe, the results for labor and hired labor imply that there is a labor constraint in Doho because of an inactive labor market. The proportion of salary earners has a negative impact on total labor input, with the peak of the U-shape relationship at a much higher value (21%) than the average (2%), which is consistent with our intuition because the more salary earners a household has, the less dependent the household is on rice farming. Puzzling results are the U-shape relationship between total labor use and the average years of schooling, and the inverted U-shape relationship between total labor use and the size of unirrigated cultivable area, for which we cannot find any good explanations.

3.4.4 Yield Function

Table 3.8 summarizes the results of the yield function in Chokwe. The OLS results of the linear approximation model and the corresponding IV results (models (1), (2)) indicate that management ability and experience do not have much impact on yield, particularly in the IV model, presumably because they do not have direct impacts but indirect impacts on yield through their effect on the change in endogenous input variables. Hence, in models (3) and (4), we remove them from our yield function and use them only in the first-stage regressions.

The test statistics for the IV approach of our final model presented in the lower part of Table 3.5 indicate that inputs may suffer from endogenity (the chi-square test for endogeneity at the 15% significance level) but they are significantly predicted by the instrumental variables (first-stage F test) that can be considered as exogenous to the model (chi-square test for overidentification), providing confidence in the validity of the model specification (Wooldridge 2002).

A key finding is that chemical fertilizer, labor, and irrigation water are the crucial factors that affect yield. Fertilizer application has a positive impact on yield. Yield is low when insufficient irrigation water is received. Labor input is also a crucial input. On the other hand, mechanization does not have much impact on yield increases. This feature is also observed in Asia as machine power can be replaced by animal power or human labor to some extent (David and Otsuka 1994).[12] The negative and significant coefficient of the size of cultivated area indicates that higher

[12] Although tractor power can be replaced by human labor, not all kinds of human labor activities can be replaced by tractor power (for example, crop establishment and harvesting). Thus, a labor shortage can still be a constraint.

Table 3.8 Determinants of paddy yield in Chokwe (structural form estimation)

	(1) OLS	(2) IV	(3) OLS	(4) IV
NPK[a]	0.005	0.020	0.006	0.022
	(1.32)	(1.90)*	(1.56)	(2.02)**
Labor[a]	0.003	0.006	0.003	0.008
	(2.19)**	(1.74)*	(2.33)**	(2.06)**
Prop of hired labor[a]	0.693	−0.595	0.712	−0.386
	(2.72)***	(0.67)	(2.89)***	(0.49)
Use of tractor[a]	−0.011	0.063	−0.000	0.207
	(0.06)	(0.10)	(0.00)	(0.29)
Use of thresher[a]	1.085	1.829	0.983	1.483
	(2.79)***	(1.59)	(2.57)**	(1.36)
Transplanting[a]	0.214	−0.480	0.217	−0.369
	(0.90)	(0.57)	(0.92)	(0.47)
Size of cultivated area in the sample plot	−0.083	−0.147	−0.078	−0.126
	(2.02)**	(2.30)**	(1.93)*	(2.06)**
Insufficient irrigation (relative freq.)	−0.810	−0.686	−0.787	−0.694
	(2.94)***	(1.93)*	(2.87)***	(1.81)*
Downstream parcels	−0.551	−0.354	−0.627	−0.459
	(2.01)**	(1.02)	(2.34)**	(1.34)
Av. schooling years of working mem	0.194	0.187		
	(1.71)*	(1.30)		
Av. schooling years of working mem sq	−0.018	−0.015		
	(1.59)	(1.10)		
Age of HH head	−0.002	−0.008		
	(0.09)	(0.38)		
Age of HH head sq	0.000	0.000		
	(0.10)	(0.22)		
Constant	1.220	1.817	1.538	1.536
	(2.14)*	(1.78)*	(5.68)***	(1.99)**
Edogeneity test (chi-sq)[b]	6.46 (0.37)		9.36 (0.15)	
First-stage F test		All significant		All significant
Overidentification test (chi-sq)[c]		4.26		6.88
		(0.89)		(0.80)
Observations	176	176	176	176

*Significant at 10%; **Significant at 5%; ***Significant at 1%

[a] Instrumented variable. Identifying instruments are the variables used in Table 3.6. Table 3.6 shows the first-stage regression results for model (5) of this table

[b] Durbin-Wu-Hausman endogeneity test

[c] Sargan's overidentification test

yield is achieved under a smaller scale operation, which is also consistent with the observation in Asia that small farmers contributed to the rice Green Revolution.

In Table 3.9, we show the estimation results of reduced-form yield functions in Doho. In model (1), we use the distance from the main channel to the intake of the strip and the distance from the intake of the strip to each plot as proxies for the availability of irrigation water. In model (2), we use water depth (cm) at the critically important stage of flowering, and treat it as an exogenous variable. The distance from the main channel to the intake of the strip has a negative and significant coefficient on yield. Water depth has a significant and positive impact on yield. Both results indicate the importance of irrigation water for rice productivity. According to model (2), a 1-cm increase in irrigation water raises paddy yield by 0.13 t/ha.

3.5 Concluding Remarks

This chapter investigated the potential of and constraints to the rice Green Revolution in SSA's large-scale irrigation schemes, using data from Uganda, Mozambique, Burkina Faso, Mali, Niger, and Senegal. The results of regression analyses for Uganda and Mozambique reveal the crucial importance of irrigation water for rice productivity. When irrigation water is available, both study sites achieve high yield. Furthermore, the availability of irrigation water may have both a direct impact on rice yield and an indirect impact through an increase in fertilizer application. Since the conditions for water access are generally good in the four Sahelian countries, farmers achieve attractive yield with sufficient application of chemical fertilizer. In many schemes, Asian varieties or varieties with an Asian origin perform well under irrigated conditions. This implies that proper management of irrigation schemes for timely and sufficient water distribution, together with variety transfer from Asia, is one of the key strategies to increase rice production in large-scale irrigation schemes.

The sufficient use of chemical fertilizer in the four Sahelian countries seems to be attributed not only to their good water access but also to the institutional support for fertilizer purchase. Unless the cost of support is unduly high, this kind of support may be effective in Uganda and Mozambique, where no such support exists yet. In addition, our regression results for Uganda and Mozambique imply that an improvement in credit access would help cash-constrained farmers purchase chemical fertilizer.

We also find that labor shortages are another critical constraint to the achievement of high productivity. The results in Uganda and Mozambique indicate that improvement in credit access could encourage hiring wage labor. The development of varieties with shorter maturity could be another solution as they would spread out the peak season's labor demand. Moreover, it is worth considering a strategy to substitute machines for labor in areas where the relative wage rate is high. A challenge is the strategy to promote this relatively expensive equipment. Further investigation is needed to see whether collective ownership (maybe through a co-op) can be a solution. In addition, it is clear that, unless local repair shops are accessible to local farmers, dissemination would be limited.

Table 3.9 Determinants of paddy yield in Doho (reduced-form estimation)

	(1)	(2)
	OLS	OLS
Water depth (cm)		0.135
		(2.15)**
Distance from main channel to the intake of the strip (km)	−0.271	
	(3.04)***	
Distance from the intake of the strip to each plot (km)	0.291	
	(0.60)	
Plot size (ha)	−0.836	−1.596
	(1.44)	(1.51)
HH size	0.012	−0.031
	(0.23)	(0.30)
Unirrig owned area	0.468	0.407
	(2.34)**	(1.08)
Unirrig owned area sq	−0.066	−0.057
	(2.26)**	(1.05)
Irrig owned area	0.276	0.412
	(0.79)	(0.38)
Irrig owned area sq	−0.049	−0.13
	(0.55)	(0.21)
Prop of salary earners	−3.016	−8.305
	(0.81)	(1.17)
Prop of salary earners sq	9.566	20.75
	(1.13)	(1.16)
Ave sch years	0.018	0.122
	(0.16)	(0.53)
Ave sch years sq	0	−0.007
	(0.04)	(0.32)
Female-headed HH	0.42	0.263
	(1.06)	(0.34)
Age of head	−0.066	−0.063
	(1.36)	(0.51)
Age of head sq	0.001	0.000
	(1.17)	(0.38)
Season (2007 2nd)	−0.701	
	(3.92)***	
Block dummy	No	No
Constant	4.768	3.531
	(4.15)***	(1.23)
Observations	288	103
R-squared	0.14	0.14

Absolute value of t statistics in parentheses
**Significant at 5%; ***Significant at 1%

Although small-scale irrigation development seems to be a current trend in SSA among aid organizations, our analyses show that large-scale irrigation schemes also have high potential under proper management and are equally important. When the Comprehensive Africa Agriculture Development Programme (CAADP) called for investment in improved water control for 15.9 million ha by 2030, the proposed share of the large-scale irrigation area (including new and rehabilitation investment) still consists of about 17%, while the proposed share of small-scale irrigation area is 14%, that of wetlands and inland valley bottoms is 23%, and that of water harvesting and rainfed areas is 45% (World Bank 2007).[13] Thus, large-scale irrigation schemes are as important as other means such as small-scale schemes and rainfed area development. The lessons drawn from our study sites are important for the development of strategies for SSA's rice Green Revolution.

References

Africa Rice Center (2006) Medium term plan 2007–2009: charting the future of rice in Africa. Africa Rice Center, Cotonou

Aw D, Diemer G (2005) Making a large irrigations scheme work: a case study from Mali. World Bank, Washington, DC

Balasubramanian V, Sie M, Hijmans R, Otsuka K (2007) Increasing rice production in Sub-Saharan Africa: challenges and opportunities. Adv Agron 94(1):55–133

David CC, Otsuka K (eds) (1994) Modern rice technology and income distribution in Asia. Lynne Rienner, Boulder

Estudillo JP, Otsuka K (2006) Lessons from three decades of Green Revolution in the Philippines. Dev Econ 44(2):123–148

Food and Agriculture Organization of the United Nations (FAO) (2012) Rice market monitor 15(3), July 2012

Hayami Y, Godo Y (2005) Development economics: from the poverty to the wealth of nations. Oxford University Press, Oxford

Hayami Y, Kikuchi M (2000) A rice village saga – three decades of Green Revolution in the Philippines. International Rice Research Institute, Los Banos

Inocencio A, Kikuchi M, Tonosaki M, Maruyama A, Merrey D, Sally H, de Jong I (2007) Costs and performance of irrigation projects: a comparison of Sub-Saharan Africa and other developing regions, vol 109, Research report. International Water Management Institute, Colombo

Kijima Y, Serunkuuma D, Otsuka K (2006) How revolutionary is the 'NERICA Revolution'? Evidence from Uganda. Dev Econ 44(2):232–251

Kijima Y, Otsuka K, Sserukuuma D (2008) Assessing the impact of NERICA on income and poverty in Central and Western Uganda. Agric Econ 38(3):327–337

Kijima Y, Otsuka K, Sserunkuuma D (2011) An inquiry into constraints on a Green Revolution in Sub-Saharan Africa: the case of NERICA rice in Uganda. World Dev 39(1):77–86

Kore H (2006) Politiques rizicoles et développement des marchés. Rapport Technique du Niger, Université de Niamey, Niger

Minten B, Randrianarisoa JC, Barrett CB (2006) Productivity in Malagasy rice systems: wealth-differentiated constraints and priorities. Invited panel paper presented at the international association of agricultural economists conference, Gold Coast, Australia

[13] Large scale refers to an irrigated area of 1,000 ha and more, whereas small scale refers to an area of more than 1 ha but less than 100 ha in the report.

Nakano Y (2009) Irrigation management in the Doho Rice Scheme in Uganda: an inquiry into the potential of a Green Revolution in Sub-Saharan Africa. Ph.D. dissertation, National Graduate Institute for Policy Studies, Tokyo, Japan

Otsuka K (2006) Why can't we transform traditional agriculture in Sub-Saharan Africa? Rev Agric Econ 28(3):332–337

Otsuka K, Kalirajan K (2005) An exploration of a Green Revolution in Sub-Saharan Africa. J Agric Dev Econ 2(1):1–7

Otsuka K, Kijima Y (2010) Technology policies for a Green Revolution and agricultural transformation in Africa. J Afr Econ 19(S2):60–96

Sakurai T (2006) Intensification of rainfed lowland rice production in West Africa: present status and potential Green Revolution. Dev Econ 44(2):232–251

Spencer DSC (1994) Infrastructure and technology constraints to agricultural development in the humid and sub-humid tropics of Africa, vol 3, Environmental and Production Technology Division (EPTD) discussion paper. International Food Policy Research Institute, Washington, DC

United States Department of Agriculture (USDA) (2008) Rice projections, 2008–2017. http://www.ers.usda.gov/Briefing/Rice/2008baseline.htm

Wooldridge JM (2002) Econometric analysis of cross section and panel data. MIT Press, Cambridge, MA

World Bank (2007) Investment in agricultural water for poverty reduction and economic growth in Sub-Saharan Africa, A synthesis report, collaborative program of AfDB, FAO, IWMI and the World Bank. World Bank, Washington, DC

World Bank (2008) World development report 2008: agriculture for development. World Bank, Washington, DC

Chapter 4
The Declining Impacts of Climate on Crop Yields During the Green Revolution in India, 1972–2002

Takuji Tsusaka and Keijiro Otsuka

Abstract One of the decisive factors determining agricultural yields is known to be climatic conditions, typically temperature and rainfall, which have a direct impact on agricultural production. To date, most of the empirical studies on how agro-climatic factors affect agricultural productivity have focused on developed countries. The purpose of this study is to demonstrate that the impacts of climatic factors on cereal yields have been mitigated during the Green Revolution period, using a district-level panel data set. Given India's diverse cropping patterns and agro-climate, it will be useful to review the experience of India's agricultural growth with particular reference to the impact of GR technology and climate on crop yields, so as to draw effective lessons for facilitating agricultural growth in other parts of the developing world, including sub-Saharan Africa.

Keywords Climate change • Green Revolution • India • Sub-Saharan Africa • Crop yields • Modern variety • Crop choice • Irrigation

4.1 Introduction

Agricultural productivity growth is an important factor to reduce the population of the extremely poor across the globe (World Bank 2008). Econometric analysis over the last two decades for 42 developing countries shows that a 1% growth in

T. Tsusaka
Social Sciences Division, International Rice Research Institute,
DAPO Box 7777, Metro Manila 1301, Philippines
e-mail: t.tsusaka@irri.org

K. Otsuka (✉)
National Graduate Institute for Policy Studies,
7-22-1 Roppongi, Minato-ku, Tokyo 106-8677, Japan
e-mail: otsuka@grips.ac.jp

K. Otsuka and D.F. Larson (eds.), *An African Green Revolution: Finding Ways to Boost Productivity on Small Farms*, DOI 10.1007/978-94-007-5760-8_4,

agricultural GDP increases the incomes of the poorest deciles in the expenditure distribution by more than 2.5% (Christiaensen and Demery 2007). Among the poor at slightly higher income levels, the effect of agricultural growth on poverty reduction is found to decrease but remain superior to that of non-farm activities.

In sub-Saharan Africa, the rate of increase in staple food production has been exceeded by its high population growth rate. In contrast, in Asia, growth in agricultural production has consistently outpaced population growth owing to the Green Revolution (GR) (Otsuka and Kalirajan 2006). This is not only because the population growth has been somewhat slower in Asia, but much more importantly because the technological innovation represented by the diffusion of improved crop varieties and other complementary production practices spurred the agricultural yields in Asia, which has led to significant reductions in rural poverty as well as the growth of nonfarm sectors (Otsuka et al. 2009; Lipton 2007).

One of the decisive factors determining agricultural yields is known to be climatic conditions, typically temperature and rainfall, which have a direct impact on agricultural production (Omamo 2003; Mwabu and Thorbecke 2004). To date, most of the empirical studies on how agro-climatic factors affect agricultural productivity have focused on the United States (Adams et al. 1995) and other developed countries (Olesen and Bindi 2002; Bruce et al. 1996; Reilly et al. 1996). As for developing countries, case studies of India and Brazil by Seo and Mendelsohn (2007), Sanghi et al. (1998), and Auffhammer et al. (2006) show significant effects of climate on agricultural income per unit of land and crop yields, particularly the positive effect of rainfall. This means that unfavorable climate endowments can be a constraint to agricultural productivity growth. The critical question is whether and to what extent the influence of climatic conditions has been alleviated or augmented by GR technology, irrigation, or other factors.

The purpose of this study is to assess the changing impacts of climatic factors on the performance of Indian agriculture during the GR period from the early 1970s to the early 2000s, using a district-level panel data set. Given India's diverse cropping patterns and agro-climate, it will be useful to review the experience of India's agricultural growth with particular reference to the impact of GR technology and climate on crop yields, so as to draw effective lessons for further facilitating agricultural growth in other parts of the developing world, including sub-Saharan Africa.

The organization of this chapter is as follows. Section 4.2 overviews the historical agricultural performance in India and discusses the relevant descriptive statistics. Section 4.3 explains our data sources and how the database is constructed, and introduces the econometric models. The regression results are carefully examined in Sect. 4.4. Lastly, Sect. 4.5 presents the concluding remarks

4.2 An Overview of the Agricultural Performance in India

4.2.1 Background

By the time the GR began in the late 1960s, irrigation was available in some parts of India, and road conditions had been considerably improved as well, which had set the stage for the adoption of GR technologies (Rosegrant and Hazell 2000; Bhalla and Singh 2001). Then, massive investments were made in rural areas during the GR, when modern varieties (MVs) were diffused and subsequently, small-scale irrigation with pumps and tubewells became popular. Public investments in rural areas grew at a rate of about 13% annually during the 1970s and increased fivefold by the end of the 1980s, which led to phenomenal rural poverty reduction in India (Fan et al. 2000; Fujita 2010). The steady growth in irrigation investments resulted in a nearly twofold increase in the area under irrigation by the 1990s (Johnson et al. 2003).

4.2.2 Cropping Patterns

India consists of 35 states (including union territories) with diverse cropping patterns reflecting its diverse agro-climates. Figure 4.1 shows the proportions of harvested areas of five major crops grown in India with reference to other Asia and SSA, over the period of 2003–2007. The crop composition in other Asia is largely skewed to rice, whereas sorghum and millet have large shares in SSA. India stands in-between the two regions, with relatively diverse cropping patterns, which reflects the similarity in agro-climate between some parts of India and SSA, and between some other parts of India and other Asia.

4.2.3 Crop Yields

How different are the crop yields between India and the other developing world? To answer this question, Fig. 4.2 compares the average cereal yields between India and SSA, as well as Southeast Asia. The most important finding is that despite the more favorable production environments, the cereal crop yield in India was not significantly superior to that in SSA up to the early 1980s. As can be seen, however, the yields in the two regions started to diverge in the mid-1980s and today the gap is approximately twofold. Therefore, it seems clear that climatic conditions can explain, at best, only a small portion of today's large gap in crop yield between the two regions.

	SSA	*India*	*asia except India*
Other	6	1	6
Wheat	3	27	32
Rice	10	44	42
Maize	34	8	18
Millet	23	12	1
Sorghum	24	9	1
Total Cereal	100	100	100
upland	80	29	20

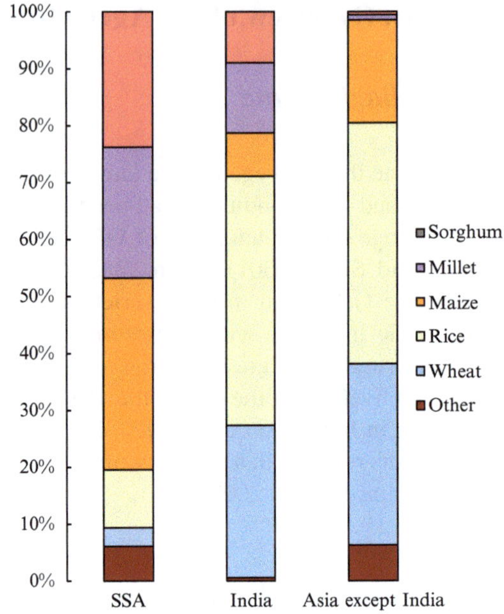

Fig. 4.1 Proportions of harvested areas by cereal crop (%), 2003–2007 average Others*: Ragi, Oats, Barley, Cassava, Teff, Potatos, etc (Source: FAOSTAT data)

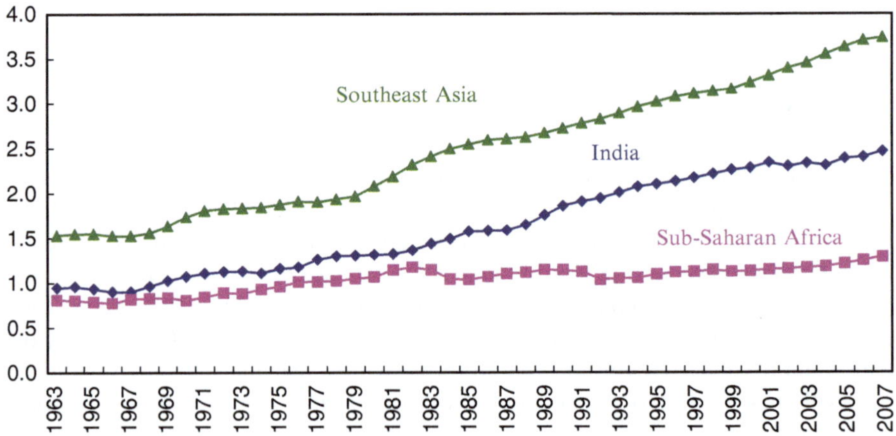

Fig. 4.2 Average cereal yields in India, Southeast Asia, and SSA, 3-year moving averages (Source: FAOSTAT data)

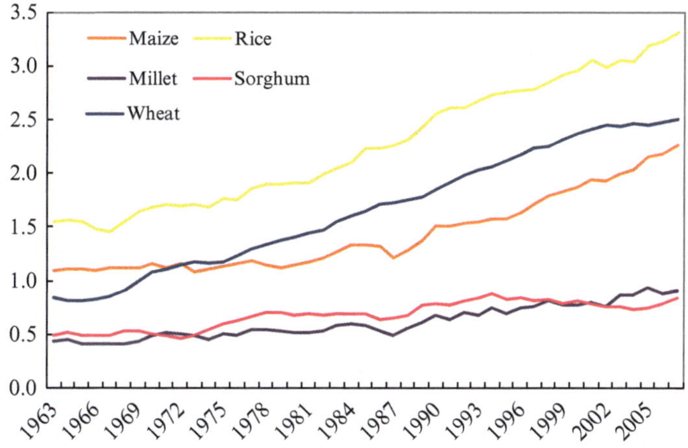

Fig. 4.3 Cereal yields in India by crop, 3-year moving averages (Source: FAOSTAT data)

India's crop-wise yield data are shown in Fig. 4.3. It is noticeable that the yields of rice, wheat and maize have soared dramatically over the last several decades. Although less dramatic, the yields of sorghum and millet have almost doubled, which is consistent with the finding by Pray and Nagarajan (2010), who argue that the production technology of these crops, including improvement of varieties, has improved significantly.

4.2.4 *Changes in Cropping Patterns*

Figure 4.4 shows how cropping patterns have been evolving in India. The average annual growth rates of harvested areas indicate that farmers in India have been increasing the areas planted to rice, wheat and maize, whose yields have been increasing. Since the total harvested area has not been growing at all, it is clear that farmers have been replacing sorghum and millet with these three crops. Thus, India has been feeding its increasing population not only by raising the yield of various crops but also by switching crops from low performers (sorghum and millet) to high performers (rice, wheat, and maize).

4.2.5 *Irrigation Expansion*

The availability of irrigation is widely considered to be vitally important for yield performance (see, e.g., David and Otsuka 1994). According to Fig. 4.5, the proportion of irrigated area in wheat and rice has been notably rising, while it has been

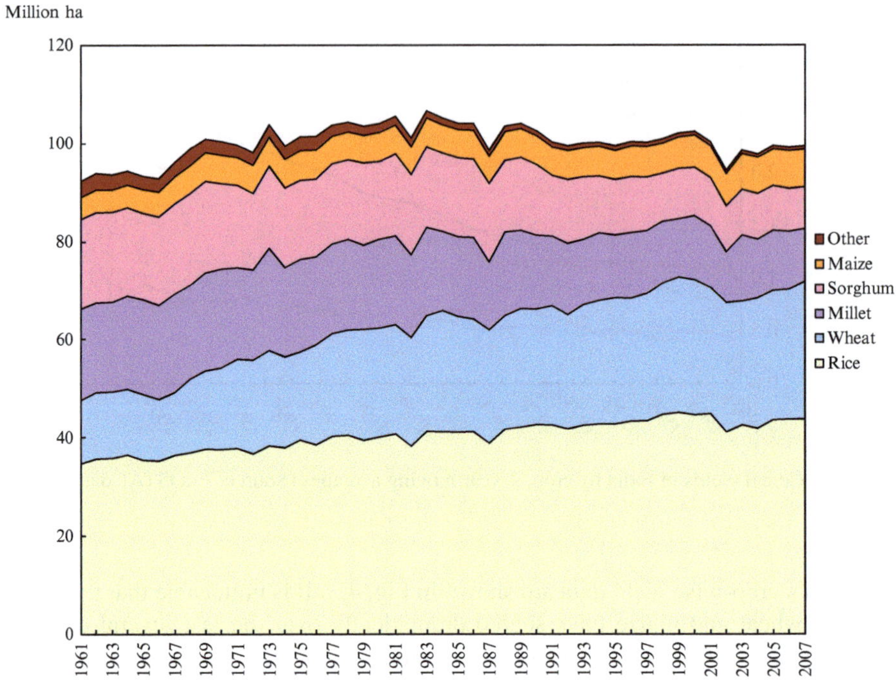

Fig. 4.4 Changes in harvested areas by cereal crop in India (Source: FAOSTAT data)

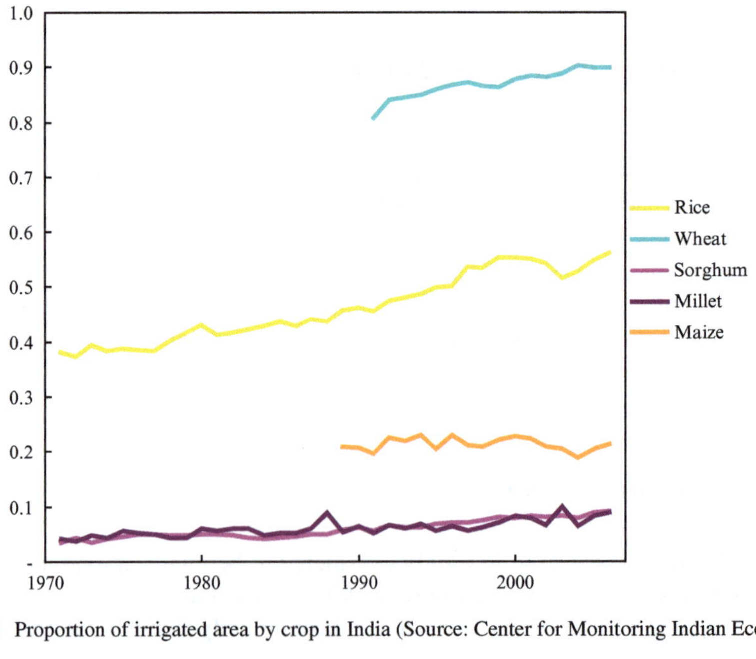

Fig. 4.5 Proportion of irrigated area by crop in India (Source: Center for Monitoring Indian Economy)

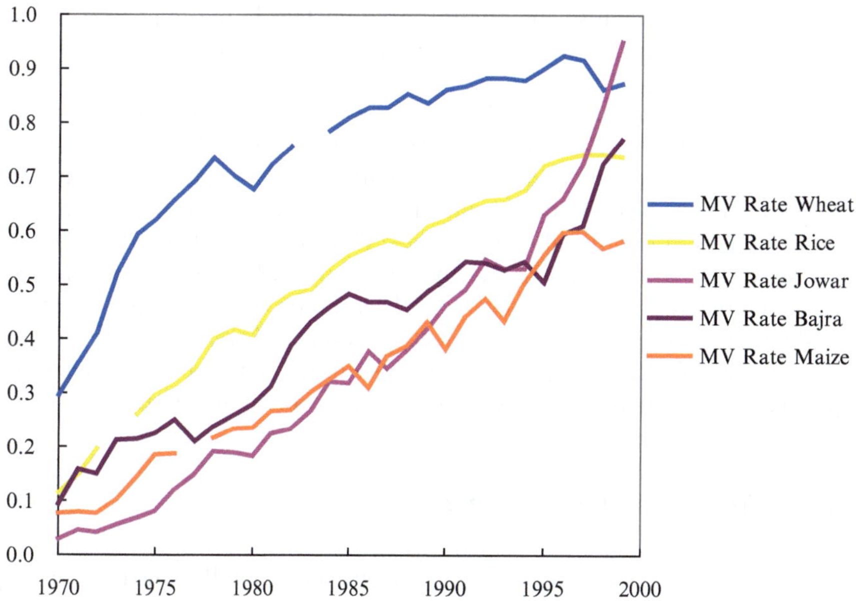

Fig. 4.6 Proportion of area sown to modern varieties by crop (Source: Center for Monitoring Indian Economy)

more or less constant at low levels for maize, millet and sorghum. Clearly the impacts of irrigation on productivity would be higher for wheat and rice than for other cereals.

A question may arise as to whether the marginal effects of irrigation on crop yields have been changing over time along with the introduction of improved technologies. If the new technologies are less dependent on stable supply of water, the effect of irrigation may decline. On the other hand, it may increase if intensification of wheat and rice farming system requires more intensive use of irrigation water. It would also be interesting to investigate how the effect of irrigation interacts with the impact of climatic conditions on crop yield.

4.2.6 *Modern Variety Adoption*

Figure 4.6 shows the changes in the adoption rate of modern varieties (MVs) by crop. What is striking is that the adoption rates of MVs have been rapidly increasing, even for sorghum and millet in recent years, which confirms the report by Pray and Nagarajan (2010). As is emphasized by Estudillo and Otsuka (Chap. 2) in this volume, the quality of MVs has improved over time. The data in Fig. 4.6, however, do not distinguish the quality. Also unclear is the impact of MVs of sorghum and millet on their yields, as it has seldom been reported in the economic literature.[1]

A potentially important trait of MVs is the short growth duration (Cavatassi et al. 2011; Khush 2001; Hossain and Fischer 1995; Lawn 1989), allowing cereal crops to grow in a shorter period during which rainfall is assured. For example, the latest rice MVs mature in 105–110 days, which is much shorter than the growth duration of 160–170 days of traditional rice varieties (TVs) in Asia (Khush 2001). If so, it seems reasonable to hypothesize that the adoption of MVs lessens the impact of precipitation on crop yields. However, our dataset only contains India's state-level MV adoption rates by crop. In the absence of proper district-level MV adoption data, we are unable to perform a direct analysis of the impact of MV adoption on crop yields. Even so, strong suggestive results are reported in this study.

4.2.7 Climate and Crop Choice

Table 4.1 shows the 5-year averages of temperature, rainfall, irrigation coverage rate, and crop yields, over the districts where each crop is grown, for the years 1998–2002. The temperature is largely the same across crops, though it is slightly higher for sorghum and millet. Rainfall varies and it is notably higher for rice, suggesting that rainfall plays an important role in rice cultivation. Considering the level of rainfall and irrigation coverage, it is understood that sorghum and millet have comparative advantages in drier environments.

In the lower section of each box in the table are the statistics calculated for two groups: high and low in irrigation ratio, where for convenience high refers to above 50% and low refers to below 50% of the total sown area for each crop. There are two major observations. First, in the regions with high irrigation coverage, rainfall is lower, and vice-versa, for all the five major crops, which is not much of surprise if irrigation is designed to help compensate for inadequate rainfall. Second, even in the regions with lower rainfall, the availability of irrigation results in higher yields than in the regions with higher rainfall, particularly for wheat and rice, followed by maize and millet, but not for sorghum. These observations are mostly consistent with the differences in irrigation coverage among the five crops shown in Fig. 4.5.

Table 4.2 demonstrates the same statistics as in Table 4.1 for the early 1970s, which corresponds to the early stage of the GR in India. We may ignore regions with high irrigation coverage for millet and sorghum, as there were only few such districts at that time. Again, irrigation coverage and crop yields are positively associated, but not as clearly as in the late 1990s. These observations are suggestive of the increasingly important role that irrigation plays in enhancing the crop yields in low rainfall environments in recent years. Thus, the comparison of Tables 4.1 and 4.2 suggests

[1] Cavatassi et al. (2011) found in eastern Ethiopia that the early-maturing sorghum MVs adopted in their study site can cope with downward yield risks associated with moderate droughts, while the TVs (traditional varieties) are more tolerant of extreme drought events.

Table 4.1 Yield, climate, and irrigation coverage in India, by crop, 1998–2002 5-year averages

	Wheat		Rice		Maize		Sorghum		Millet	
Grouping by irrigation	*Average*		*Average*		*Average*		*Average*		*Average*	
coverage rate[a, b]	*Low*	*High*	*Low*	*High*	*Low*	*High*	*Low*	*High*	*Low*	*High*
Temperature (°C)[c]	25.3		25.5		25.6		26.3		26.2	
	23.4	25.7	24.9	26.0	25.4	26.0	26.3	25.9	26.2	26.4
Rainfall (mm)[c]	852		1,007		863		848		794	
	1045	809	1127	920	895	802	877	658	811	701
Irrigation coverage rate[a] (%)	79		58		34		14		18	
	23	92	17	90	10	81	4	82	6	83
Yield (kg/ha)	2,153		2,007		1,825		821		1,001	
	1,203	2,365	1,418	2,455	1,630	2,198	836	720	964	1,207
Num. of districts	356		412		327		258		269	
	65	291	178	234	215	112	224	34	228	41

Source: Author's calculation with India Water Portal's data and Center for Monitoring Indian Economy's database

[a] Irrigation coverage = Irrigated area's percentage of area sown to each crop

[b] *High* referes to the coverage of more than 50%, and *Low* refers to the coverage of less than 50%

[c] In India, wheat is cultivated in the winter season when temperature and rainfall are lower than in the other seasons. Thus, note that the presented annual climate overstates the actual wheat growing environment. The discussion applies, in particular, to rainfall because most of the rainfall is observed in the summer monsoon season whereas temperature is ralatively stable across seasons

Table 4.2 Yield, climate, and irrigation coverage in India, by crop, 1972–1976 5-year averages

Grouping by irrigation coverage rate[a, b]	Wheat		Rice		Maize		Sorghum		Millet	
	Average		*Average*		*Average*		*Average*		*Average*	
	Low	*High*	*Low*	*High*	*Low*	*High*	*Low*	*High*	*Low*	*High*
Temperature (°C)[c]	25.2		25.5		25.5		25.8		25.8	
	25.2	25.3	25.4	25.7	25.5	25.6	25.8	24.6	25.7	26.5
Rainfall (mm)[c]	996		1,072		1,003		950		911	
	1,134	868	1,138	963	1,032	898	961	735	916	801
Irrigation coverage rate[a] (%)	52		41		25		8		6	
	23	78	15	86	10	78	3	73	3	66
Yield (kg/ha)	1,243		1,045		1,093		570		532	
	1,036	1,430	858	1,368	1,041	1,276	568	601	522	752
Num. of districts	211		205		187		169		152	
	100	111	130	75	146	41	159	10	145	7

Source: Author's calculation with India Water Portal's data and Center for Monitoring Indian Economy's database

[a] Irrigation coverage = Irrigated area's percentage of area sown to each crop

[b] *High* referes to the coverage of more than 50%, and *Low* refers to the coverage of less than 50%

[c] In India, wheat is cultivated in the winter season when temperature and rainfall are lower than in the other seasons. Thus, note that the presented annual climate overstates the actual wheat growing environment. The discussion applies, in particular, to rainfall because most of the rainfall is observed in the summer monsoon season whereas temperature is ralatively stable across seasons

the hypothesis that the impact of irrigation on crop yields has increased over time with the adoption of GR technologies. However, it is too soon to come to conclusion before performing a formal testing of the changing impacts and interactive effects of irrigation in the later section of this chapter.

4.3 Data Source and Empirical Methodology

4.3.1 Data Source

One unique aspect of this research is the use of district-level panel data to quantify the effects of various factors on yields by crop and their changes over time. The data set covers 270 main districts and many small districts over a period of 31 years from 1972 to 2002. The important variables are crop yields, i.e., the quantity of agricultural output divided by the respective planted areas for five major crops (millet, sorghum, rice, wheat, maize), the output-fertilizer price ratio (i.e., farm harvest price by crop divided by fertilizer price), climate represented by temperature and rainfall, irrigation ratio by crop, and such district specific characteristics as population density and literacy rate. The database is composed of several different data sources, both public and private. Data on agricultural outputs, inputs, and rainfall are obtained from Center for Monitoring Indian Economy Limited (CMIE). Temperature data are collected through Water Portal Service powered by India's Meteorological Department (IMD). Some district characteristic data are obtained from Indiastat Service provided by Datanet India. Other district characteristics data are cordially provided by Drs. Zhang and Fan of the International Food Policy Research Institute. Detailed agricultural data for years 1958–1986 are available through the database assembled by Dr. Kumar of the World Bank (http://ipl.econ.duke.edu/dthomas/dev_data/).

There are currently as many as 600 districts in India; this number has been steadily increasing as more and more districts declare independence. Hence, combining the data from different sources in a consistent manner over the long period has been a major challenge.[2]

4.3.2 Crop Choice Function by Crop

Once the database is constructed, the next task is to assess the effects of climate (in particular temperature and rainfall), markets (as revealed in output-input price ratio), and irrigation on crop yields, and their changes over time. We also considered the

[2] As far as simple mergers and separations are concerned, the data can be adjusted by arithmetic operations, in which post-merger or pre-separation district bordering is adopted so as to construct a long-term panel. Otherwise, we used our best guess, including temporal interpolation and extrapolation.

effects of other factors such as population density and literacy rate. It is important to note that each crop is grown in some districts but not all. Therefore, the estimation procedure consists of two steps:

In the first step, we use a Probit regression to estimate the crop choice functions in order to eliminate the sample selection bias (Heckman 1979). The set of explanatory variables in the first-step estimation includes normal climate represented by 3-year moving averages of rainfall and temperature in the preceding years.[3] The crop choice function can be specified as follows:

$$y_{ijt} = \begin{cases} 1 \, if \, y_{ijt}^* > 0, \\ 0 \text{ otherwise,} \end{cases}$$

and

$$y_{ijt}^* = \beta_{1i}\overline{temp}_{jt} + \beta_{2i}\overline{temp}_{jt}^2 + \beta_{3i}\overline{rain}_{jt} + \beta_{4i}\overline{rain}_{jt}^2 + \beta_{5i}stdev(rain)_{jt} + \beta_{6i}irri_{ijt} + \beta_{7i}P_{ijt} + \beta_{8i}lit_{jt} + \beta_{9i}popden_{jt} + \varepsilon_{ijt}$$

where y_{ijt} is 1 if crop *i* is grown in district *j* in year *t*, and 0 otherwise; $\overline{temp}_{jt}$ and $\overline{rain}_{jt}$ represent normal climate expressed by the 3-year moving averages of temperature (in °C) and rainfall (in meter), respectively, over the preceding years *t-1, t-2 and t-3*; $stdev(rain)_{jt}$ is the standard deviation of rainfall over the preceding 3 years; $irri_{jt}$ is the irrigation coverage rate for district *j*; lit_j is the literacy rate; $popden_{jt}$ is the population density; ε_{ijt} is the error term such that $\varepsilon \sim N(0,1)$. Many alternative specifications were tested and it is found that use of the interaction terms, year dummy variables, or district specific effects leads to failure in the MLE (maximum likelihood estimation) associated with the Probit regression. Therefore, such variables are dispensed with.

4.3.3 Yield Function Estimation

The second step is to estimate the yield function by crop, for which the basic estimation model can be specified as follows:

$$Y_{ijt} = \alpha_{1i}C_{jt} + \alpha_{2i}C_{jt}\cdot t' + \alpha_{3i}C_{jt}\cdot t'^2 + \beta_{1i}P_{ijt} + \beta_{2i}P_{ijt}\cdot t' + \beta_{3i}P_{ijt}\cdot t'^2 + \gamma_{1i}irri_{ijt} + \gamma_{2i}irri_{ijt}\cdot t' + \gamma_{3i}irri_{ijt}\cdot t'^2 + \zeta_{1i}X_{jt} + \zeta_{2i}X_{jt}\cdot t' + \zeta_{3i}X_{jt}\cdot t'^2 + \theta_i\tau_t + \rho_i\lambda_{ijt} + \nu_{ij} + \upsilon_{ijt}$$

[3] As for normal climate, we also tried using 5-year moving averages to check the robustness. The results are largely the same. The shortcoming of using a longer period is that it leads to a reduced number of observations.

where Y_{ijt} is the yield of crop *i*, district *j*, year *t*; C_{jt} is a vector of climate variables (temperature and rainfall); X_{jt} is a vector of district characteristics (population density and literacy rate); τ_t is a vector of year dummies; λ_{ijt} is the inverse Mill's ratio obtained from the first-step Probit estimation; ν_{ij} is the unobservable time-invariant district specific effect; ε_{ijt} is the error term; t' is such that $t' = t - 1972$, thus ranging from 0 to 30. Finally and very importantly, interaction terms with t' and t'^2 are included to examine whether there have been over-time changes in the impacts of the explanatory variables due to technological changes, such as the introduction of short-maturity and drought-tolerant MVs. The yield functions are estimated separately for each crop.

The time trend (t') and its squared term (t'^2) themselves are also included with an aim to capture the effect of general technology improvement over time and its acceleration (or deceleration) that are not picked up by the over-time changes in the effects of the explanatory variables. In this way, we hope to be able to observe whether there has been a positive trend in yield improvement due to technological changes, among other things.

4.3.4 Model Specification

The specification I employ is what is known as the two-way fixed effect model. That is, in addition to the district fixed (or random) effect, the model includes the year dummies for all the available years except the base year, in order to control for the yearly change in yield that is not explained by the time trend and the explanatory variables, e.g., aggregate macroeconomic and climatic shocks. For robustness, I also tried performing regressions (a) without the year dummies but with the time trend, and (b) without the time trend but with the year dummies. The estimated coefficients of the explanatory variables remain largely unchanged for both (a) and (b), indicating that the year dummies mostly capture the random shocks.

The endogeneity of irrigation coverage may need to be considered. However, gravity irrigation is the most popular method of irrigation (Ostrom 1990), particularly in the early stage of the GR.[4] However, as far as gravity irrigation is concerned, it is constructed by public sectors, so that it is basically exogenous for farmers. Also, as the construction of such irrigation is likely to be influenced by some district specific geographical and climatic characteristics which are essentially constant over time, the district-level fixed (or random) effect model, which controls for unobserved time-invariant district-specific effects, can fairly mitigate the potential endogeneity bias in the irrigation effect. Therefore, we assume that the endogeneity of the irrigation variable is not serious in the model we use.

Regarding the sample selection model, in some cases Heckman's ρ is estimated to be insignificant ($p > 0.05$), presumably because the district-level fixed (or random) effect model can mitigate, if not solve, the sample selection bias.

---

[4] Recently, tubewell and pump irrigation, which can be installed by farmers, have become increasingly common.

The Hausman test is conducted to compare the fixed effect and random effect estimations. When the GLS random effect estimators are diagnosed as inconsistent ($p<0.05$), the fixed effect estimation is adopted.[5]

To estimate the elasticities, the logarithm of the variables are taken whenever applicable, the exceptions being the ratio variables (viz., irrigation ratio and literacy rate), the time trend variables, the interval-scale variable (temperature in Celsius), and the year dummy variables (Stevens 1946; Rozeboom 1966).

4.4 Regression Results

4.4.1 First-Step Selection Estimations

Table 4.3 presents the estimation results of the first-step Probit estimations for wheat, rice, maize, sorghum, and millet. Shown on the table are the estimated coefficients with the standard errors in parentheses. Note that the estimated coefficients do not represent the marginal effects since it is not a linear probability model.

The first major finding is that normal climate significantly affects which crop is chosen in the districts. The probability of rice selection increases as normal temperature decreases, while the probabilities of maize and sorghum selection increase as normal temperature increases. As for wheat selection, the coefficient on normal temperature appears to be positive and significant. However, since wheat is a winter crop in India, it is important to note that the actual wheat growing environments have lower temperature than for the other crops.[6] Normal rainfall also seems to affect the crop choice. Rice is the crop of which the selection is most affected by the availability of rainfall, while millet tends to be chosen in relatively dry districts. Maize selection is positively affected by rainfall but to much lesser extent than for rice selection. Sorghum selection is unaffected by the availability of rainfall. Since wheat is a dry season crop, the negative impact of rainfall on wheat selection is expected. The standard deviation of rainfall has a larger positive effect on sorghum and millet selections than for the other crops, which means that sorghum and millet are relatively resistant to volatile rainfall.

Apart from climatic factors, district's irrigation coverage seems to be an important variable for crop selection. The probability of rice selection increases when district's irrigation coverage is higher. The availability of irrigation does not affect

[5] When the Hausman test fails and returns a negative probability statistic, we opt for the fixed effect model for safety as the fixed effect estimators are always consistent even if not efficient.

[6] In India, wheat is mostly grown in the dry winter season, while most of rice, maize, and millet, and more than a half of sorghum are cultivated in the summer monsoon season. The difference between the average summer temperature and the average winter temperature is 7–8°C according to IMD.

Table 4.3 First-step probit regression, by crop, 1972–2002

	Wheat selection	Rice selection	Maize selection	Sorghum selection	Millet selection
Normal temperature	0.4968***	–0.1316***	0.1317***	0.3833**	0.0606
	(0.0470)	(0.0507)	(0.0505)	(0.1622)	(0.0696)
Normal temperature squared	–0.0142***	0.0052***	–0.0029***	–0.0020	0.0019
	(0.0011)	(0.0012)	(0.0012)	(0.0033)	(0.0015)
Normal rainfall	–2.0262***	1.7436***	0.2499***	–0.1319	–1.7970***
	(0.1570)	(0.1000)	(0.0923)	(0.1280)	(0.1242)
Normal rainfall squared	0.3428***	–0.3722***	–0.1122***	–0.2753***	0.1599***
	(0.0469)	(0.0364)	(0.0314)	(0.0444)	(0.0426)
Standard deviation of rainfall	–0.0857	0.5676***	0.3758***	2.0134***	1.2459***
	(0.1586)	(0.1261)	(0.1069)	(0.1513)	(0.1357)
Irrigation coverage for district	–1.0211***	0.5565***	0.0231	–0.3564***	–0.4889***
	(0.0742)	(0.0822)	(0.0661)	(0.0655)	(0.0639)
Literacy rate	–0.7286***	–0.1014	–1.5724***	–0.8689***	–0.1756
	(0.1483)	(0.1581)	0.1322	(0.1293)	(0.1193)
Ln population density	0.2023***	0.0929***	–0.0597**	–0.3842***	0.0330
	(0.0278)	(0.0317)	(0.0260)	(0.0267)	(0.0247)
Constant term	–0.6732	–0.7076	0.5554	–4.9045***	0.7707
	(0.5092)	(0.5323)	(0.5656)	(1.9576)	(0.8218)
Number of observations	7,080	7,080	7,080	7,080	7,080
Likelihood ratio χ^2	887.05	580.75	232.14	1979.40	1751.10
statistic (degree of freedom)	(8)	(8)	(8)	(8)	(8)
Prob>χ^2	0.0000	0.0000	0.0000	0.0000	0.0000
Pseudo R^2	0.1719	0.1329	0.0400	0.2458	0.1968

In the first step, rainfall is expressed in meters

***, **, and * indicate 1, 5, and 10% statistical significance levels, respectively

maize selection while sorghum and millet are chosen in districts with low irrigation coverage, confirming that these crops have a comparative advantage in rain-fed areas. It is interesting to find that the probability of choosing wheat is higher when district's irrigation coverage is lower. This result indicates that even though wheat fields are highly irrigated (Fig. 4.5), the wheat-growing district's overall irrigation coverage including irrigation for other crops and non-cereal agricultural produce is not necessarily high. In other words, it is indicated that wheat tends to be grown in irrigated area in districts where the overall irrigation coverage is low.

The coefficients on the population density variable show that wheat and rice are chosen in relatively densely populated districts, presumably because the production of these crops is labor intensive.

4.4.2 Estimated Yield Functions

Table 4.4 presents the estimation results of the yield functions for wheat, rice, and maize, and Table 4.5 shows those for sorghum, and millet. To keep the table succinct, the estimated coefficients on the year dummy variables are not presented.

Table 4.4 Second-step outcome regression result for wheat, rice, and maize, 1972–2002

Dependent variable: Ln yield	Wheat	Rice	Maize	
District effect	Fixed effect	Fixed effect	Fixed effect	Random effect
Temperature	–0.0016	0.1091***	–0.0600**	–0.0288*
	(0.0192)	(0.0249)	(0.0263)	(0.0168)
Temperature × Time trend	–0.0054***	–0.0044***	0.0080***	0.0079***
	(0.0013)	(0.0016)	(0.0020)	(0.0019)
Temperature × Time trend squared	0.0001***	0.0001**	–0.0002***	–0.0002***
	(0.0000)	(0.0000)	(0.0001)	(0.0001)
Temperature × Irrigation coverage	0.0361***	–0.0584***	0.0405*	0.0165
	(0.0145)	(0.0237)	(0.0228)	(0.0165)
Ln rainfall	0.3149***	0.5713***	0.0610	0.0881**
	(0.0338)	(0.0395)	(0.0425)	(0.0420)
Ln rainfall × Time trend	–0.0173***	–0.0223***	–0.0024	–0.0032
	(0.0029)	(0.0040)	(0.0044)	(0.0044)
Ln rainfall × Time trend squared	0.0004***	0.0001	0.0000	0.0000
	(0.0001)	(0.0001)	(0.0001)	(0.0001)
Ln rainfall × Irrigation coverage	–0.1228***	–0.1509***	0.0051	0.0431*
	(0.0289)	(0.0262)	(0.0275)	(0.0261)
Irrigation coverage for each crop	0.1366	2.8437***	–0.9642	–0.4740
	(0.4338)	(0.6575)	(0.6431)	(0.4784)
Irrigation coverage × Time trend	–0.0001	0.0155***	0.0030	0.0040
	(0.0060)	(0.0056)	(0.0076)	(0.0076)
Irrigation coverage × Time trend squared	–0.0001	–0.0006***	–0.0003	–0.0001
	(0.0002)	(0.0002)	(0.0002)	(0.0002)

(continued)

Table 4.4 (continued)

Dependent variable: Ln yield	Wheat	Rice	Maize	
District effect	Fixed effect	Fixed effect	Fixed effect	Random effect
Literacy rate	0.2318	0.0645	1.3287***	1.5071***
	(0.2294)	(0.3017)	(0.3779)	(0.2655)
Literacy rate × Time trend	–0.0290*	–0.0258	–0.0544**	–0.0671***
	(0.0158)	(0.0204)	(0.0248)	(0.0245)
Literacy rate × Time trend squared	0.0006	–0.0002	–0.0001	0.0005
	(0.0005)	(0.0006)	(0.0008)	(0.0007)
Ln population density	0.1067**	0.1497***	–0.2211***	–0.2251***
	(0.0437)	(0.0457)	(0.0614)	(0.0315)
Ln population density × Time trend	–0.0023	–0.0019	0.0124***	0.0144***
	(0.0023)	(0.0028)	(0.0036)	(0.0036)
Ln population density × Time trend squared	0.0001*	0.0001	–0.0001	–0.0002**
	(0.0001)	(0.0001)	(0.0001)	(0.0001)
Inverse Mills ratio	–0.0765	0.4889***	0.7461**	0.6788**
	(0.0640)	(0.0852)	(0.3127)	(0.2915)
Time trend	0.3142***	0.2550***	–0.2679**	–0.3343***
	0.0401	(0.0505)	(0.1251)	(0.1229)
Time trend squared	–0.0065***	–0.0023	–0.1114	–0.0480
	(0.0012)	(0.0015)	(0.1095)	(0.1072)
Constant term	3.9938***	–0.9224**	8.8777***	7.7295***
	(0.5848)	(0.7262)	(0.8114)	(0.5758)
Number of observations	5,809	5,840	5,317	
R-squared: Overall	0.3175	0.4169	0.2250	0.2855
Hausman test				
χ^2 statistics (degree of freedom)	764.24	100.54	13.53	
	(43)	(43)	(7)	
Prob>χ^2	0.0000	0.0000	0.0602	

In the second step, rainfall is expressed in millimeters
***, **, and * indicate 1, 5, and 10% statistical significance levels, respectively

Based on the result of the Hausman test, the GLS random effect estimators are diagnosed as inconsistent in all cases except for maize. Therefore, the fixed effect is adopted in the estimation of the yield functions of wheat, rice, sorghum, and millet. In the case of maize, both fixed and random effect estimation results are presented since the diagnosis of the Hausman test is somewhat ambiguous ($p=0.06$). The sample selection bias is diagnosed as statistically significant and is duly treated in the case of rice and maize, shown by the estimated coefficients on the inverse Mills ratios.

4.4.2.1 Results for Wheat

Temperature by itself does not pose a statistically significant effect on wheat yield while rainfall has a positive and significant effect. Somewhat unexpectedly the

Table 4.5 Second-step outcome regression result for sorghum and millet, 1972–2002

Dependent variable: Ln yield	Sorghum	Millet
District effect	Fixed effect	Fixed effect
Temperature	0.0640**	0.0659**
	(0.0328)	(0.0316)
Temperature × Time trend	–0.0025	0.0029
	(0.0025)	(0.0024)
Temperature × Time trend squared	0.0001	–0.0001
	(0.0001)	(0.0001)
Temperature × Irrigation coverage	–0.1147**	–0.1287***
	(0.0547)	(0.0506)
Ln rainfall	0.4167***	0.2581***
	(0.0494)	(0.0484)
Ln rainfall × Time trend	–0.0288***	–0.0021
	(0.0052)	(0.0051)
Ln rainfall × Time trend squared	0.0006***	–0.0002
	(0.0001)	(0.0001)
Ln rainfall × Irrigation coverage	0.1538***	0.0649
	(0.0548)	(0.0697)
Irrigation coverage for each crop	2.4500*	3.2677***
	(1.4449)	(1.3315)
Irrigation coverage × Time trend	–0.0636***	–0.0221
	(0.0207)	(0.0153)
Irrigation coverage × Time trend squared	0.0020***	0.0004
	(0.0006)	(0.0005)
Literacy rate	–0.5039	–1.2734***
	(0.4283)	(0.4194)
Literacy rate × Time trend	0.0498	0.0834***
	(0.0311)	(0.0297)
Literacy rate × Time trend squared	–0.0011	–0.0015*
	(0.0010)	(0.0009)
Ln population density	0.1501**	0.1096
	(0.0744)	(0.0747)
Ln population density × Time trend	0.0004	0.0088**
	(0.0046)	(0.0045)
Ln population density × Time trend squared	0.0000	–0.0002*
	(0.0002)	(0.0001)
Inverse Mills ratio	–0.0588	–0.0578
	(0.0657)	(0.0673)
Time trend	0.2330***	–0.1620**
	(0.0752)	(0.0708)
Time trend squared	–0.0050**	0.0070***
	(0.0025)	(0.0024)
Constant term	1.0884	2.7625***
	(0.9656)	(0.9298)
Number of observations	4,809	4,408
R-squared: Overall	0.1982	0.2759
Hausman test		
χ^2statistics (degree of freedom)	87.72 (42)	–133.50 (43)
Prob>χ^2	0.0000	n/a

In the second step, rainfall is expressed in millimeters
***, **, and * indicate 1, 5, and 10% statistical significance levels, respectively

independent effect of irrigation on wheat yield is not significant. This would be partly because most wheat fields are irrigated (Singh and Jain 2000), so that it is difficult to identify its marginal effect on yield and partly because wheat is a dry season crop (Fujisaka et al. 1994) and, hence, irrigation provides water only occasionally. However, since irrigation has interactive effects with the climate variables as shown by the coefficients on the temperature-irrigation and rainfall-irrigation interaction terms, the average total irrigation effect on wheat yield appears largely positive.[7] The signs of these interaction effects indicate that higher irrigation coverage leads to a positive and significant effect of higher temperature, and to a reduced dependence on rainfall. The latter result suggests that the role of rainfall can be substituted for by irrigation. The coefficients on the time trend variables indicate that the impact of general technological advancement is positive on average, holding other variables constant.

The remarkable result is the declining dependence of yield on rainfall over time. The rainfall elasticity of wheat yield is 0.3149 at the beginning of the period under study (i.e., 1972), but it is likely to decrease to nearly 0.16 toward the end (i.e., 2002), as predicted by the coefficients on the time interaction terms: i.e., since the coefficient on the rainfall-time trend interaction term and the one on the rainfall-time trend squared interaction term are –0.0173 and 0.0004, respectively, the rainfall elasticity of wheat yield in each year is predicted by $0.3149-0.0173*t+0.0004*t^2$, where $t=0$ in 1972 and $t=30$ in 2002. It is important to note that this over-time change in the impact of rainfall is distinct from the influence of the availability of irrigation, since that influence is controlled for by the rainfall-irrigation interaction term. The result clearly supports our hypothesis that the rainfall effect is lessened, not augmented, by technological change over time.

The effect of population density is significant and positive, which is supportive of the induced innovation hypothesis of Hayami and Ruttan (1985) which states that the increasing scarcity of land induces the development and diffusion of land-saving and yield-enhancing technologies.[8]

4.4.2.2 Results for Rice

Rice is known as a crop that grows well on adequate heat and water, which is confirmed by the positive and significant coefficients on temperature, rainfall, and irrigation, which are 0.109, 0.571, and 2.844, respectively. In other words, in the early 1970s, a 1°C rise in temperature leads to an 11% increase in yield on average; the rainfall elasticity of yield is about 0.6. In contrast to wheat, a positive independent effect of irrigation on rice yield is confirmed, in which a one percentage point increase in irrigation coverage improves the yield by 2.8%, holding other variables constant.

---

[7] It is fairly easy to see this by substituting the climate variables in the interaction terms by the average climate shown in Tables 4.1 and 4.2.

[8] It is assumed that the marginal product of labor is sufficiently low.

The estimated coefficients on the interaction terms between the climate variables and the time trend or its squared term indicate that the dependence of rice yield on climate is alleviated over time, since the coefficients on the climate variables and the ones on the climate-time interaction terms have the opposite signs. This result supports our hypothesis that the rice yield has been affected less by climatic conditions over time, due probably to the adoption of improved rice varieties. Again, these over-time changes in the impact of climate are net of the influence of irrigation diffusion, since that influence is controlled for by the climate-irrigation interaction terms. Therefore, the critically important finding is that the dependence of rice yield on climate is mitigated over time regardless of the availability of irrigation, which cannot be understood without considering the impact of the adoption of the MVs with shorter maturity and drought-tolerant traits. Also, the negative coefficients on the rainfall-irrigation and the temperature-irrigation interaction terms indicate that irrigation can substitute the role of climate, to some extent.

One other highly interesting point is that the predicted irrigation effect increases in the early period, from 2.84 (percent per percentage point) in the early 1970s to 2.94 in the mid 1980s, but then decreases in the late period and becomes 2.77 in the early 2000s, which can be arithmetically calculated by the positive coefficient (0.0155) on the irrigation-time trend interaction term and the negative coefficient (–0.0006) on the irrigation-time trend squared interaction term. This result clearly indicates that the early generations of rice technology require more irrigation than do the recent ones, which is actually consistent with the suggestions by Janaiah et al. (2005), and Byerlee (1996).

The induced innovation hypothesis is supported, to greater extent than for wheat, with the elasticity of about 0.15. The coefficient on the time trend variable indicates that the impact of general technological advancement through the study period is positive and significant on average.

4.4.2.3 Results for Maize

The independent temperature effect is initially negative and significant in both fixed effect and random effect specifications: A 1°C rise in temperature results in a 3–6% drop in maize yield in the early 1970s, shown by the coefficients (–0.060 and –0.029). This adverse effect of higher temperature is found to be mitigated over time at a diminishing rate, since the coefficients on the temperature-time trend interaction terms and the ones on the temperature-time trend squared interaction terms are positive (0.008) and negative (–0.0002), respectively. The spreadsheet computation using these figures indicates that the adverse effect of temperature on maize yield actually approaches zero by the mid 1980s in both specifications. Even though the signs of the coefficients are the opposite to the case of rice, the two cases show essentially the same phenomenon of declining adverse temperature effect over time.

The unique result for maize may be the rainfall effect being weakly significant, with the elasticity of 0.09 which does not evolve over time. The effects of irrigation and its interactions with the time trend variables are all statistically insignificant. Also, the substitution relationship between climate and irrigation is weak. The absence of both rainfall effects and irrigation effects throughout the study period indicates that maize, whether it is TVs or MVs, is likely to be a drought-tolerant crop that can well cope with the lack of stable water supply.

For a one percentage point increase in literacy rate, the maize yield rises by 1.3–1.5% in the early 1970s, and this marginal effect decreases over time. The population density elasticity of yield turns out to be negative (–0.22 to –0.23) in the early 1970s, but it increases over time to become positive by the mid 1990s, which is supportive of the induced innovation hypothesis for maize farming in recent years.

Somewhat unexpectedly, the coefficient on the time trend variable is negative and significant, which may indicate that the positive impacts of the adoption of improved maize technology are captured, for the most part, by the changing coefficients on the explanatory variables over time.

4.4.2.4 Results for Sorghum

A time-invariant positive and significant effect of higher temperature is found, indicating that sorghum may be a heat-preferring crop. Particularly noteworthy is the decreasing coefficient on rainfall, which again suggests the declining impacts of rainfall on crop yield: The independent effect of rainfall on sorghum yield is positive and significant, with the initial elasticity of 0.42, but this dependence on rainfall decreases over time and accordingly the elasticity becomes 0.09 in the early 2000s. Although more concrete evidence must be obtained, it appears that a major effect of the improved traits of the sorghum MVs is the reduced downward yield risk associated with drought, rather than an enhancement in the maximum yield potential.[9]

The independent impact of irrigation is weakly significant, which seems to decrease over time. Thus, as for sorghum, the impacts of both rainfall and irrigation decline over time. In addition, irrigation seems to have a substitution effect for temperature, shown by the negative and significant coefficient on the interaction term (–0.115), indicating that when irrigation availability increases, the impact of temperature on sorghum yield is lessened.

As in the case of wheat and rice, the induced innovation hypothesis seems to apply to sorghum farming, which is shown by the positive and significant impact of population density.

4.4.2.5 Results for Millet

The estimate of the impact of temperature on millet yield is similar to that for sorghum, including the interactive effect of temperature through irrigation. The major difference from the case of sorghum is that the effects of both rainfall and irrigation,

[9] According to Cavatassi et al. (2011), the sorghum MVs adopted in eastern Ethiopia have early-maturing traits and, thus, can better cope with downward yield risks associated with moderate droughts.

though being positive and highly significant on average over the study period, do not exhibit an over-time decrease. Similar to sorghum, irrigation seems to have a substitution effect for temperature, shown by the negative and significant coefficient on the interaction term (−0.129), indicating that when irrigation availability increases, the impact of temperature on millet yield is lessened. Although it should be fully understood that the absolute level of irrigation coverage is rather low for millet (Fig. 4.5), it could be the case that the availability of irrigation is positively linked to the yield performance of millet.

The induced innovation hypothesis may be applicable in an increasing manner, as the predicted elasticity, being insignificant in the early 1970s, increases over time and becomes 0.08 in the early 2000s.

4.5 Concluding Remarks

While it is well-known that the GR enormously contributed to the growth of crop yields in Asia, it is much less known whether it mitigated or aggravated the impacts of adverse climate on crop yields. This study demonstrates, based on district level data in India for the three decades from 1972 to 2002, that the impact of climate represented by temperature and rainfall has been reduced over time. As for wheat, rice, and sorghum, the dependence on rainfall has evidently decreased and even disappeared in the recent years. For wheat, rice, and maize, which are the three major GR crops, the influence of temperature is also alleviated over time. These findings suggest that improved varieties and production practices can cope with unfavorable climatic conditions, due presumably to the shorter growth duration of those varieties and to the careful crop care. Furthermore, biotechnology offers considerable potential for strengthening such traits (Johnson et al. 2003), which may help improve agricultural productivity in some regions under unfavorable climate around the world.

As would be expected, irrigation plays an essential role in achieving higher yields for wheat, rice, sorghum, and millet in India. Moreover, irrigation contributes not only to increasing the yields but also to mitigating the impacts of harsh climatic conditions on crop yields.

A critically important policy implication of this study for agriculture in SSA may be a focus on rice as a strategic crop, if we admit that wheat can be grown only in relatively cool regions as in the temperate climate zone. In contrast, sorghum and millet do not appear to be attractive crops at this moment, as the small difference in yield between Asia and SSA indicates the absence of the opportunity to transfer technology for these crops from Asia to SSA. It is also important to note that although the transferability of Asian rice technology seems to be high (see Chaps. 2 and 3 in this volume), investments in irrigation and other water management may be required in SSA for this transfer to be successful.

References

Adams RM, Fleming RA, Chang C-C, McCarl BA, Rosenweig C (1995) A reassessment of the economic effects of global climate change on US agriculture. Clim Change 30(1):147–167

Auffhammer M, Ramanathan V, Vincent JR (2006) Integrated model shows that atmospheric brown clouds and greenhouse gases have reduced rice harvests in India. Proc Natl Acad Sci USA 103(52):19668–19672

Bhalla GS, Singh G (2001) Indian agriculture: four decades of development. Sage, New Delhi

Bruce JP, H-s Y, Haites EF (1996) Climate change 1995 – economic and social dimensions of climate change. Cambridge University Press, Cambridge

Byerlee D (1996) Modern varieties, productivity, and sustainability: recent experience and emerging challenges. World Dev 24(4):697–718

Cavatassi R, Lipper L, Barloch U (2011) Modern variety adoption and risk management in drought prone areas: insights from the sorghum farmers of Eastern Ethiopia. Agric Econ 42(3):279–292

Christiaensen L, Demery L (2007) Down to earth: agriculture and poverty reduction in Africa, directions in development. World Bank, Washington, DC

David CC, Otsuka K (eds) (1994) Modern rice technology and income distribution in Asia. Lynne Rienner, Boulder

Fan S, Hazell P, Thorat S (2000) Government spending, growth, and poverty in rural India. Am J Agric Econ 82(4):1038–1051

Fujisaka S, Larry H, Peter H (1994) Rice-wheat in South Asia: systems and long-term priorities established through diagnostic research. Agric Syst 46(2):169–187

Fujita K (2010) The Green Revolution and its significance for economic development: the Indian experience and its implications for Sub-Saharan Africa. Working Paper 17, Japan International Cooperation Agency Research Institute

Hayami Y, Ruttan VW (1985) Agricultural development: an international perspective. Johns Hopkins University Press, Baltimore

Heckman JJ (1979) Sample selection bias as a specification error. Econometrica 47(1):153–161

Hossain M, Fischer KS (1995) Rice research for food security and sustainable agricultural development in Asia: achievements and future challenges. Geo J 35(3):286–298

Janaiah A, Achoth L, Bantilan C (2005) Has the Green Revolution bypassed coarse cereals? The Indian experience. J Agric Dev Econ 2(1):20–31

Johnson ME, Hazell P, Gulati A (2003) The role of intermediate factor markets in Asia's Green Revolution: lessons for Africa? Am J Agric Econ 85(5):1211–1216

Khush GS (2001) Green Revolution: the way forward. Nat Rev Genet 2(10):815–822

Lawn RJ (1989) Agronomic and physiological constraints to the productivity of tropical grain legumes and prospects for improvement. Exp Agric 25(4):509–528

Lipton M (2007) Plant breeding and poverty: can transgenic seeds replicate the 'Green Revolution' as a source of gains for the poor? J Dev Stud 43(1):31–62

Mwabu G, Thorbecke E (2004) Rural development, growth, and poverty in Africa. J Afr Econ 13(Supplement):16–65

Olesen JE, Bindi M (2002) Consequences of climate change for European agricultural productivity, land use and policy. Eur J Agron 16(4):239–262

Omamo SW (2003) Policy research on African agriculture: trends gaps, and challenges. International Service for National Agricultural Research of International Food Policy Research Institute, The Hague

Ostrom E (1990) Governing the commons the evolution of institutions for collective action. Cambridge University Press, Cambridge

Otsuka K, Kalirajan K (2006) Rice Green Revolution in Asia and its transferability to Africa: an introduction. Dev Econ 44(2):107–122

Otsuka K, Estudillo JP, Sawada Y (2009) Rural poverty and income dynamics in Asia and Africa. Routledge, London

Pray CE, Nagarajan L (2010) Pearl millet and sorghum improvement in India. In: Spielman DJ, Pandya-Lorch R (eds) Proven successes in agricultural development: a technical compendium to millions fed, Chapter 11, International Food Policy Research Institute, Washington, DC

Reilly J, Baethgen W, Chege FE, Iglesias A, Kenny G, Patterson D, Rogasik J, Rötter R, Rosenzweig C, Sombroek W, Westbrook J, Bachelet D, Brllacich M, Dämmgen U, Howden M (1996) Agriculture in a changing climate: impacts and adaptation. In: Sedjo R, Brown S (eds) Climate change 1995: impacts, adaptations and mitigation of climate change: scientific–technical analysis, management of forests for mitigation of greenhouse gas emissions. Res Future Press, Washington, DC

Rosegrant MW, Hazell P (2000) Transforming the rural Asian economy: the unfinished revolution. Oxford University Press, Oxford

Rozeboom WW (1966) Scaling theory and the nature of measurement. Synthesis 16(2):170–233

Sanghi A, Mendelsohn R, Dinar A (1998) The climate sensitivity of Indian agriculture: in measuring the impact of climate change on Indian agriculture, vol 402, World Bank technical paper. World Bank, Washington, DC

Seo N, Mendelsohn R (2007) A Ricardian analysis of the impact of climate change on Latin American farms, vol 4163, World Bank policy research working paper. World Bank, Washington, DC

Singh AK, Jain GL (2000) Effect of sowing time, irrigation and nitrogen on grain yield and quality of durum wheat (triticum durum). Indian J Agric Sci 70(8):532–533

Stevens SS (1946) On the theory of scales of measurement. Science 103(2684):677–680

World Bank (2008) World development report 2008: agriculture for development. World Bank, Washington, DC

Chapter 5
The Impact of Technological Changes on Crop Yields in Sub-Saharan Africa, 1967–2004

Takuji Tsusaka and Keijiro Otsuka

Abstract Many specialists in African agriculture doubt whether a Green Revolution similar to the one achieved in Asia is possible in Sub-Saharan Africa (SSA). The major reasons why SSA has failed to realize a Green Revolution are considered to be its unfavorable, dry and diverse climate. The purpose of this study is to assess the impacts of climate, as well as population pressure on the agricultural crop yields in SSA from the late 1960s to the early 2000s. Using a country-level panel data set, we found evidence that technology advancements in SSA have mitigated the adverse effects of climatic factors on wheat, rice, and maize yields.

Keywords Technological change • Technology adoption • Crop yield • Cereal yield • Sub-Saharan Africa • Green Revolution • Irrigation • Cropping patterns

5.1 Introduction

Sub-Saharan Africa's slow economic growth and widespread poverty are a human tragedy that poses a challenge to the development profession. In sub-Saharan Africa (SSA), agriculture accounts for as much as 70% of its employment and about one-third of its economic growth from 1990 to 2005. Though urban slums gather most

T. Tsusaka
Social Sciences Division, International Rice Research Institute,
DAPO Box 7777, Metro Manila 1301, Philippines
e-mail: t.tsusaka@irri.org

K. Otsuka (✉)
National Graduate Institute for Policy Studies, 7-22-1 Roppongi, Minato-ku,
Tokyo 106-8677, Japan
e-mail: otsuka@grips.ac.jp

K. Otsuka and D.F. Larson (eds.), *An African Green Revolution: Finding Ways to Boost Productivity on Small Farms*, DOI 10.1007/978-94-007-5760-8_5,

of the attention due to their conspicuousness, over 70% of the poor in SSA actually live in rural areas and depend on agriculture for their livelihood (World Bank 2008). According to Ligon and Sadoulet (2007), in SSA, a 1% decrease in the agricultural gross domestic product (GDP) leads to a decrease in the consumption of the three poorest decile groups by as much as 4–6%. Thus, agriculture has to develop in SSA in order to reduce the persistent poverty.

Although staple food production has been increasing in SSA, the rate of increase has been exceeded by its high population growth rate. Consequently, the per-capita agricultural production in SSA has declined by about 10% since 1960 (Food and Agriculture Organization of the United Nations (FAO) 2008). The cultivated land per farm population has also declined in this region by about 40% since the 1960s (World Bank 2008). Furthermore, investments in new agricultural technologies have declined in recent years. Public spending on African agriculture, including investment in R&D, stands at an all-time low of less than 7% of the agricultural GDP, which is lower than the corresponding rates of 11% in Asia and almost 13% in Latin America. Donor support to agriculture in Africa has shrunk from 3 to 4 billion US dollars in the late 1980s to 1 billion US dollars today (Kuyvenhoven 2008).

In contrast, in Asia, growth in agricultural production has consistently outpaced population growth owing to the Green Revolution (GR) since the late 1960s (Otsuka and Kalirajan 2006). This is not only because the population growth has been somewhat slower in Asia, but much more importantly because the technological innovation represented by the diffusion of improved crop varieties and other complementary production practices spurred the agricultural yields in Asia. The GR has led to significant reductions in rural poverty as well as the growth of nonfarm sectors (Otsuka et al. 2009). At present, the adoption rate of modern crop varieties (MVs) is on average 78% in South Asia and 84% in East Asia, compared with 22% in SSA (Evenson and Gollin 2003).

Many specialists in African agriculture, however, doubt whether a GR similar to the one achieved in Asia is possible in SSA. One of the major reasons why SSA has failed to realize a GR is considered to be its unfavorable (e.g., dry) and diverse climate. Although direct statistical evidence is scanty for SSA, case studies of India and Brazil by Seo and Mendelsohn (2007), Sanghi et al. (1998), and Auffhammer et al. (2006) show significant effects of climate on crop yields and on agricultural income per unit of land, particularly the positive effect of rainfall, indicating that the relatively dry climate in SSA can be a major constraint on agricultural productivity growth.[1] The diversity of the climate in SSA results in a wide range of rain-fed farming systems producing a broad range of staple crops, which creates disadvantages for the Asian-type GR technologies that rely on standard technical packages

[1] To date, most of the empirical studies on how agro-climate factors affect agricultural productivity have focused on the United States (Adams et al. 1995; Mendelsohn et al. 1994) and other developed countries (Olesen and Bindi 2002; Bruce et al. 1996; Reilly et al. 1996).

for mono-cropping under controlled water management (Omamo 2003; Mwabu and Thorbecke 2004). Since the Asian GR technologies focuses on irrigated rice and wheat, the technology transfer to Africa may require location-specific adjustments so as to adapt to its adverse agro-climate (Otsuka and Kalirajan 2006).

One other reason for the failure of a GR in SSA is that irrigation and improved water management system have not been widely introduced in SSA. The availability of irrigation is considered to be vitally important for yield performance (see e.g., David and Otsuka 1994). As is confirmed by Tsusaka and Otsuka in Chap. 4, the Asian GR crops, i.e., rice and wheat, require the availability of sufficient water. Consequently, the adoption of improved rice and wheat technologies in SSA has been confined to the limited regions with high irrigation ratios or reliable rainfall. What is worse, only less than 5% of the favorable wetlands are planted with rice because of various constraints (Balasubramanian et al. 2007).

Yet, many recent studies show that there is high potential for new technology adoption and improved crop yields in SSA, which is yet to be fully exploited. Examples are the case studies of Diagne (2006), Sakurai (2006), Goufo (2008), Kajisa and Payongayong (2011), and Kijima et al. (2006), all of which point out that the rice yields will significantly increase once the constraints are properly addressed.[2] Otsuka and Kijima (2010) also argue that the GR in Asia has been technology-led, and thus investments in agricultural research and extension would open the door to an African GR. In fact, some analysis of the current state of agriculture shows that the adverse effects of the unfavorable production environments can be lessened by technological change. For example, in India, though modern varieties of cereal crops were introduced in irrigated areas at the initial stage of technology adoption, the adoption rate in non-irrigated areas started to rise at the later stage as technology gradually improved (Byerlee 1996). Similarly, in rural India, the rain-fed areas, including many less-favored areas, have recently exhibited higher agricultural growth for an additional unit of public investment than irrigated areas have (Fan and Hazell 1999).

Drawing on the above arguments, the purpose of this study is to assess the impacts of climate and population pressure on the agricultural crop yields in SSA from the late 1960s to the early 2000s. One feature of this analysis is the use of a country-level panel dataset. Thus, the over-time changes in those impacts can be examined in order to determine whether any technology advancements have mitigated the effects of those factors on crop specific yields.

The organization of this chapter is as follows. Section 5.2 overviews the basic agricultural statistics in SSA. Section 5.3 explains our data sources and how the database is constructed, followed by the introduction of the econometric methodology in Sect. 5.4. The regression results are carefully examined in Sect. 5.5. Lastly, Sect. 5.6 presents the concluding remarks.

[2] See the case studies of Côte d'Ivoire by Diagne (2006) and Sakurai (2006), Cameroon by Goufo (2008), Mozambique by Kajisa and Payongayong (2011), and Uganda by Kijima et al. (2006).

5.2 An Overview of Agricultural Performance in SSA

5.2.1 Cropping Patterns

SSA consists of 48 countries with diverse cropping patterns reflecting its diverse agro-climates. Figure 5.1 shows the proportions of areas harvested to major crops grown in SSA over the period of 2003–2007 in comparison with Asia. The crop composition in Asia is notably skewed to rice, accounting for 43% of the total cereal crop area vs. 10% in SSA. On the other hand, sorghum and millet have considerable shares in SSA, accounting for 24 and 23%, respectively, vs. 3 and 4% in Asia. This discrepancy is attributed to the relatively drought-tolerant trait of sorghum and millet, and the water-demanding nature of rice varieties (see e.g., Table 4.1 in Chap. 4).

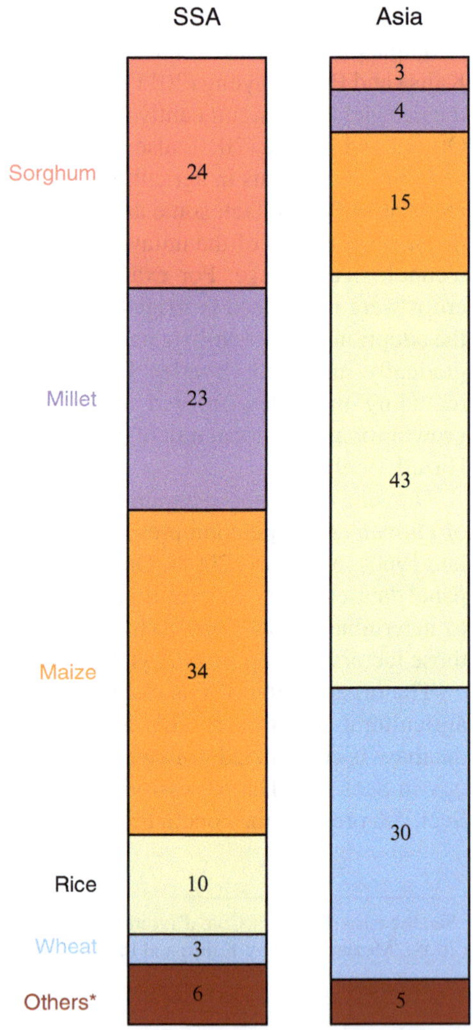

Fig. 5.1 Proportions of harvested areas by cereal crop (%), 2003–2007 average (Source: Author's calculation with FAOSTAT data).
* Others include ragi, oats, barley

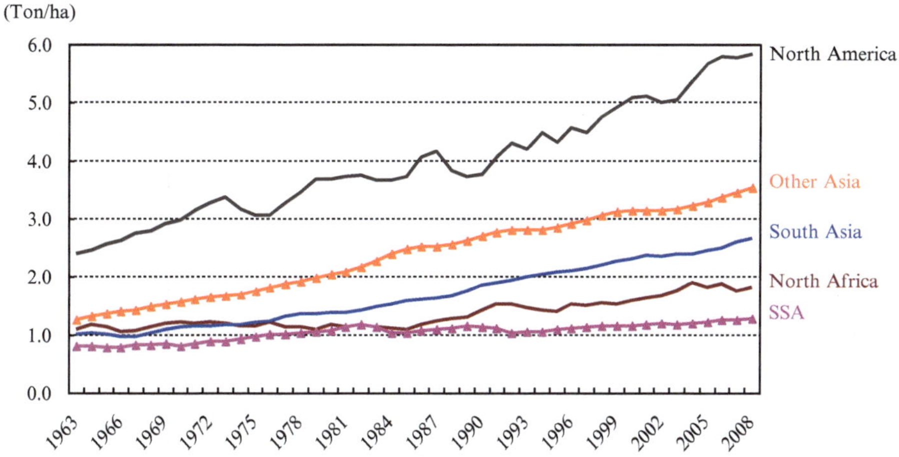

Fig. 5.2 Average cereal yields in SSA, Asia, North America, South Asia, and North Africa 3-year moving averages (Source: Author's calculation with FAOSTAT data)

5.2.2 Average Cereal Yields

Given the substantial difference in cropping patterns, how different are the crop yields between SSA and Asia? To answer this question, Fig. 5.2 compares the average cereal yields among SSA, South Asia, North Africa, other Asia, and North America. The GR has boosted the Asian cereal yield 2.8 times compared with the level in the early 1960s vs. 1.6 times in SSA. The most interesting finding from Fig. 5.2 is that despite the less favorable production environments, the cereal crop yield in SSA was not significantly inferior to that in other Asia, particularly to that in South Asia, until the early 1980s. The yield in North Africa was nearly the same as that in SSA until the late 1980s. These observations strongly indicate that before the GR, the productivity of cereal crops was not significantly different between SSA and other developing countries. As can be seen, however, the yields in SSA and Asia have diverged since the mid-1980s, and today the gap is approximately threefold. Therefore, it seems clear that the difference in climatic factors between SSA and Asia can explain only a small portion of the large gap in crop yields between the two regions.

5.2.3 Crop Yields

SSA's crop-wise yield statistics are shown in Fig. 5.3, for which the key figures are summarized in Table 5.1. It is noticeable that the yield of wheat has soared dramatically in SSA over the last several decades, increasing to 3.1 times the level in the early 1960s, which is actually comparable to the growth rate of the wheat yield in South Asia, 3.0 times. The yields of the other four crops have increased 1.4–1.5

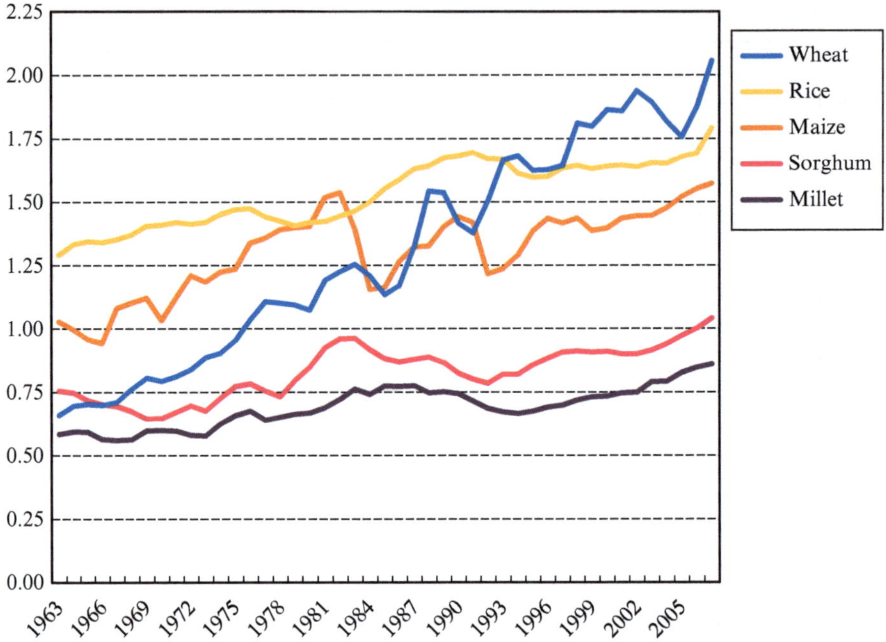

Fig. 5.3 Cereal yields in SSA by crop, 3-year moving averages (Source: Author's calculation with FAOSTAT data)

Table 5.1 Crop yields in SSA and South Asia and their growth

		Crop yield (t/ha)		Growth (times)
		1961–1963 Average	2005–2007 Average	1963–2007
SSA	Wheat	0.7	2.1	3.1
	Rice	1.3	1.8	1.4
	Maize	1.0	1.6	1.5
	Sorghum	0.8	1.0	1.4
	Millet	0.6	0.9	1.5
South Asia	Wheat	0.8	2.5	3.0
	Rice	1.5	3.3	2.1
	Maize	1.1	2.3	2.1
	Sorghum	0.5	0.8	1.7
	Millet	0.4	0.9	2.1
South Asian yield/ SSA yield	Wheat	1.3	1.2	
	Rice	1.2	1.9	
	Maize	1.1	1.4	
	Sorghum	0.6	0.8	
	Millet	0.7	1.0	

Source: Author's calculation with FAOSTAT data

times compared with the levels in the early 1960s, which is inferior to the growth rates of those in South Asia, 1.7–2.1 times.[3] Hence, as far as the yield growth rate is concerned, the GR seems to be occurring in wheat in SSA, but not really so in the other crops. Nonetheless, it is remarkable that the yields of the other four crops have also recorded considerable growth despite the much less favorable economic and climatic conditions in SSA.

If we were to consider the possibility of technology transfer from Asia to SSA, a direct comparison of the current yield levels between SSA and some parts of Asia, such as South Asia, whose climates are not so different from those in SSA, may be appropriate. The bottom section of Table 5.1 shows how much higher the yields in South Asia are than those in SSA. It is clear that the difference in the current rice yield is huge, followed by maize and then wheat, which seems to leave plenty of room for the transfer of rice technology to take place. On the other hand, when it comes to sorghum and millet, there would be limited transferability of technology from Asia to SSA.

5.2.4 Limitation of Wheat

The above discussions on the crop yields may suggest the expansion of wheat planted area within SSA, and the promotion of the technology transfer for rice and maize from Asia. However, it must be pointed out that the feasibility of the former is limited because wheat can be grown well only under a cool climate, which is associated with the temperate climate zone. In the African continent, the temperate climate zone is found only in some parts of North Africa and South Africa, and wheat is thus grown only in the Republic of South Africa, the highlands in Ethiopia, and a few other regions in SSA (Fig. 5.4), which explains why a mere 3% of the total crop area is planted to wheat (Fig. 5.1). It is also known that, during the GR in India, wheat crop was more likely to be chosen to be grown in districts with lower temperature (Table 4.4 in Chap. 4). Therefore, it appears that agricultural growth potential in SSA rests in the intensification of rice and maize farming by new technology adoption.[4]

5.2.5 Changes in Crop Areas

Figure 5.5 shows how the crop areas have been changing in SSA. The average annual growth rates of harvested areas indicate that farmers in SSA have been expanding the areas planted to rice, maize, sorghum, and millet, for which the growth of the yields

[3] The reason why we compare the yields in SSA with those in South Asia instead of with the whole of Asia is that South Asia is the only region in Asia where sorghum and millet, the two major crops in SSA, are grown (see Fig. 4.1 in Chap. 4 in this volume).

[4] Tsusaka and Otsuka also found for India that wheat crop has been becoming more heat-tolerant over time (Table 4.5 in Chap. 4). Therefore, the potential of wheat expansion in SSA may not be totally excluded.

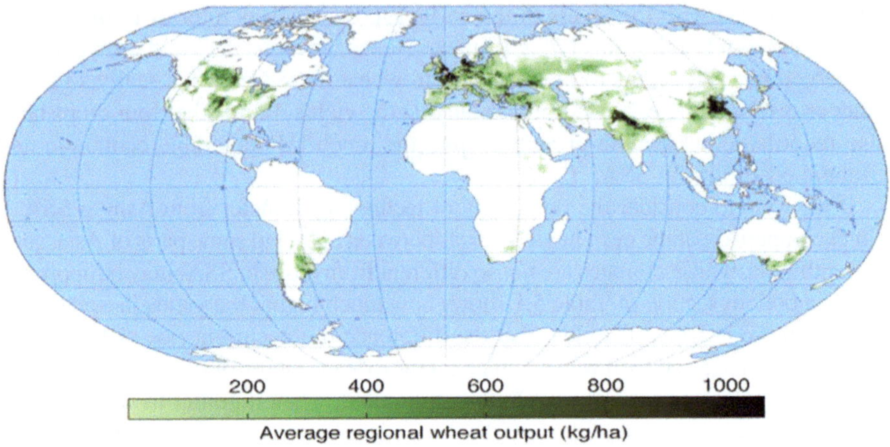

Fig. 5.4 Global wheat map: amount of wheat output per hectare of total agricultural land (Source: Compiled by the University of Minnesota Institute on the Environment with data from: Monfreda et al. 2008)

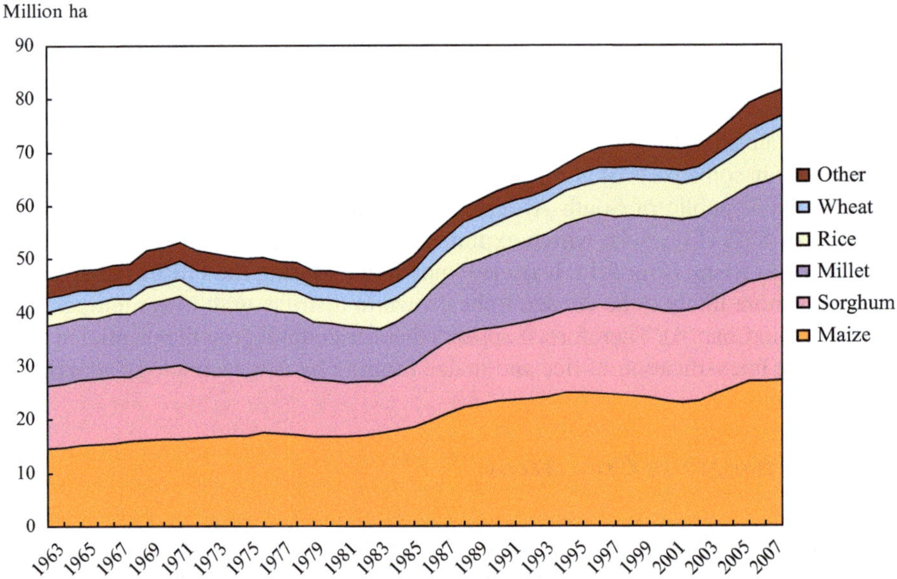

Fig. 5.5 Changes in harvested areas by cereal crop in SSA, 3-year moving averages (Source: Author's calculation with FAOSTAT data)

has been relatively slow, but have been decreasing the area planted to wheat, whose yield has been growing dramatically, which may be related, in part, to the limitation of the wheat growing environment previously mentioned. This observation is in sharp contrast with India's case, where farmers have been gradually shifting from sorghum and millet to wheat, rice, and maize (Fig. 4.4 of Chap. 4).

5.2.6 Technology Adoption

Unfortunately, direct statistics to show the level of the adoption of new technology are scanty in SSA, including the adoption of modern varieties (MV) and other improved farming techniques. Nevertheless, the yield statistics discussed above suggest that technology improvement has been taking place, particularly in wheat and also in other crops to lesser extent. A potentially important trait of MVs is the short growth duration (Khush 2001; Hossain and Fischer 1995; Lawn 1989), so that the cereal crops can be grown in a shorter period during which rainfall is much assured. For example, the latest rice MVs mature in 105–110 days, which is much shorter than the growth duration of 160–170 days of traditional varieties in Asia (Khush 2001). If so, it seems reasonable to hypothesize that the adoption of the Asian-type MVs lessens the impact of precipitation on crop yields in SSA.

5.3 Database Construction

One unique feature of this study is the use of country-level panel data to quantify the effects of various factors on crop specific yields and their changes over time. The data set covers 48 countries in SSA over an extended period of 38 years from 1967 to 2004. The important variables are crop yields, i.e., the quantity of agricultural output per area for five major crops (wheat, rice, maize, sorghum, and millet), climate represented by temperature and rainfall, real output price deflated by consumer price index,[5] population density, and literacy rate. The database is composed of several different data sources. Data on agricultural outputs are obtained from FAOSTAT provided by FAO. Meteorological station-level data on temperature and rainfall are collected through the Global Observing Systems Information Center. There are some limitations with these climate data: quite a few meteorological stations have many missing data points over the period of study. Fourteen out of the 48 countries have no station,[6] while some of the rest have more than one station, in which cases we chose the most representative station in terms of the data availability and the agro-climatic perspective rather than manipulating to reflect incomplete data from all the existing stations. Regarding the output price, FAO offers the nominal prices in local currencies from 1966 to 1990 and the international prices in US dollars from 1991 to 2004. From 1966 to 1990, we deflated the nominal prices by the consumer price indices (CPIs) provided by the World Development Indicator (WDI), though

[5] It is desirable to use the output-input price ratio to account for the practical impact on the output. However, we cannot do so because the input prices such as fertilizer prices are unavailable.

[6] The 14 countries with no meteorological stations, which are hence excluded from the regression analyses, are Burundi, Central African Republic, Comoros, Djibouti, Gambia, Ghana, Guinea, Guinea-Bissau, Lesotho, Liberia, Rwanda, Sao Tome and Principe, Somalia, and Uganda.

the CPIs are not available for all the countries.[7] The data on population density and literacy rate are respectively obtained from WDI and the United Nation's Educational, Scientific and Cultural Organization (UNESCO).

5.4 Empirical Methodology

5.4.1 Estimated Yield Functions

Once the database is constructed, the next step is to assess the effects of climate (temperature and rainfall), population density, and literacy rate on crop yields, and their changes over time. To estimate the yield functions by crop, the basic estimation model can be specified as follows:

$$Y_{ijt} = \alpha_{1i}C_{jt} + \alpha_{2i}C_{jt} \cdot t' + \alpha_{3i}C_{jt}t'^2 + \beta_{1i}P_{ijt} + \beta_{2i}P_{ijt} \cdot t' + \beta_{3i}P_{ijt} \cdot t'^2$$

$$+\gamma_{1i}X_{jt} + \gamma_{2i}X_{jt} \cdot t' + \gamma_{3i}X_{jt} \cdot t'^2 + \theta_i\tau_t + v_{ij} + \varepsilon_{ijt}$$

where Y_{ijt} is the yield of crop *i* in country *j* in year *t*, C_{jt} is a vector of climate variables (temperature and rainfall); P_{ijt} is the real price of the output; X_{jt} is a vector of country characteristics (population density and literacy rate); τ_t is a vector of year dummies; v_{ij} is the unobservable time-invariant country specific effect; ε_{ijt} is the error term; t' is time trend ranging from 0 to 37. Finally and very importantly, the interaction terms with t' and t'^2 are included to examine whether there have been over-time changes in the impacts of the explanatory variables due to any technological change, such as the introduction of short-maturity MVs and better crop care. The yield functions are estimated separately for the different crops.

The coefficients on population density are particularly important for our purpose. If the coefficient γ_{1i} is significantly positive, it means that population density has a positive effect on the crop yield, which is supportive of the induced innovation hypothesis of Hayami and Ruttan (1985) which states that the increasing scarcity of land induces the development and diffusion of land-saving and yield-enhancing technologies. γ_{2i} and γ_{3i} capture the over-time change in the impact of population density. If their combined effect is positive, it may suggest that the inducement of innovation takes place with a lag.

The reason why the country-level fixed (or random) effect model is employed is that there must be several omitted variables such as irrigation coverage and road density. We began with panel regressions and accordingly conducted the Hausman test, to compare the fixed effect and random effect estimations (Hausman 1978). In all cases, the random effect estimation is diagnosed as inconsistent (i.e., $p<0.10$). Therefore, we decided to

[7] As for the deflator, it is ideal to use the producer price indices because producer prices and consumer prices may exhibit different behaviors. We cannot do so, however, because of the data unavailability.

employ the country-level fixed-effect model by explicitly using the country dummies, which are expected to absorb unobservable time-invariant country-specific effects.

If district-level or household-level data were used, prices would be somewhat exogenous. The use of country-level data suggests that the endogeneity of the price variable must be considered. It is desirable to endogenize it using instruments such as distance to major markets and the development of infrastructure. Because of the unavailability of such variables, however, we cannot do so. Also, the application of two-stage least squares regressions using the population density as an instrument severely worsens the statistical significance of the estimators and the overall fit of the regressions, which may be taken to imply that proper instruments are absent. Therefore, we do not employ such corrective techniques but perform the regressions with and without the price variable to check the robustness.

Another issue to be taken into account is that each crop is grown in some countries but not in others, which means that to be formal, the sample selection bias needs to be treated (Heckman 1979). However, it turns out either that we do not succeed in performing the first step probit regressions, or that we cannot obtain any statistically significant coefficients on the inverse Mills ratios in the second step outcome estimations.[8] This is due presumably to the limited number of cross-sectional observations, the absence of proper instrumental variables in the first step, and the use of the country specific effect model that can mitigate, if not solve, the sample selection bias. Therefore, we directly perform the outcome estimations, by assuming that the sample selection biases are negligible in the model we use.

To obtain the elasticities, the logarithm of the variables are taken whenever applicable, the exceptions being the ratio variable (viz., literacy rate), the time trend variables, the interval-scale variable (temperature in Celsius), and the year and country dummies (Stevens 1946; Rozeboom 1966).

5.4.2 *Model Specifications*

Two model specifications are employed as follows:

Model 1 with Year Dummies: This specification is what is known as the two-way fixed effect model; that is, it includes the year dummies for all the available years (1968–2004 with 1967 as the base year), in order to capture the average yearly change in yield that is not explained by the explanatory variables, e.g., aggregate macroeconomic and climatic shocks as well as major and minor technological improvements. A major defect of this methodology is that we cannot directly observe whether there has been a positive trend of yield improvement due to technological change, among other things.

[8] The instruments we attempted to use in the first step probit estimations are the longer-term temperature and rainfall, which are likely to affect the crop choice decisions. For maize, the probit regression does not succeed because maize is grown in all the countries included in the regression. For the other crops, the coefficients on the instrumental variables are estimated to be highly significant. However, in the second step outcome estimations, the sample selection bias associated with this first step estimation is identified to be insignificant in all cases.

Model 2 with Time Trend Variables ($t' = 0$ for 1967): Instead of the year dummies, this specification includes the time trend (t') and its squared term (t'^2), in order to capture the effect of general technology improvement over time and its acceleration (or deceleration) that is not picked up by the interaction terms. Since the year-specific effects are not controlled for in this model, the estimated coefficients may be biased. Thus, care must be taken in comparing the estimation results of Models 1 and 2.

5.5 Regression Results

Tables 5.2, 5.3, 5.4, 5.5, and 5.6 present the estimation results of the yield functions for wheat, rice, maize, sorghum, and millet, respectively. For simplicity, the estimated coefficients on the year dummies are not presented in the regression tables. Shown in the tables are the estimated coefficients with the standard errors in parentheses. The results with and without the price variable are shown for comparison. As explained in the previous section, the price data are not available for all countries and years. For meaningful comparison, the countries with missing price data, which are expected to be relatively low-profile countries, are excluded from the without-price regressions. Therefore, the same set of countries is included in both the with- and without-price regressions, although the years included are not necessarily the same between the two because the with-price regression requires the price data which are missing in some years. This is why the numbers of observations are generally not the same between the two cases.

5.5.1 Results for Wheat

Table 5.2 shows the estimation results of the yield function of wheat, in which ten countries are included. They are Cameroon, Ethiopia, Kenya, Madagascar, Malawi, Mali, Mozambique, Niger, Rep. of South Africa, and Zimbabwe, with Cameroon as the base for the country dummies. The differences in the estimated coefficients and their significance levels between the with- and without-price specifications are generally not so conspicuous, indicating that the price variable is not causing a serious problem to the estimation of the coefficients on the other variables. On the other hand, the estimated coefficients are somewhat different between Models 1 and 2, indicating that the year dummies capture the impacts of various shocks in climate and economy, so that the estimated coefficients on the explanatory variables change to some extent when the year dummies are replaced by the time trend variables.

According to the estimation results, temperature seems to have a positive effect on wheat yield at the beginning, but the effect is mitigated over time and finally disappears around 1990. The originally positive effect of temperature on wheat yield contrasts with what we found in the Indian case (Table 4.5 in Chap. 4). Our interpretation is that because in SSA wheat is grown in limited regions with relatively low temperature, the temperature effect is positive in such a low temperature range. The effects of rainfall are almost insignificant, whereas price has a

Table 5.2 Country fixed-effect regression results for yield of wheat, 1967–2004

Dependent variable: Ln wheat yield	Without price		With price	
	Model 1	Model 2	Model 1	Model 2
Estimation				
Temperature	0.2699***	0.0670	0.4518***	0.1445*
	(0.0971)	(0.0752)	(0.1078)	(0.0789)
Temperature × Time trend	−0.0229***	−0.0077*	−0.0277***	−0.0091**
	(0.0062)	(0.0047)	(0.0066)	(0.0047)
Temperature × Time trend squared	0.0004***	0.0001	0.0004***	0.0001**
	(0.0000)	(0.0001)	(0.0000)	(0.0001)
Ln rain	−0.2229	−0.1300	−0.4205	−0.1841
	(0.2412)	(0.2153)	(0.3118)	(0.2856)
Ln rain × Time trend	0.0149	0.0054	0.0426*	0.0199
	(0.0040)	(0.0154)	(0.0040)	(0.0211)
Ln rain × Time trend squared	−0.0001	0.0001	−0.0007*	−0.0003
	(0.0003)	(0.0003)	(0.0004)	(0.0004)
Ln real price			−1.7140***	−0.9755**
			(0.4785)	(0.4192)
Ln real price × Time trend			0.1015***	0.0600**
			(0.0336)	(0.0304)
Ln real price × Time trend squared			−0.0017***	−0.0011**
			(0.0006)	(0.0005)
Ln population density	1.6184**	1.3718**	1.5865***	1.1113***
	(0.6584)	(0.6110)	(0.2664)	(0.2366)
Ln population density × Time trend	−0.1279***	−0.0913***	−0.1295***	−0.0864***
	(0.0212)	(0.0179)	(0.0192)	(0.0168)
Ln population density × Time trend squared	0.0017***	0.0010***	0.0018***	0.0010***
	(0.0004)	(0.0003)	(0.0004)	(0.0090)
Literacy rate	−0.3458	−2.2333**	4.9869***	0.7397
	(1.2243)	(1.0966)	(1.8329)	(1.4724)
Literacy rate × Time trend	−0.0464	0.1492**	−0.2691**	0.0376
	(0.0889)	(0.0707)	(0.1157)	(0.0846)
Literacy rate × Time trend squared	0.0012	−0.0020*	0.0044***	0.0002
	(0.0015)	(0.0090)	(0.0017)	(0.0013)
Time trend		0.3088		−0.0514
		(0.1771)		(0.2366)
Time trend squared		−0.0045		0.0022
		(0.0030)		(0.0042)
Ethiopia	0.1451	0.3807	0.9412***	1.3222***
	(0.5190)	(0.4912)	(0.3647)	(0.332)
Kenya	0.7741***	0.5646***	0.7979***	0.6156***
	(0.2323)	(0.2143)	(0.2215)	(0.2087)
Madagascar	0.6749***	0.6559***	0.4786***	0.5392***
	(0.0000)	(0.0000)	(0.0000)	(0.0000)
Malawi	0.4275	0.3084	0.4948***	0.5482***
	(0.5432)	(0.5139)	(0.1786)	(0.1668)

(continued)

Table 5.2 (continued)

Dependent variable: Ln wheat yield	Without price		With price	
	Model 1	Model 2	Model 1	Model 2
Mali	0.1102	0.5859	0.3044	0.7171*
	(0.0040)	(0.0040)	(0.0040)	(0.0040)
Mozambique	–0.1437	0.0780	0.0816	0.3996*
	(0.2385)	(0.2134)	(0.2436)	(0.2227)
Niger	0.0526	0.5841	0.5397	1.1113***
	(0.6829)	(0.6324)	(0.4927)	(0.4457)
Republic of South Africa	0.4060	0.1734	0.3562	0.1034
	(0.2807)	(0.2528)	(0.2739)	(0.2582)
Zimbabwe	1.6285***	1.4210***	1.5566***	1.3391***
	(0.3006)	(0.2700)	(0.2655)	(0.2496)
Constant term	0.6661	1.0912	2.0161	4.5411
	(1.8474)	(2.3101)	(1.9311)	(2.9599)
Number of observations	312	312	279	279
R-squared	0.7992	0.7775	0.8600	0.8314
Adjusted R-squared	0.7532	0.7597	0.8206	0.8140

***, **, and * indicate 1, 5, and 10% statistical significance levels, respectively

negative impact, even though it diminishes over time.[9] Population density has a positive effect on wheat yield initially, though the effect decreases over time. The positive impact of population density on the yield is supportive of the induced innovation hypothesis as mentioned earlier. The decreasing effect of the population density over time may be explained by the exhaustion of yield-enhancing technological possibilities. The time trend variables in Model 2 are not significant, which suggests that the impact of technological change is captured by the changing coefficients of some explanatory variables such as temperature and population density. Another reason may be that there are a large number of small wheat-producing countries which have not experienced appreciable growth in wheat yield. The coefficients on the country dummies indicate that the wheat yield is significantly higher in Kenya, Madagascar, and Zimbabwe than in Cameroon, on average, during the period under study.

5.5.2 *Results for Rice*

Table 5.3 shows the estimation results of the yield function of rice, in which nine countries are included (Burkina Faso, Ethiopia, Kenya, Madagascar, Malawi, Mali, Mozambique, Niger, and Zimbabwe, with Burkina Faso as the base for the country

[9] The negative effect of the real output price is unexpected. A possible reason is that the wheat price is lower (higher) in major (minor) wheat growing and exporting (importing) regions, where the wheat yield is higher (lower).

Table 5.3 Country fixed-effect regression results for yield of rice, 1967–2004

Dependent variable: Ln rice yield	Without price		With price	
	Model 1	Model 2	Model 1	Model 2
Estimation				
Temperature	−0.1485**	−0.1339**	−0.1425*	−0.1154*
	(0.0759)	(0.0635)	(0.0799)	(0.0641)
Temperature × Time trend	0.0136***	0.0111***	0.0136***	0.0103***
	(0.0044)	(0.0037)	(0.0049)	(0.0038)
Temperature × Time trend squared	−0.0002***	−0.0002***	−0.0003***	−0.0002***
	(0.0000)	(0.0001)	(0.0000)	(0.0001)
Ln rain	0.6256***	0.5375***	0.3229	0.1102
	(0.2247)	(0.2051)	(0.3332)	(0.2628)
Ln rain × Time trend	−0.0413**	−0.0310*	−0.0200	0.0005
	(0.0040)	(0.0162)	(0.0040)	(0.0204)
Ln rain × Time trend squared	0.0006*	0.0005	0.0003	0.0000
	(0.0003)	(0.0003)	(0.0005)	(0.0004)
Ln real price			−0.3171	−0.4602
			(0.3598)	(0.31)
Ln real price × Time trend			0.0320	0.0462**
			(0.0257)	(0.0223)
Ln real price × Time trend squared			−0.0005	−0.0008*
			(0.0005)	(0.0004)
Ln population density	1.4199**	1.6446**	1.9960**	2.0197***
	(0.7159)	(0.6807)	(0.8116)	(0.7483)
Ln population density × Time trend	0.0531***	0.0413***	0.0460**	0.0334**
	(0.0167)	(0.0144)	(0.0191)	(0.0148)
Ln Population density × Time trend squared	−0.0012***	−0.0010***	−0.0011***	−0.0009***
	(0.0003)	(0.0003)	(0.0003)	(0.0090)
Literacy rate	−6.2579***	−5.8521***	−5.5703***	−4.2862***
	(1.149)	(1.0603)	(1.6821)	(1.3834)
Literacy rate × Time trend	0.2094***	0.1572**	0.1721	0.0813
	(0.0776)	(0.0641)	(0.1074)	(0.0748)
Literacy rate × Time trend squared	−0.0035***	−0.0025**	−0.0033*	−0.0019
	(0.0014)	(0.0090)	(0.0018)	(0.0013)
Time trend		−0.2704*		−0.6729***
		(0.1607)		(0.2352)
Time trend squared		0.0061**		0.0137***
		(0.003)		(0.0044)
Ethiopia	0.0316	−0.2115	−0.3308	−0.6891
	(0.6441)	(0.5491)	(0.6807)	(0.5733)
Kenya	2.4974***	2.4260***	2.5058***	2.3163***
	(0.6915)	(0.6065)	(0.7451)	(0.6773)
Madagascar	2.5875***	2.6277***	2.9004***	2.7228***
	(0.0000)	(0.0000)	(0.0000)	(0.0000)
Malawi	−0.2090	−0.3388	−0.5495	−0.6182
	(0.4934)	(0.4361)	(0.6118)	(0.5588)

(continued)

Table 5.3 (continued)

Dependent variable: Ln rice yield	Without price		With price	
	Model 1	Model 2	Model 1	Model 2
Mali	3.0795***	3.2533***	3.8097***	3.6329***
	(0.0040)	(0.0040)	(0.0040)	(0.0040)
Mozambique	1.0980**	1.0697**	1.3312**	1.0659**
	(0.511)	(0.4773)	(0.5686)	(0.5179)
Niger	2.8772***	2.8359***	2.8809***	2.5913***
	(0.9021)	(0.8138)	(0.9487)	(0.868)
Zimbabwe	2.6191***	2.5895***	2.7181***	2.4295***
	(0.9501)	(0.8608)	(0.9937)	(0.9109)
Constant term	−0.7005	0.7871	−0.7003	4.1911
	(2.5)	(2.6841)	(2.7355)	(3.3172)
Number of observations	305	305	291	291
R-squared	0.7547	0.7340	0.7695	0.7541
Adjusted R-squared	0.6980	0.7133	0.7093	0.7309

***, **, and * indicate 1, 5, and 10% statistical significance levels, respectively

dummies). The estimated coefficients and their significance levels remain largely the same regardless of whether the price variable is added. Also note that, in the case of rice, the estimated coefficients and their statistical significance are not so different between Models 1 and 2, indicating that the year dummies capture the pure random shocks in addition to the trend effects, so that the estimated coefficients on the explanatory variables remain largely unchanged when the year dummies are replaced by the time trend variables.

As for the impacts of climate on rice yield, a one-degree Celsius increase in temperature leads to a 12–15% decrease in the yield in the late 1960s, but this negative impact of higher temperature is alleviated over time at a diminishing rate. According to the without-price estimations, the elasticity of rainfall ranges from 0.5 to 0.6 in the early 1970s, but around 1990 it becomes almost zero. These results support the hypothesis that the climate effect is lessened by technological change.

The price effects on rice yield are much weaker than in the case of wheat. The elasticity of population density turns out to be 1.6–2.1 in the early period, and it increases to 2.0–2.5, and then decreases to 1.9–2.3, experiencing an inverted U-shape like change over time. The positive impact of population density throughout the period under study is, again, supportive of the induced innovation hypothesis. What may be odd is that the time trend has negative coefficients. However, when we look into the year dummies one by one, it turns out that the coefficients are significantly negative in the early years but become insignificant and positive in the recent years, which may indicate that the new technology adoption for rice began recently rather than earlier on. The coefficients on the country dummies show that the rice yield is considerably higher in Kenya, Madagascar, Mali, Niger, and Zimbabwe, than in Burkina Faso, on average, during the period under study. It is interesting to note that irrigation coverage for rice harvested areas is reported to be high in Kenya (100%),

Madagascar (52%), and Niger (80%) (Balasubramanian et al. 2007), so that the positive coefficients of some country dummies would capture the positive effects of irrigation on rice yields.

5.5.3 *Results for Maize*

Table 5.4 shows the estimation results of the yield function of maize, in which 12 countries are included (Burkina Faso, Cameroon, Ethiopia, Kenya, Madagascar, Malawi, Mali, Mozambique, Namibia, Rep of South Africa, Togo and Zimbabwe,

Table 5.4 Country fixed-effect regression results for yield of maize, 1967–2004

Dependent variable: Ln maize yield	Without price		With price	
	Model 1	Model 2	Model 1	Model 2
Estimation				
Temperature	−0.0237	−0.0001	0.0815	0.0594
	(0.0584)	(0.0528)	(0.0566)	(0.0509)
Temperature × Time trend	−0.0015	−0.0051*	−0.0093***	−0.0095***
	(0.0033)	(0.0031)	(0.0031)	(0.0029)
Temperature × Time trend squared	0.0000	0.0001	0.0001*	0.0001**
	(0.0000)	(0.0001)	(0.0000)	(0.0001)
Ln rain	0.6656***	0.5151***	0.1644	0.1574
	(0.1782)	(0.1728)	(0.1931)	(0.1832)
Ln rain × Time trend	−0.0560***	−0.0371***	−0.0136	−0.0073
	(0.0040)	(0.0144)	(0.0040)	(0.0143)
Ln Rain × Time trend squared	0.0011***	0.0007***	0.0004	0.0002
	(0.0003)	(0.0003)	(0.0003)	(0.0003)
Ln real price			−0.5721***	−0.4181**
			(0.2169)	(0.2163)
Ln real price × Time trend			0.0384**	0.0266
			(0.0191)	(0.0190)
Ln real price × Time trend squared			−0.0007*	−0.0005
			(0.0004)	(0.0004)
Ln population density	0.1530	0.3365	−0.3977	−0.7004
	(0.5194)	(0.5288)	(0.5903)	(0.5937)
Ln population density × Time trend	0.0225**	0.0101	−0.0076	−0.0076
	(0.0100)	(0.0092)	(0.0123)	(0.0117)
Ln population density × Time trend squared	−0.0004*	−0.0001	0.0002	0.0002
	(0.0002)	(0.0002)	(0.0002)	(0.0090)
Literacy rate	−0.5547	0.1493	3.3870***	3.6863***
	(0.9386)	(0.9205)	(0.9726)	(0.9289)
Literacy rate × Time trend	0.0269	−0.0679	−0.2016***	−0.2179***
	(0.0633)	(0.0556)	(0.0633)	(0.0548)
Literacy rate × Time trend squared	−0.0014	0.0003	0.0019*	0.0022**
	(0.0012)	(0.0090)	(0.0011)	(0.0010)

(continued)

Table 5.4 (continued)

Dependent variable: Ln maize yield	Without price		With price	
	Model 1	Model 2	Model 1	Model 2
Time trend		0.3449***		0.2561*
		(0.1334)		(0.1419)
Time trend squared		−0.0059**		−0.0025
		(0.0025)		(0.0027)
Burkinafaso	−0.2262	−0.1732	0.2737	0.6321
	(0.4542)	(0.4354)	(0.4764)	(0.4505)
Ethiopia	−0.8172*	−1.0229**	−0.8466*	−0.7423
	(0.4789)	(0.4817)	(0.4784)	(0.4799)
Kenya	0.1294	0.1568	0.2404	0.3292
	(0.0000)	(0.0000)	(0.0000)	(0.0000)
Madagascar	0.1886	0.2967	−0.1702	−0.1391
	(0.2163)	(0.2075)	(0.2083)	(0.205)
Malawi	−0.4444	−0.4697	0.3269	0.6754
	(0.0040)	(0.0040)	(0.0040)	(0.0040)
Mali	0.7301	0.9418	−0.2621	−0.3373
	(0.6193)	(0.6136)	(0.6357)	(0.6356)
Mozambique	−0.5414***	−0.5546***	−1.2266***	−1.2475***
	(0.1901)	(0.1820)	(0.2118)	(0.2056)
Namibia	1.3229	1.5989	−1.5300	−2.3096
	(1.4093)	(1.4424)	(1.525)	(1.5332)
Republic of South Africa	0.5202**	0.5908***	0.2424	0.2040
	(0.2452)	(0.2390)	(0.2308)	(0.2241)
Togo	−0.3834	−0.3505	0.6299	1.0822*
	(0.559)	(0.5593)	(0.6246)	(0.6083)
Zimbabwe	0.2110	0.2865	−0.1897	−0.2638
	(0.2651)	(0.2587)	(0.2566)	(0.2499)
Constant term	2.5314*	0.2626	5.6548***	4.4663**
	(1.5511)	(1.9269)	(1.5106)	(1.8666)
Number of observations	391	391	364	364
R-squared	0.6266	0.5564	0.7193	0.6645
Adjusted R-squared	0.5587	0.5260	0.6603	0.6365

***, **, and * indicate 1, 5, and 10% statistical significance levels, respectively

with Cameroon as the base for the country dummies). In the case of maize, the estimated coefficients and their significance levels are substantially different between the with- and without-price specifications, indicating that the price variable is correlated with the other variables. Considering the problematic nature of the price variable, it will probably be safe to pay more attention to the without-price specifications. On the other hand, the use of the year dummies and that of the time trend variables do not alter the estimated coefficients of the other variables substantially, indicating that the year dummies capture the pure random shocks on top of the trend effects.

The without-price estimation shows that the impact of temperature on maize yield is almost insignificant, which suggests that maize has a heat tolerant trait.

The remarkable result is that the dependence of maize yield on rainfall decreases over time. That is, the initial elasticities are 0.4–0.5 but diminish and disappear around 1990. This result clearly supports the hypothesis that the maize yield has been less and less dependent on rainfall due probably to more drought tolerant traits of improved maize varieties.

Compared with wheat and rice, the impact of population density on maize yield is much less significant. According to Model 1, the elasticity of population density is insignificant initially and then becomes significantly positive, reaching 0.3 in the 1990s, which indicates that the population pressure has stimulated efforts to increase maize yield in recent years, and that the regional diffusion of maize technology may be difficult. The last result is supportive of Chap. 8 in this volume, which implies the difficulty in the regional diffusion of maize technology. The unique result for maize is that the time trend variables exhibit a positive effect at a diminishing rate, which suggests that there is some technological improvement, e.g., increased pest-resistance, of which the impact is not captured by the changing coefficients on the explanatory variables such as rainfall and population density. According to Smale et al. (Chap. 8), the maize MV adoption rate is currently 44% in South and East Africa and 60% in West and Central Africa. The coefficients on the country dummies show that, compared with the case in Cameroon, the maize yield is significantly higher in Rep of South Africa, and significantly lower in Ethiopia, and Mozambique, on average, during the period under study.

5.5.4 *Results for Sorghum*

Table 5.5 shows the estimation results of the yield function of sorghum, in which 13 countries are included (Burkina Faso, Cameroon, Ethiopia, Kenya, Madagascar, Malawi, Mali, Mozambique, Namibia, Niger, Rep of South Africa, Togo and Zimbabwe, with Cameroon as the base for the country dummies). Similar to maize, the estimated coefficients and their significance levels are considerably different between the with- and without-price specifications, indicating that the price variable is correlated with the other variables. Thus, again, it is probably preferable to pay more attention to the without-price specifications. On the other hand, it is important to note that the time trend and its squared term are insignificant, whereas the year dummies are significant, indicating that the latter capture the random shocks rather than the long-term trend. The without-price estimation, as well as the with-price estimation, shows that the impact of temperature on sorghum yield is negative and almost constant over time, whereas the rainfall has no impact at all. This means that sorghum is highly drought-tolerant, as is expected, but not heat-resistant within the temperature range in SSA, and these traits have not been changing over time. The non-significant effects of the time trend variables and their interaction terms with temperature and rainfall strongly suggest that there has been no major technological progress in sorghum farming in SSA.

Table 5.5 Country fixed-effect regression results for yield of sorghum, 1967–2004

Dependent variable: Ln sorghum yield	Without price		With price	
	Model 1	Model 2	Model 1	Model 2
Estimation				
Temperature	−0.1902***	−0.1392***	−0.1138*	−0.1041*
	(0.0586)	(0.0525)	(0.0667)	(0.0565)
Temperature × Time trend	0.0060*	0.0017	0.0012	−0.0002
	(0.0035)	(0.0031)	(0.0037)	(0.0031)
Temperature × Time trend squared	−0.0001	0.0000	0.0000	0.0000
	(0.0000)	(0.0001)	(0.0000)	(0.0001)
Ln rain	0.1685	0.0415	0.2404	0.1634
	(0.1492)	(0.1422)	(0.2058)	(0.2032)
Ln rain × Time trend	−0.0081	0.0064	−0.0065	0.0077
	(0.0040)	(0.011)	(0.0040)	(0.0153)
Ln rain × Time trend squared	0.0000	−0.0002	0.0000	−0.0003
	(0.0002)	(0.0002)	(0.0003)	(0.0003)
Ln real price			−0.7406***	−0.6912***
			(0.1857)	(0.1910)
Ln real price × Time trend			0.0544***	0.0506***
			(0.0150)	(0.0154)
Ln real price × Time trend squared			−0.0009***	−0.0009***
			(0.0003)	(0.0003)
Ln population density	−1.7672***	−1.5738***	0.5120	0.4091
	(0.4154)	(0.4239)	(0.6403)	(0.6534)
Ln population density × Time trend	0.0023	−0.0122	−0.0486***	−0.0529***
	(0.0093)	(0.0086)	(0.0180)	(0.0165)
Ln population density × Time trend squared	0.0001	0.0004**	0.0009***	0.0010***
	(0.0002)	(0.0002)	(0.0003)	(0.0090)
Literacy rate	−4.1966***	−3.3752***	−3.5472***	−3.0604***
	(0.8648)	(0.8539)	(0.9293)	(0.9171)
Literacy rate × Time trend	0.2419***	0.1453***	0.0855	0.0474
	(0.0612)	(0.0526)	(0.0682)	(0.0572)
Literacy rate × Time trend squared	−0.0034***	−0.0017*	−0.0014	−0.0008
	(0.0011)	(0.0090)	(0.0011)	(0.0010)
Time trend		−0.0317		−0.1244
		(0.1177)		(0.1732)
Time trend squared		0.0003		0.0033
		(0.0021)		(0.0030)
Burkinafaso	0.8586**	0.7716**	−0.1724	0.0119
	(0.4034)	(0.3819)	(0.5137)	(0.4883)
Ethiopia	1.1848***	1.0157***	−0.5932	−0.5690
	(0.3981)	(0.4036)	(0.4872)	(0.4993)
Kenya	0.6632***	0.7118***	0.2032	0.2825
	(0.0000)	(0.0000)	(0.0000)	(0.0000)
Madagascar	−0.4008**	−0.3747**	−0.0338	−0.0300
	(0.1800)	(0.1769)	(0.2062)	(0.2067)

(continued)

Table 5.5 (continued)

Dependent variable: Ln sorghum yield	Without price		With price	
	Model 1	Model 2	Model 1	Model 2
Malawi	1.6264***	1.5821***	−0.0967	0.0728
	(0.0040)	(0.0040)	(0.0040)	(0.0040)
Mali	−1.6292***	−1.6028***	−0.2087	−0.1680
	(0.5104)	(0.5132)	(0.6505)	(0.6467)
Mozambique	−0.4635***	−0.5323***	−0.6186***	−0.6659***
	(0.1764)	(0.1708)	(0.2156)	(0.201)
Namibia	−5.7669***	−5.5475***	−1.1916	−1.4508
	(1.0972)	(1.1289)	(1.4778)	(1.5323)
Niger	−2.7795***	−2.7475***	−1.8849***	−1.8273***
	(0.5182)	(0.5199)	(0.6265)	(0.6147)
Republic of South Africa	1.0646***	1.155***	1.2603***	1.2783***
	(0.2260)	(0.2203)	(0.2504)	(0.2389)
Togo	2.0900***	2.0526***	0.4544	0.6522
	(0.4574)	(0.4567)	(0.6237)	(0.6284)
Zimbabwe	−0.5669**	−0.4694**	−0.1719	−0.1652
	(0.2349)	(0.2279)	(0.2743)	(0.2632)
Constant term	12.2914***	11.8313***	7.6921***	8.3003***
	(1.2650)	(1.6463)	(1.9117)	(2.5715)
Number of observations	418	418	345	345
R-squared	0.8355	0.8016	0.8567	0.8196
Adjusted R-squared	0.8074	0.7884	0.8240	0.8030

***, **, and * indicate 1, 5, and 10% statistical significance levels, respectively

Curiously, the impact of population density on sorghum yield is found to be negative, i.e., the induced innovation hypothesis does not apply to sorghum farming. It may well be that the opportunity for technological change for sorghum in response to the increasing scarcity of land is limited in SSA, so that the increasing population pressure did not lead to a significant increase in yields.

5.5.5 *Results for Millet*

Lastly, Table 5.6 shows the estimation results of the yield function of millet, in which 11 countries are included (Burkina Faso, Cameroon, Ethiopia, Kenya, Madagascar, Malawi, Mali, Mozambique, Namibia, Niger, and Rep of South Africa, with Cameroon as the base for the country dummies). Similar to maize and sorghum, the estimated coefficients and their significance levels are somewhat different between the with- and without-price specifications, indicating that the price variable is correlated with the other variables. The coefficients of the year dummies and the time trend variables are insignificant and only marginally affect the estimated coefficients, indicating that the year dummies capture the random shocks. It is also important to note that the without-price estimation shows that the impact of climate

Table 5.6 Country fixed-effect regression results for yield of millet, 1967–2004

Dependent variable:	Without price		With price	
Ln millet yield	Model 1	Model 2	Model 1	Model 2
Estimation				
Temperature	−0.1822**	−0.0986	0.0405	0.0818
	(0.0805)	(0.0721)	(0.0970)	(0.0855)
Temperature × Time trend	0.0057	0.0024	−0.0108**	−0.0102**
	(0.0043)	(0.0041)	(0.0050)	(0.0048)
Temperature × Time trend squared	−0.0001	0.0000	0.0001*	0.0001
	(0.0000)	(0.0001)	(0.0000)	(0.0001)
Ln rain	0.2547	0.1610	0.1788	0.0360
	(0.1783)	(0.1697)	(0.2859)	(0.2587)
Ln rain × Time trend	−0.0217	−0.0132	−0.0266	−0.0101
	(0.0040)	(0.0128)	(0.0040)	(0.0191)
Ln rain × Time trend squared	0.0005**	0.0004	0.0007*	0.0004
	(0.0003)	(0.0002)	(0.0004)	(0.0003)
Ln real price			−0.3688	−0.4111*
			(0.2570)	(0.2506)
Ln real price × Time trend			0.0321	0.0382**
			(0.0206)	(0.0199)
Ln real price × Time trend squared			−0.0005	−0.0006*
			(0.0004)	(0.0004)
Ln population density	−0.1933	−0.0358	−0.4396	−0.2293
	(0.4834)	(0.4746)	(0.6769)	(0.6658)
Ln population density × Time trend	−0.0092	−0.0172*	0.0084	−0.0032
	(0.0105)	(0.0098)	(0.0167)	(0.0154)
Ln population density × Time trend squared	0.0002	0.0004*	−0.0001	0.0002
	(0.0002)	(0.0002)	(0.0003)	(0.0090)
Literacy rate	−3.3231***	−2.1490*	1.0088	1.1193
	(1.3141)	(1.2502)	(1.4740)	(1.4076)
Literacy rate × Time trend	0.2309***	0.1522**	−0.1150	−0.1124
	(0.0743)	(0.0684)	(0.0937)	(0.0882)
Literacy rate × Time trend squared	−0.0049***	−0.0038***	0.0000	−0.0002
	(0.0013)	(0.0090)	(0.0015)	(0.0014)
Time trend		0.0082		0.1504
		(0.1520)		(0.1913)
Time trend squared		−0.0006		−0.0019
		(0.0027)		(0.0034)
Burkinafaso	0.2132	−0.0566	−0.0998	−0.3695
	(0.4710)	(0.4361)	(0.6091)	(0.5631)
Ethiopia	−0.5572	−0.5725	−1.5687***	−1.5004**
	(0.4544)	(0.4400)	(0.6258)	(0.6167)
Kenya	−0.0111	−0.1386	−0.3426	−0.4938
	(0.0000)	(0.0000)	(0.0000)	(0.0000)
Madagascar	0.1423	0.0215	−0.2145	−0.2932
	(0.5658)	(0.5530)	(0.7046)	(0.7049)

(continued)

Table 5.6 (continued)

Dependent variable:	Without price		With price	
Ln millet yield	Model 1	Model 2	Model 1	Model 2
Malawi	−0.7101***	−0.8033***	−1.1594***	−1.2462***
	(0.0040)	(0.0040)	(0.0040)	(0.0040)
Mali	−1.8075	−1.5792	−1.7616	−1.4194
	(1.3147)	(1.2923)	(1.8618)	(1.8397)
Mozambique	−0.5091	−0.6057	−1.0355	−1.1016*
	(0.5720)	(0.5612)	(0.6621)	(0.6623)
Namibia	−0.1068	−0.0585	−0.0948	0.0351
	(0.2547)	(0.2488)	(0.2623)	(0.2549)
Niger	0.5831	0.2797	0.5557	0.2173
	(0.5611)	(0.5339)	(0.7867)	(0.7472)
Republic of South Africa	−0.6864***	−0.5994**	−0.6465**	−0.4680*
	(0.2657)	(0.2557)	(0.2967)	(0.2849)
Constant term	7.9895***	6.4312***	6.3902**	4.3952
	(1.8858)	(2.4273)	(2.6216)	(3.2512)
Number of observations	342	342	281	281
R-squared	0.7073	0.6537	0.7622	0.7056
Adjusted R-squared	0.6461	0.6274	0.6945	0.6742

***, **, and * indicate 1, 5, and 10% statistical significance levels, respectively

on millet yield is generally insignificant with no over-time change, even though heat may worsen the yield to some extent. These results imply not only that millet is relatively resistant to climate stresses in SSA but also that similar to sorghum, there have been no major technological changes in millet farming in SSA.

The impact of population density on millet yield is found to be generally insignificant, i.e., the induced innovation hypothesis does not apply to millet farming. Again, similar to the case of sorghum, the potential for technological change in millet farming seems limited in SSA.

5.6 Concluding Remarks

While it is well-known that the GR enormously contributed to the growth of crop yields in Asia, no GR has been taking place in SSA due to various constraints. The fact that the crop yields were not significantly different between Asia and SSA in the early 1960s indicates the difference in agro-climate alone cannot explain the huge yield gap between the two continents today. This study began by pointing out that among the five major crops, wheat yield is growing faster in SSA than in Asia. We also pointed out that the yields of maize, rice, sorghum, and millet also increased appreciably, even though the yield levels are still low as of now. These observations suggest that technological changes might have taken place in SSA.

The econometric analysis using the country-level macroeconomic data in SSA for the 38 years from 1967 to 2004 revealed the following points. First, the impacts of climate represented by temperature and rainfall on wheat, rice, and maize yields have been mitigated over time due presumably to some technological improvements. This finding is consistent with our finding in India (Chap. 4), supporting our hypothesis that improved varieties along with better crop care can cope with unfavorable climatic conditions in SSA. Second, for wheat, rice, and maize, population density is found to play an important role in increasing the yields. That is, continued population pressure is likely to have increased the relative profitability of land-saving and yield-enhancing technologies along the lines of the induced innovation hypothesis. Since these three crops are known as the GR crops in Asia, this result may be a strong sign that an African GR akin to the Asian GR is actually starting to take off. Third, we found no evidence that technological changes took place in sorghum and millet. Furthermore, there is no indication of the inducement effects of population pressure on the yields of these two crops. In all likelihood, the opportunities for technological breakthroughs for these crops are limited.

In policy formulation, however, the limitation of wheat area expansion in SSA requires due attention. It is also important to note that the current yield levels are different between SSA and Asia for these GR crops, but not so for sorghum and millet, even though infrastructure, markets, and policy environments are much more favorable in Asia than in SSA. In other words, the potential to increase crop yields in SSA seems high for rice and maize presumably because there is ample room to transfer technology from Asia to SSA for rice and maize (see Chap. 2 for rice and Chap. 8 for maize). We, thus, agree with Otsuka and Kijima (2010), in order to realize a GR in SSA, the limited budget for research and extension should focus on the target crops, i.e., rice and maize.

References

Adams RM, Fleming RA, Chang C-C, McCarl BA, Rosenweig C (1995) A reassessment of the economic effects of global climate change on US agriculture. Clim Change 30(1):147–167

Auffhammer M, Ramanathan V, Vincent JR (2006) Integrated model shows that atmospheric brown clouds and greenhouse gases have reduced rice harvests in India. Proc Natl Acad Sci USA 103(52):19668–19672

Balasubramanian V, Sie M, Hijmans R, Otsuka K (2007) Increasing rice production in Sub-Saharan Africa: challenges and opportunities. Adv Agron 94(1):55–133

Bruce JP, H-s Y, Haites EF (1996) Climate change 1995 – economic and social dimensions of climate change. Cambridge University Press, Cambridge

Byerlee D (1996) Modern varieties, productivity, and sustainability: recent experience and emerging challenges. World Dev 24(4):697–718

David CC, Otsuka K (eds) (1994) Modern rice technology and income distribution in Asia. Lynne Rienner, Boulder

Diagne A (2006) The diffusion and adoption of NERICA rice varieties in Cote d'Ivoire. Dev Econ 44(2):208–231

Evenson RE, Gollin D (2003) Crop variety improvement and its effect on productivity: the impact of international agricultural research. CAB International, Wallingford, Oxford

Fan S, Hazell P (1999) Are returns to public investment lower in less-favored rural areas? An empirical analysis of India, vol 43, Environment and production technology division discussion paper. International Food Policy Research Institute, Washington, DC

Food and Agriculture Organization of the United Nations (FAO) (2008) FAOSTAT. FAO, Rome

Goufo P (2008) Evaluating the constraints and opportunities for sustainable rice production in Cameroon. Res J Agric Biol Sci 4(6):734–744

Hausman JA (1978) Specification tests in econometrics. Econometrica 46(6):1251–1271

Hayami Y, Ruttan VW (1985) Agricultural development: an international perspective. Johns Hopkins University Press, Baltimore

Heckman JJ (1979) Sample selection bias as a specification error. Econometrica 47(1):153–161

Hossain M, Fischer KS (1995) Rice research for food security and sustainable agricultural development in Asia: achievements and future challenges. Geo J 35(3):286–298

Kajisa K, Payongayong E (2011) Potential and constraints to the rice Green Revolution in Mozambique: a case study of the Chokwe irrigation scheme. Food Policy 35(5):614–625

Khush GS (2001) Green Revolution: the way forward. Nat Rev Genet 2(10):815–822

Kijima Y, Serunkuuma D, Otsuka K (2006) How revolutionary is the 'NERICA Revolution'? Evidence from Uganda. Dev Econ 44(2):232–251

Kuyvenhoven A (2008) Africa, agriculture, aid. Natl J Agric Sci 55(2):93–112

Lawn RJ (1989) Agronomic and physiological constraints to the productivity of tropical grain legumes and prospects for improvement. Exp Agric 25(4):509–528

Ligon E, Sadoulet E (2007) Estimating the effects of aggregate agricultural growth on the distribution of expenditures, Background paper for the world development report 2008. World Bank, Washington, DC

Mendelsohn R, Nordhaus WD, Shaw D (1994) The impact of global warming on agriculture: a Ricardian analysis. Am Econ Rev 84(4):753–771

Monfreda C, Ramankutty N, Foley JA (2008) Farming the planet: 2. geographic distribution of crop areas, yields, physiological types, and net primary production in the year 2000. Global Biogeochem Cycl 22:GB1022

Mwabu G, Thorbecke E (2004) Rural development, growth, and poverty in Africa. J Afr Econ 13(supplement):16–65

Olesen JE, Bindi M (2002) Consequences of climate change for European agricultural productivity, land use and policy. Eur J Agron 16(4):239–262

Omamo SW (2003) Policy research on African agriculture: trends gaps, and challenges. International Service for National Agricultural Research of International Food Policy Research Institute, The Hague

Otsuka K, Kalirajan K (2006) Rice Green Revolution in Asia and its transferability to Africa: an introduction. Dev Econ 44(2):107–122

Otsuka K, Kijima Y (2010) Technology policies for a Green Revolution and agricultural transformation in Africa. J Afr Econ 19(S2):60–96

Otsuka K, Estudillo JP, Sawada Y (2009) Rural poverty and income dynamics in Asia and Africa. Routledge, London

Reilly J, Baethgen W, Chege FE, Iglesias A, Kenny G, Patterson D, Rogasik J, Rötter R, Rosenzweig C, Sombroek W, Westbrook J, Bachelet D, Brllacich M, Dämmgen U, Howden M (1996) Agriculture in a changing climate: impacts and adaptation. In: Sedjo R, Brown S (eds) Climate change 1995: impacts, adaptations and mitigation of climate change: scientific–technical analysis, management of forests for mitigation of greenhouse gas emissions. Res Future Press, Washington, DC

Rozeboom WW (1966) Scaling theory and the nature of measurement. Synthesis 16(2): 170–233

Sakurai T (2006) Intensification of rainfed lowland rice production in West Africa: present status and potential Green Revolution. Dev Econ 44(2):232–251

Sanghi A, Mendelsohn R, Dinar A (1998) The climate sensitivity of Indian agriculture: in measuring the impact of climate change on Indian agriculture, vol 402, World Bank technical paper. World Bank, Washington, DC

Seo N, Mendelsohn R (2007) A Ricardian analysis of the impact of climate change on Latin American farms, vol 4163, World Bank policy research working paper. World Bank, Washington, DC

Stevens SS (1946) On the theory of scales of measurement. Science 103(2684):677–680

World Bank (2008) World development report 2008: agriculture for development. World Bank, Washington, DC

Part II
Prospects for Upland Rice and Maize Green Revolution in Sub-Saharan Africa

Chapter 6
Causes and Consequences of NERICA Adoption in Uganda

Yoko Kijima and Keijiro Otsuka

Abstract In Uganda, a new upland rice variety, namely New Rice for Africa (NERICA), was introduced in 2003 as one of its poverty eradication strategies essentially because it is high-yielding, which can results in both increased cash income and food security. In addition, NERICA is considered to be cultivable in most parts of Uganda thanks to its short maturity and drought tolerant trait. The major question is whether the "NERICA Revolution" is sustainable and extendable to wide areas. The major purposes of this study are to identify the determinants of NERICA adoption with a special focus on the high incidence of dropouts and to assess the consequences of NERICA adoption in terms of changes in crop income using the panel data collected in 2004 and 2006.

Keywords NERICA • Upland rice • NERICA yield • Uganda • Sub-Saharan Africa • Population pressure • Agricultural productivity • High-yielding variety

6.1 Introduction

In Sub-Sahara Africa (SSA), agriculture is still basis for pro-poor economic growth, as the most poor depend on agriculture for their livelihood. According to Ligon and Sadoulet (2007), GDP growth from agriculture benefits the poorest half at least

Y. Kijima (✉)
Graduate School of Systems and Engineering, University of Tsukuba,
1-1-1 Tennodai, Tsukuba, Ibaraki 305-8577, Japan
e-mail: kijima@sk.tsukuba.ac.jp

K. Otsuka
National Graduate Institute for Policy Studies, 7-22-1 Roppongi, Minato-ku,
Tokyo 106-8677, Japan
e-mail: otsuka@grips.ac.jp

K. Otsuka and D.F. Larson (eds.), *An African Green Revolution: Finding Ways to Boost Productivity on Small Farms*, DOI 10.1007/978-94-007-5760-8_6,

twice as much as that from non-agriculture. Growth rates of agriculture have been accelerating in SSA since the 1980s but productivity growth has lagged behind. Public agricultural investment in R&D fell in half of countries in the 1990s and the donor support for agriculture had been decreasing since 1990 until the "food crisis" in 2008 (World Bank 2008). Moreover, crop land per agricultural population in SSA decreased to 60% of 1960 level due to population pressure. In Asia, agricultural productivity growth has driven poverty reduction by increasing crop income and investment in schooling of children who later found lucrative non-farm jobs (Otsuka et al. 2009). In order to persistently decrease poverty in SSA, development of profitable agricultural technology and its diffusion is urgently needed.

In Uganda, a new upland rice variety, namely New Rice for Africa (NERICA), was introduced by vice-president initiative development program in 2003 as one of its poverty eradication strategies essentially because it is high-yielding, which can results in both increased cash income and food security. In addition, NERICA is considered to be cultivable in most parts of Uganda thanks to its short maturity and drought tolerant trait. Indeed, the yield of NERICA is much higher than that of traditional upland rice (Kijima et al. 2006). Due to its relatively high price and the increasing demand, rice production in Uganda has been increasing. According to the survey covering the major production areas conducted by the authors, the rice production more than doubled between 2004 and 2008 and the share of NERICA in total rice production in Uganda reached 10% in 2004 and 30% in 2008. Therefore, by far the most important reason for the increasing rice production in Uganda is introduction of NERICA.[1]

The major question is whether the "NERICA Revolution" is sustainable and extendable to wide areas. Our panel data shows that more than 50% of farmers who planted NERICA in 2004 did not cultivate NERICA in 2006. Why is the dropout rate so high in such a short period of time? There are several reasons to explain this high dropout rate. First, it is possible that NERICA did not increase income compared with the alternative crops in some areas, which discourages households to grow rice. In other words, NERICA might have been disseminated to inappropriate areas. Second, since rice is a new crop to many areas in Uganda, farmers might have difficulty in selling rice due to the absence of rice millers located nearby. Third, the supply of NERICA seeds by commercial seed suppliers is limited and, hence, farmers might have used mistreated low-quality self-produced seeds. Fourth, the higher labor and seed costs of NERICA rice can make it difficult for poor farmers to cultivate NERICA especially with inadequate availability of credit and without any support from program.

[1] Figure 6.1 come from different survey (rice miller survey). We asked total amount of processed rice from 2004 to 2008. This figure does not mean that the adoption rate of NERICA is 30%. The NERICA's production share out of total rice produced increased from 10 to 30%. These seemingly contradicting figures with the high dropout rate discussed in this chapter suggest that NERICA adoption areas have been expanding but the adoption rate would have been much higher if there were no problems mentioned in this study.

In order to achieve major increase in rice production in Uganda, it is critically important to explore the causes and consequences of changing NERICA adoption behaviors. The major purposes of this study are to identify the determinants of NERICA adoption with a special focus on the high incidence of dropouts and to assess the consequences of NERICA adoption in terms of changes in crop income using the panel data collected in 2004 and 2006.

The rest of this chapter is structured as follows. Section 6.2 presents the descriptive data used in this study and examines their major characteristics. Section 6.3 explains the empirical models and variables used in this study, while the results of the estimated NERICA adoption and yield functions are examined in Sect. 6.4. Section 6.5 discusses the conclusions and policy implications.

6.2 Data and Sample

The data used in this chapter were collected in 2004 and 2006. Since the dissemination of NERICA started in 2001, farm households growing NERICA rice were found mainly in areas with NERICA seed dissemination programs. Thus, we intentionally selected ten NERICA growing areas covering the Central and Western regions (Kijima et al. 2008).[2] In each sample area, we drew a random sample of 25 households that grew NERICA rice in the second cropping season of 2004, and 15 households that did not. In the second survey (2006), we had to drop one district due to budget constraint and there was attrition due to out-migration from the sample areas, the dissolution of households, and the absence of household members during the data collection period. In total, we used a panel sample of 347 households in nine districts for this analysis.

There have been different kinds of interventions in Uganda. The major one is the Vice-President Program in which NERICA seeds are distributed through NGOs such as APEP and SG2000.[3] Under this program, farmers can obtain seeds at no charge in some areas while seeds are distributed on credit in other areas (farmers have to return the seeds or money to the micro-finance institution after harvest). Normally, seeds are distributed to local farmers' groups by agricultural extension workers, and then group members allocate the seeds to individual farmers. Since the NERICA dissemination program under vice-president initiative was introduced for his political motive in 2004, it is possible that seeds were distributed to areas which are not suitable to upland rice production and to households who are not seriously engaged in farming. Since most of the districts where NERICA was introduced in 2004 received NERICA seeds under vice-president initiative, we call these districts "program districts" and examine whether there is difference in adoption behavior between program and non-program districts.

[2] Two to three local council 1s (the lowest administrative unit in Uganda called LC1) constitute each NERICA growing area and our sample covers 29 LC1s.

[3] Another way is contract farming with a private seed company and an agricultural research organization from which farmers obtain NERICA seeds and to which farmers sell outputs.

Based on differences in the NERICA adoption behavior, we stratified the sample into four groups. The first group, non-adopters, has never adopted NERICA; the second group, dropouts, includes households that grew NERICA in 2004 but not in 2006; the third group, continuous adopters, consists of households that grew NERICA in both 2004 and 2006; and the last group, late adopters, refers to households that adopted NERICA in 2006 but not in 2004. We call dropouts and continuous adopters the early adopters since they grew NERICA in 2004.

It is possible that geographical variations make the adoption patterns and average yields of NERICA different. Table 6.1 shows the year when NERICA was introduced, whether NERICA was introduced by the vice-president initiative, the adoption rate in 2004 and 2006, the percent of dropout out of early adopters, the percent of late adopters out of NERICA non-adopters in 2004, the rainfall, the market access measured by the distance to the nearest town, and the average cultivated land size per household in the sample areas, including the average area planted to NERICA. As shown in column 2, in the sample areas where NERICA was recently introduced, NERICA was introduced by the vice-president initiative. Since the program could have disseminated NERICA seeds partly based on political motives, we differentiate the sample areas into the program districts and the non-program districts. The estimated adoption rate in 2004 varies across areas from 5 to 63%. Except in Hoima and Masindi districts, the adoption rate decreased between 2004 and 2006. The dropout rate, as shown in column 5, is quite high, especially in the program districts (73–83%). This may be the sign of mis-targeting of NERICA seed dissemination program under vice-president initiative. Though the dropout rate is high even in non-program districts (more than 20%), the proportion of late adopters is also high, thereby partially offsetting the reduction in the adoption rate.

Most of the sample areas have more than 1,200 mm of rainfall annually. Market access varies considerably across the sample areas. The areas in Wakiso and Mpigi districts have good access not only to nearby towns but also to the capital city, Kampala. Except for these areas, the sample areas are located far from rural towns and have some of the available farmlands fallowed or uncultivated.

Table 6.2 shows the average NERICA yield and the proportion of rice growers who obtained zero harvest by districts. In the program districts, the yield tends to be lower than the non-program districts. Moreover, the yield significantly dropped in 2005 only in the program districts.[4] The table also shows the proportion of rice plots with zero yields, which indicates that more than half of NERICA growers did not obtain any harvest in some of the program districts, especially in 2005. The high probability of getting zero yield can be the reason for high dropout rate in the short period of time.

Table 6.3 indicates the characteristics of sample households and communities by type of adopters. Household characteristics pertain to 2004. The early adopters, who include the dropouts and the continuous adopters, are more experienced with rice

[4] In Mpigi and Mubende districts, the yield turns to be higher in the second season of 2006 than 2005. This is probably because only "better" farmers remain in rice cultivation, which has a positive effect on average yields.

Table 6.1 Rainfall, access to nearby town, and land use in sample areas of NERICA production in Uganda

District	Year NERICA introduced[a]	Introduced by vice-president program	Adoption rate in 2004[c,f]	Adoption rate in 2006[c,f]	% of dropout out of adopters in 2004[d,f]	% of late adopters out of non-adopters in 2004[d,f]	Total rainfall in 2003 (mm)[e]	Distance to nearby town (km)[a]	Average area cultivated 2004 (ha)[a]	Average area fallowed 2004 (ha)[a]
Masindi	n.a.[b]	No	5.4	11.6	41.7	7.1	1,294	21	3.42	1.65
Kibaale	2001	No	63.0	59.3	54.2	42.9	1,602	17	1.88	4.04
Kamwenge	2001	No	35.8	33.2	20.0	44.4	643	16	1.57	1.16
Hoima	2002	No	26.0	67.1	21.7	60.0	1,530	9	1.36	2.12
Luwero	2003	Yes	19.8	11.3	53.3	0.0	1,373	17	1.95	0.99
Wakiso	2004	Yes	8.9	5.0	75.9	0.0	1,460	7	1.08	0.97
Mpigi	2004	Yes	5.3	0.2	82.8	13.3	1454	2	1.70	1.71
Mubende	2004	Yes	6.0	0.1	73.3	0.0	1,248	25	2.11	1.60
Kiboga	2004	Yes	4.8	2.9	76.2	21.4	818	8	2.63	3.29

[a] NERICA community survey data
[b] The year in which NERICA was introduced for the first time is not known
[c] Since we collected data based on the stratified sampling by adoption status in 2004 (the numbers of adopters and non-adopters in 2004 are 25 and 15, respectively, in each sample area), the adoption rate is estimated using population weights
[d] The percent of dropouts and late adopters are calculated without using population weight (i.e., if there are ten dropouts in a sample area, the dropout rate out of the adopters in 2004 is 10/25)
[e] Data is obtained from Department of Meteorology, Uganda
[f] Calculated from NERICA household survey data

Table 6.2 NERICA rice yield and seed use by district in Uganda

	Average yield t/ha			% of zero yields out of rice plots		
District	2004	2005	2006	2004	2005	2006
Masindi	1.96	2.13	3.11	0.0	22.2	7.7
Kibale	2.21	2.30	2.11	0.0	9.5	6.3
Kamwenge	2.63	2.48	2.60	0.0	0.0	6.7
Hoima	2.90	2.55	2.85	0.0	0.0	3.7
Luwero	1.97	1.70	1.84	0.0	0.0	7.1
Wakiso	1.74	0.97	1.60	3.6	11.8	12.5
Mpigi	1.42	0.65	(1.12)[a]	21.4	50.0	(0.0)[a]
Mubende	1.67	1.38	(3.04)[a]	10.3	5.9	(20.0)[a]
Kiboga	1.78	1.10	1.14	3.5	58.3	25.0

All the information is calculated using the data in the second cropping season
[a] The number of observation is only 4

Table 6.3 Household and community characteristics by type of adopters of NERICA in Uganda[a]

	Dropouts	Continuous adopters	Late adopters	Non- adopters
Number of households	129	99	25	94
Household characteristics				
Rice cultivation experience (years)	1.49	1.88	0.50	0.07
Household head's age	48.2	43.9	49.4	48.7
Household head's years of schooling	7.0	7.7	4.5	4.9
Number of adult male aged 15–59	1.80	1.81	0.99	1.43
Number of adult female aged 15–59	1.96	1.68	1.34	1.50
% Female headed households	10.2	8.0	29.7	32.8
% Bakiga tribe	8.3	9.0	20.3	0.4
Land area per capita (ha)[b]	0.38	0.47	0.38	0.24
Land area per household (ha)[b]	4.52	4.23	2.53	2.56
Land cultivated in 2nd season (ha)	1.19	1.30	0.78	0.92
Household asset (USD)	149	172	54	160
Value of livestock (USD)	371	390	80	307
Community characteristics				
Availability of Seed Program in 2004 (%)	37.4	33.7	17.3	18.6
Availability of Seed Program in 2006 (%)	20.4	28.6	23.9	10.7
Distance to rice miller in 2004 (km)	15.4	26.9	28.9	19.0
Distance to rice miller in 2006 (km)	11.1	6.2	5.5	14.1
Traveling time to town (h)	0.62	0.77	0.66	0.42
Community area per household (mile^2)	0.020	0.022	0.019	0.023
Relative price of maize to rice	0.512	0.465	0.448	0.385
Annual rainfall in 2004 (mm)	424.0	429.6	393.0	368.5
Annual rainfall in 2006 (mm)	435.8	432.0	450.2	446.2
Average annual rainfall[c]	435.5	441.3	421.6	409.4
C.V. of rainfall[c]	0.17	0.15	0.17	0.20

[a] The data pertain to 2004 unless stated otherwise
[b] Land area refers to owned land and tenanted land under the mailo regime
[c] Average of 6 years from 2001 to 2006

cultivation than the late adopters and non-adopters. Among the early adopters, the rice cultivation experience before the adoption of 2004 is significantly higher for the continuous adopters, suggesting that learning from own experience enhances the profitability of rice cultivation, which affects the dynamic decision of NERICA adoption, as shown in Moser and Barrett (2006). It is also clear that the continuous adopters tend to be younger than the other groups. The significant differences in the household head's age between the continuous adopters and the dropouts suggest that younger households are less likely to abandon rice cultivation in a short period of time.

The early adopters are significantly more educated than the non-adopters and the late adopters. These findings are consistent with those in the large body of literature on the adoption of agricultural technologies (Sunding and Zilberman 2007). The early adopters are also endowed with a larger number of adult household members than the non-adopters and the late adopters. The early adopters also cultivate lager areas of land than the others.[5] It is, however, important to recognize that the early adopters (both the dropouts and the continuous adopters) are almost equally educated and endowed with family labor and land, indicating that the reason for abandoning NERICA rice cannot be attributed to the lack of human capital, family labor and land. Regarding the endowment of household assets (such as furniture, bicycles, and electrical appliances) and livestock, a major difference is found between the late adopters and the other categories, suggesting that poverty (or the lack of assets) compelled some farmers to adopt NERICA later, even though they may have been aware that it is potentially more profitable than other crops. This may suggest a failure of the credit market. Female headed households are less likely to be the early adopters, which is consistent with the common belief that women are disadvantaged, particularly in the rural African setting. The proportion of the Bakiga tribe, which is well recognized as a tribe of hard-working people because they originate from land-scarce hilly areas, is markedly high among the late adopters.

As NERICA rice is a new crop to many farmers in Uganda, the seed distribution programs are intended to provide not only seed but also training on rice production in areas where NERICA was newly introduced. As a consequence, the availability of seed distribution programs is expected to affect the NERICA adoption significantly. In fact, Table 6.3 clearly indicates that the early adopters had better access to seed distribution programs in 2004. However, the availability of such programs decreased in 2006 among all NERICA adopter categories, except in areas where the late adoption took place. Therefore, it seems reasonable to hypothesize that the seed distribution programs influenced the NERICA adoption by providing valuable information on the new technology as well as seed. Because NERICA rice seed can be produced by farmers, unlike some improved seeds for other crops, such as hybrid maize, it is also reasonable to postulate that the productivity or the quality of seed, originally distributed by the program, is moderated over time as farmers produce their own seeds.

---

[5] Since mailo tenants have secure land rights according to the government Land Act of 1998, we include mailo tenanted land in the land area along with the owned land. The difference between the sum of the owned area and the mailo tenanted area and the cultivated area is the fallow area.

One of the major constraints on NERICA adoption in 2004 was identified to be the absence of rice millers in nearby towns to mill or sell the paddy rice. Common transportation means from homestead to rice miller is bicycle in sample areas in 2004. As is indicated in Table 6.3 a typical farmer had to transport rice 15–30 km by bicycle to the nearest rice millers. However, the number of rice millers, who are sometimes buyers as well, has increased rapidly in Uganda as a whole, presumably responding to the increasing demand for rice milling services that followed the increase in NERICA rice production. In all the four sample categories, access to rice millers has improved in the past 2 years, which is clearly reflected in the considerably shortened distance to rice millers from 15 to 30 km in 2004 to 6 to 11 km in 2006, particularly among the continuous adopters and the late adopters. These observations indicate that improved access to the market for paddy is a critical factor promoting NERICA adoption.

The relative profitability of NERICA can be an important determinant of its adoption. Since NERICA is highly labor intensive compared with the alternative crops (Kijima et al. 2008), its adoption may be higher in communities where labor is abundant relative to land. However, data on community (village) area per household suggest that the availability of land is similar across the four adoption categories. Since maize is a major alternative crop to rice that is less labor intensive, the relative price of maize to rice can be an important factor affecting the relative profitability and, thus, the adoption of NERICA rice.

Table 6.4 shows the NERICA yield on sample plots and the sources of NERICA seed in 2004 and 2006 by the different adoption categories. The average yield for the continuous adopters in 2004 is 3.0 t/ha, which attests to the high yield potential of NERICA.[6] It is also significantly higher among the continuous adopters than among the dropouts in 2004. On average, the yield for the late adopters is much lower than that for the continuous adopters in 2006. This low yield among the late adopters may be attributed to their lower human and physical capital (as shown earlier in Table 6.1) and lower experience in rice cultivation than the continuous adopters.[7]

There are four possible ways through which farmers obtain NERICA seeds: (1) participating in the seed distribution program, (2) directly purchasing seed from seeds companies, (3) using own seeds saved from the previous harvests, and (4) purchasing seeds from other farmers. Note that the seed distribution programs procured certified seed from seed companies and distributed it to farmers via NGOs and extension workers. However, the direct purchase of seed from seed companies by individual farmers is observed mainly in areas close to the seed companies where the farmers engage in contract farming for the seed companies. This requires farmers to use the treated seeds from the company for maintaining the high quality of their

[6] The average yield of upland rice in SSA is about 1 t/ha (Balasubramanian et al. 2007).

[7] Sserunkuuma (2008) shows that farmers trained by JICA in 2007 had significantly lower NERICA yield than those trained in 2005 who started growing NERICA earlier and accumulated experience over the years.

Table 6.4 NERICA yields and sources of seeds by type of adopters in Uganda

	2004		2006	
	Dropouts	Continuous adopters	Continuous adopters	Late adopters
Number of plots	129	107	100	24
Yield (t/ha)	2.01	2.97	2.54	1.49
% Self-produced seeds	5.2	7.7	41.5	5.8
% Purchased seed from neighbors	3.8	11.7	12.0	46.5
% Program seeds (NGO, VP)	53.8	42.9	15.0	10.2
% Other seeds (purchased from traders, contract farming)	37.2	37.7	31.5	37.5
Yield in plots from self-produced seeds	0.99	3.41	2.06	1.15
Yield in plots from purchased seeds from neighbors	1.35	2.72	2.97	1.17
Yield in plots using program seeds	1.76	2.96	2.95	3.46
Yields in plots using seeds from seed companies	2.49	2.99	2.75	1.39
% plots in low lying location	11.8	14.1	21.7	21.8
Yield in plots in low lying location	2.04	2.83	2.09	1.22
Yield in plots not in low lying location	2.01	3.00	2.67	1.56
% Late plantation	8.5	2.8	11.1	6.1
Yield in plots without late plantation	2.20	3.05	2.56	1.43
Yield in plots with late plantation	0.68	0.63	2.44	2.31
Profit from rice production (USD/ha) 2nd cropping season	40.54	202.21	164.88	−164.51
Profit from alternative crop (USD/ha)[a] 2nd cropping season	86.83	−72.54	NA	NA
Size of NERICA plot (ha)	0.423	0.377	0.471	0.241

[a] Profit from alternative crop is only available in 2004. Alternative crops to rice are mainly maize and beans (Kijima et al. 2008)

seeds. In 2004, the proportion of sample plots planted to seed obtained either from the seed companies or from the seed distribution programs reached 80–90%. The use of own seed and seed purchased from neighbors was rare in 2004, but the proportion of farmers using farmer-produced seed increased to 54% among the continuous adopters and to 52% among the late adopters in 2006. While NERICA seed can be produced by farmers themselves like other rice varieties, farmers have to remove undesirable plants to obtain "pure" or high-quality rice seed. Once farmers learn how to produce high-quality seed, the farmer-produced seed can be as good as that sold by seed companies.[8] Given that rice production was only recently started by many of our sample farmers, there is a possibility that the quality of the farmer-produced seed is not as good as that of the purchased seeds, unless the extension

[8] This is why seed suppliers cannot make large profits in Asia, where farmers are adept at producing high-quality seeds. In Uganda, certified seeds are actually produced by farmers under contract with seed companies, which provide detailed instructions for seed production.

service provides the required information for seed production. If pure seed is not produced, the quality of seed is expected to deteriorate gradually over time, thereby lowering the rice yield. As shown in Table 6.4, the average yield of the self-produced seeds tends to be lower than the other seed types (except for the continuous adopters in 2004). In particular, the yield from the self-produced seeds among the continuous adopter in 2006 is far lower than the other seed types, which suggests the possibility of mistreatment of seeds and seed-quality deterioration.

Even controlling for the effects of using different types of seeds, however, the yields of the continuous adopters are higher than the other groups. Therefore, it may be the case that there are differences in the management of rice cultivation other than the selection of seed type. Though we observe that farmers growing NERICA in plots with higher moisture content (such as the lower parts of hills) could obtain high yields, there are no statistical differences in the plot choice across the adopter types and in the yields between the plot types.

Similarly, it is widely believed that farmers who planted rice late tended to experience crop failures when they lacked sufficient rainfall. We constructed a dummy variable for whether farmers planted too late according to their planting schedule, in order to examine if the differences in the timing of planting can explain the higher yields of the continuous adopters than the dropouts and the late adopters. In 2004, the continuous adopters are less likely to plant late than the dropouts, while in 2006, the proportion of late planning is higher among the continuous adopters than the late adopter. In 2004, the plots with late planting ended up with low yields (many have zero yield) among both the dropouts and the continuous adopters. In 2006, however, there is no such trend. Significant difference between the continuous adopters and the late adopters is the proportion of plots with zero yields, which is responsible for lowering the average yield of the late adopters.

For the analysis of the impact of the profitability of crop cultivation, the profit should be a better measure since the income does not take imputed costs of owned resources such as family labor into account. Due to the data availability, the plot-level profits from NERICA plot and the alternative crop plot are calculated only in 2004. As shown in the bottom of Table 6.4, the continuous adopters earned higher profit (USD 202/ha per season) from rice plot than the dropouts (USD 40/ha per season). This result is to be expected because the poor performance of NERICA production likely discourages NERICA adoption in subsequent periods.[9] It is important to note that the profit from rice plot is lower than that from alternative crop plot among the dropouts, while for the continuous adopters, the profit from rice exceeds that from alternative crops. This may suggest that the relative profitability between rice and the alternative crop, not the absolute one, affects the decision on the adoption of NERICA for the following years.

[9] A recent study conducted in Uganda (Sserunkuuma 2008) shows that the occurrence of severe drought conditions during the cropping season significantly reduced the rice yield and negatively affected the number of Japan International Cooperation Agency (JICA)-trained households growing NERICA in the subsequent cropping seasons.

6.3 Estimation Model

In order to rigorously identify the critical factors affecting the changing behavior of NERICA adoption, we conduct a regression analysis in this section. In particular, we are interested in the effects of rainfall, the availability of seed programs, access to rice millers, and the effect of using farmer-produced seeds on the adoption and yield performance of NERICA rice.

6.3.1 *NERICA Adoption Function*

We use a multinomial logit model to identify the major factors underlying the four types of adoption decisions. Although we estimate the multinomial regressions with cross-sectional data, we use variables in 2004 and 2006 for the time-variant characteristics, such as the availability of seed distribution programs and the distance to rice millers. By comparing the estimated coefficients of these variables in 2004 and 2006, we can infer the changing impacts of seed distribution programs and access to rice millers.

To investigate the determinants of NERICA adoption decisions, we use the adoption category (dropouts, continuous adopters, late adopters, or non-adopters) as the dependent variable, while the amount of rice harvest per hectare is used as a dependent variable in the yield function. The possible determinants of the NERICA adoption and yield include (1) a set of variables indicating the suitability of rice production and the costs of acquiring seeds and selling paddy rice such as the average rainfall, rainfall variation, availability of seed program, and distance to rice millers, and (2) household characteristics such as the rice cultivation experience,[10] the household head's education, the number of adult male and female household members aged between 15 and 59 years old, and asset holdings. We use the average rainfall and its variations over the 6-year period in the estimation of the adoption function as proxies for expected rainfall patterns. It must also be noted that the availability of seed distribution programs in a village, which can be considered as exogenous, is used to explain the NERICA adoption.

6.3.2 *NERICA Yield Function*

In order to assess the factors affecting NERICA yield in general, and the sustainability of self-produced seeds in particular, another regression model is estimated by following equation:

$$y_{sit} = \beta X_{sit} + \gamma Z_{it} + \delta Y06 + \varepsilon_{it}, \tag{6.1}$$

[10] The rice cultivation experience for 2006 captures the experience in 2004 and 2005.

where ysit is rice yield per hectare from plot s of household i at time t, Xsit is the time-variant plot characteristics, Zit is the household characteristics, Y06 is year 2006 dummy variable, and it is error term. We pool all sample data except non-adopters and differences in adoption category (continuous, dropout, land ate adopter) are also controlled for. Since the adoption category is determined endogenously, we estimate equation (1) by instrumental variables regression model.

Note that we use the rainfall in a single year (90 days after planting) to explain the NERICA yield. The seed type (i.e., the use of program seeds, farmer-produced seed saved from the previous harvest, or purchased from neighbors) is also used to explain the NERICA yield.[11] As for the impact of plot characteristics, slope of the plot, cultivation practices (fertilizer application, seed used, pesticide and herbicide use, labor inputs, straight-row planting, and timing of planting) are examined.

6.4 Empirical Results

6.4.1 NERICA Adoption

Table 6.5 shows the relative risk ratio compared with the continuous adopters. If this ratio of an explanatory variable X1 in adoption category 1 (dropout) is 1.5, we can interpret that given a one unit increase in X1, the relative risk of being in the adoption category (dropout) would be 1.5 times more likely compared with the reference adoption category (continuous adopter) when the other variables in the model are held constant. Therefore, if the relative risk ratio is less than unity, we can say that if a subject were to increase X1, they would be expected to fall into the reference adoption category (continuous adopter) as compared to the adoption category 1 (dropouts).

As may be expected, lower average rainfall and higher rainfall variation, which are unfavorable for the NERICA adoption, significantly raised the probability of being dropouts compared with continuous adopters. The effects of average rainfall and rainfall variation are opposite for the probability of being late adopters. These results suggest that the new adoption in 2006 took place in areas that are more suitable for NERICA production, while some early adopters in less suitable areas for NERICA production (manifested in the higher rainfall variation) stopped growing NERICA by 2006. In other words, some of the areas that received NERICA dissemination program support at the start of the NERICA campaign were likely to have been mis-targeted.

The availability of seed distribution programs in 2004 increases the probability of being continuous adopters compared with late adopters and non-adopters. In contrast,

[11] One may wonder if the use of self-produced and purchased seeds is endogenous. It is not necessarily so, however, because farmers are almost all forced to use such seeds, if the seed distribution programs are unavailable.

Table 6.5 Determinants of NERICA adoption in Uganda (multinomial logit model, relative risk ratio)

	Dropouts	Late adopters	Non-adopters
Community characteristics			
Average Annual Rainfall, mm (2001–2006)	0.998	1.009	0.999
	(0.062)+	(0.000)**	(0.840)
C.V. of rainfall (2001–2006)	2.467	0.0424	37.935
	(0.000)**	(0.054)*	(0.017)*
Availability of seed program in 2004 dummy	0.721	0.0832	0.157
	(0.324)	(0.000)**	(0.000)**
Availability of seed program in 2006 dummy	1.744	2.665	1.548
	(0.136)	(0.013)**	(0.172)
Distance to rice miller (km) in 2004	0.973	1.065	0.994
	(0.001)**	(0.000)**	(0.401)
Distance to rice miller (km) in 2006	1.070	0.998	1.071
	(0.000)**	(0.865)	(0.000)**
Relative price of maize to rice in 2004	3.530	20.544	1.764
	(0.005)**	(0.000)**	(0.124)
Community area per household in 2004	0.012	7.790	8.640
(squared mile)	(0.700)	(0.105)	(0.000)**
Traveling time to town (hours) in 2004	0.548	0.005	0.050
	(0.230)	(0.000)**	(0.000)**
Household characteristics in 2004			
Rice cultivation experience (year)	1.060	0.836	0.396
	(0.112)	(0.000)**	(0.000)**
Female headed household dummy	0.604	2.606	1.687
	(0.199)	(0.007)**	(0.104)+
Bakiga tribe dummy	2.616	30.491	0.240
	(0.029)*	(0.000)**	(0.005)**
Household head's age	1.026	1.000	1.011
	(0.003)**	(0.989)	(0.162)
Household head's education (years of	0.948	0.896	0.823
schooling)	(0.047)*	(0.000)**	(0.000)**
Per capita land owned (mailo included) (ha)	0.867	1.546	0.929
	(0.471)	(0.010)**	(0.654)
Household asset (US$1,000)	0.813	0.443	1.014
	(0.489)	(0.088)+	(0.960)
Value of livestock (US$1,000)	1.027	0.174	0.782
	(0.892)	(0.000)**	(0.220)

The numbers in parentheses are *t*-statistics
**, *, and + indicate significance at 1, 5, and 10%, respectively

the availability of seed programs in 2006 raised the probability of being late adopters compared with continuous adopters. These results indicate the importance of seed distribution programs as a driver of NERICA adoption.

The effects of the distance to rice millers are as expected: the increase in distance to rice miller raises the probability of being dropout in 2006 and that of being late adopters in 2004. These results strongly suggest that improvements in market access in the 2-year period partly explain the changing behavior of NERICA adoption. The increase in traveling time to the nearest town in 2004 raises the probability of being the continuous adopters compared with late adopters and non-adopters. This suggests that proximity to towns was not originally considered to be an important factor in efforts to promote NERICA adoption through seed distribution.

Increase in the community area per household raises the probability of being non-adopters compared with continuous adopters, suggesting that in communities where land is abundant relative to population (labor), there is a lower incentive to adopt labor-intensive crops like NERICA. The higher price of maize relative to rice increases the probability of being dropouts compared with continuous adopters.

The estimation results clearly show that rice cultivation experience does not explain the difference in adoption behavior between dropouts and continuous adopters. It is also important to note that higher household head's education increases the probability of being continuous adopters compared with dropouts. It appears that the ability to decode new information and the rice production knowledge acquired through experience do matter in NERICA adoption as predicted by Schultz (1975). The households belonging to the Bakiga tribe tend to adopt NERICA later or not to adopt it at all compared with continuous adopters. Because the Bakiga is a predominantly migrant tribe in NERICA growing areas, having limited access to suitable land and seed for NERICA production may partly explain this.

It is important to note that, asset ownership (landholdings, household assets, and livestock) does not significantly affect NERICA dropouts compared with continuous adoption. The relative risk ratios for late adopters show that larger per capita land and lower household and livestock assets increase the probability of being late adopters compared with continuous adopters.

In conclusion, the suitability of agro-ecological conditions for NERICA production (measured by rainfall patterns and relative land abundance) and transaction costs (measured by accessibility to rice millers, and the availability of seed distribution programs) are critical factors explaining the changing behavior of NERICA adoption. Our analysis also suggests that the inappropriate targeting of areas for NERICA promotion by the seed distribution programs is responsible, at least partly, for the massive dropouts.

6.4.2 NERICA Yield

Table 6.6 reports the results of the estimated NERICA yield function. As was explained in the previous section, the yield function is estimated by instrumental variables regression. To control for the quality of seed by type, we use self-produced seed dummy.[12] Although the seed type dummy and other input variables (labor use and other input use) can be endogenous in the yield function, we failed to find suitable instruments for these variables. Thus, we have to interpret the estimated

[12] This variable takes unity when seed is farmer's own produced seed or seeds purchased from neighboring farms.

Table 6.6 Yield function (ton/ha) estimation by instrumental variables (2SLS) regression[a]

	(1)	(2)	(3)
Self-produced seed dummy	0.230	0.271	0.443
	(0.62)	(0.70)	(1.11)
Self-produced seed dummy × Year 2006 dummy	−0.781	−0.655	−0.817
	(1.91)+	(1.51)	(1.87)+
Straight-row planting dummy	0.681	0.801	0.776
	(2.28)*	(2.48)*	(2.39)*
Rainfall for 90 days in the second cropping season (mm)	0.001	0.002	0.002
	(1.17)	(1.72)+	(1.68)+
Steep slope plot dummy	0.039	0.015	−0.036
	(0.11)	(0.04)	(0.10)
Late planting dummy	−0.527	−0.428	−0.321
	(2.25)*	(1.80)+	(1.34)
Traveling time to town (hour)	−0.325	−0.145	−0.121
	(1.46)	(0.63)	(0.52)
Seed program dummy	0.361	0.318	0.290
	(2.04)*	(1.69)+	(1.52)
Distance to rice miller (km)	0.003	0.001	−0.002
	(0.48)	(0.18)	(0.25)
Number of adult male aged 15–59	−0.028	−0.052	−0.080
	(0.42)	(0.74)	(1.11)
Number of adult female aged 15–59	−0.101	−0.114	−0.115
	(1.15)	(1.23)	(1.23)
% Female headed households	−0.623	−0.514	−0.564
	(2.15)*	(1.94)+	(2.06)*
Household head's age	−0.019	−0.022	−0.023
	(3.03)**	(3.29)**	(3.35)**
Household head's years of schooling	0.003	0.004	0.003
	(0.66)	(0.87)	(0.79)
Land area per capita (ha)	0.197	0.112	0.214
	(1.03)	(0.57)	(1.04)
Household asset (1,000 USD)	0.271	0.150	0.351
	(0.96)	(0.52)	(1.16)
Value of livestock (1,000 USD)	0.327	0.398	0.355
	(1.44)	(1.58)	(1.36)
Chemical fertilizer application per ha (kg)	0.020	0.023	
	(2.23)*	(2.33)*	
Organic fertilizer application per ha (kg)	−0.001	−0.001	
	(0.68)	(0.72)	
Seed used per ha (kg)	0.006	0.007	
	(1.54)	(1.61)	
Pesticide use dummy	1.216	1.140	
	(2.94)**	(2.59)**	
Herbicide use dummy	−0.269	−0.334	
	(0.51)	(0.60)	

(continued)

Table 6.6 (continued)

	(1)	(2)	(3)
Male labor hours spent per ha	0.000		
	(0.89)		
Female labor hours spent per ha	0.001		
	(3.57)**		
Child labor hours spent per ha	−0.001		
	(1.22)		
Dropout dummy[b]	−1.297	−1.545	−1.915
	(1.48)	(1.70)+	(2.05)*
Late adopter dummy[b]	0.334	0.412	0.192
	(0.38)	(0.41)	(0.20)
Year 2006 dummy	−1.011	−1.152	−1.272
	(1.61)	(1.69)+	(1.73)+
Constant	2.491	2.725	3.268
	(2.43)*	(2.62)**	(3.23)**
	0.31	0.23	0.18
Observations	378	378	378

**, *, and + indicates significance at 1, 5, and 10%, respectively

[a] The numbers in parentheses are *t*-statistics

[b] Endogenous variables: instruments = coefficient of variation of rainfall, relative price of maize to rice, vice-president program district dummy

coefficients with caution. For checking the robustness of the results, we show the results with three specifications: with all inputs (column 1), without labor use variables (column 2), and without labor use and the other input variables (column 3). The results in these three specifications are similar.

Self-produced seed dummy is not significant, suggesting that there is no difference between self-produced seeds and non-self-produced seeds in 2004. However, the use of self-produced seeds had negative impact on yield in 2006. This may be an indication of mistreatment of seeds and deterioration of seed quality. Land area per capita is not significant, suggesting that large and small farmers achieve similar yield.

The results show that cultivation practices such as straight-row planting, timing of the planting, chemical fertilizer application, pesticide use, are important factors to increase the NERICA yield. Higher rainfall seems to have increased the yield, which is expected in Uganda where upland NERICA rice is grown under rainfed conditions. Thus, NERICA rice production ought to be promoted in areas with sufficient and reliable rainfall to enable farmers to enjoy its high production potential without exposing them to high production risk. In order to fully exploit NERICA's high-yielding traits, there is an urgent need to provide appropriate information to farmers on how to cultivate upland rice as well as to produce high-quality seed.

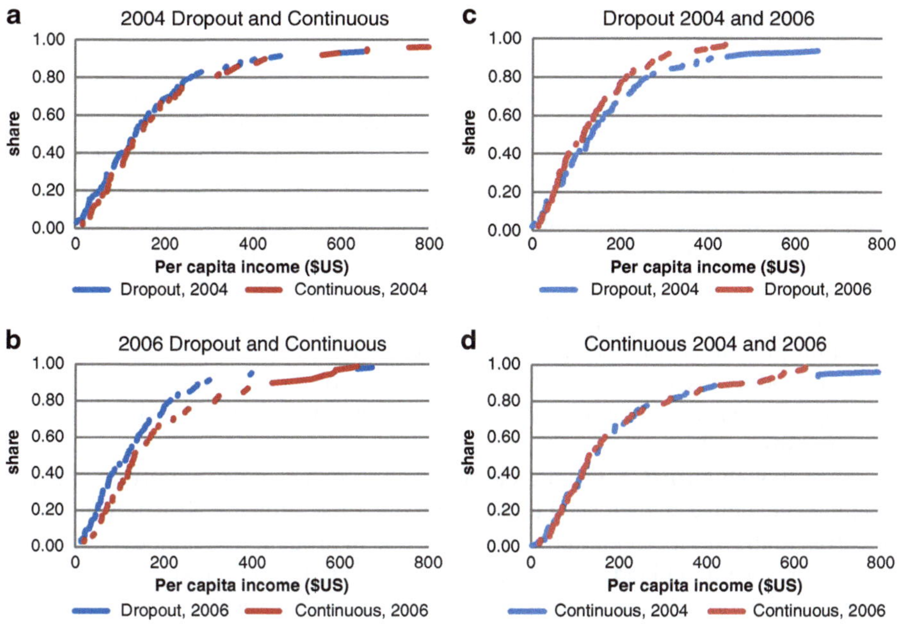

Fig. 6.1 Cumulative distribution of per capita income

6.4.3 *Effect of NERICA Adoption on Household per Capita Income*

Lastly, we examine the impact of NERICA adoption on household income. Since we did not collect income data for non-adopters in 2004, we cannot measure the impact of NERICA adoption on household income. What we can do is to examine whether continuous adoption of NERICA increases the total household income or not by using only the early adopter sample (dropout and continuous adopters) by comparing the cumulative distribution functions of the continuous adopters and dropouts' per capita total household income in 2004 and 2006 (Fig. 6.1). Panel A shows the per capita income of dropouts and continuous adopters in 2004 and there is no significant difference between two lines.[13] In contrast, per capita income of dropout in 2006 is lower than that of continuous adopters (Panel B). Comparison between Panel C and Panel D suggests that adoption of NERICA has positive impact on total household income.

[13] Note that rainfall was not particularly low in 2004 in areas where dropouts reside.

6.5 Conclusion

In this chapter, we analyzed the determinants of NERICA adoption and its effects on rice yield and crop income per hectare using 2-year panel data from rural Uganda. Using panel data of 347 households collected in 2004 and 2006 in rural Uganda, we identified four types of NERICA adoption behaviors: continuous adoption in the 2 years, dropout, late adoption, and non-adoption. A major determinant of dropouts, which account for 37% of our sample households, is the large variation in rainfall, indicating that some farmers adopted NERICA in 2004 in areas unsuitable for its production. We found that the availability of seed distribution programs was a critical determinant of NERICA adoption. The use of farmer-produced seed led to a substantial reduction in rice yields in 2006, suggesting that farmers do not have appropriate knowledge on the production of high quality seed.

Even though NERICA has a great potential to reduce poverty in rural Uganda, there are challenges which need to be addressed. First, growing NERICA in upland involves high risk of crop failure, especially due to drought. Even though NERICA is drought-tolerant upland rice, rice is a crop which requires relatively large amount of water. It is observed that many farmers lost all the harvests of NERICA due to severe drought. It is therefore important to keep in mind that growing rice in upland without irrigation has such a risk. In all likelihood, the high dropout rate was caused by the dissemination of NERICA to inappropriate areas. In fact, agricultural field officers have witnessed that farmers tend to switch upland crops after they experienced harvest loss due to drought.

Second not only the absolute profitability of NERICA but also its relative profitability compared with cooking banana, maize, cassava, and potatoes, which are major traditional staples, as well as labor intensity, matters. These crops require relatively less labor inputs than NERICA rice. According to Kijima et al. (2008), the profit of rice is five times higher than that of maize, even though the labor used in rice cultivation is more than twice as much as maize. Once the land becomes scarcer relative to labor, it is likely that NERICA rice is adopted more widely, as it is land-saving and labor-using.

Third, there seems to have been mistreatment of seeds in self-produced seeds. Therefore, the training of farmers on how to produce high-quality self-produced seeds is urgently needed. Otherwise, the prevalence of low-quality seeds can become a major constraint on the further expansion of NERICA area and upland rice production in Uganda.

In short, carefully targeted extension program to disseminate NERICA rice to areas with sufficient rainfall, few profitable alternative crops, and scarce land endowment must be implemented with a special focus on the training of farmers on the self-production of high-quality seeds as well as upland rice cultivation.

References

Balasubramanian V, Sie M, Hijmans R, Otsuka K (2007) Increasing rice production in Sub-Saharan Africa: challenges and opportunities. Adv Agron 94(1):55–133
Kijima Y, Serunkuuma D, Otsuka K (2006) How revolutionary is the 'NERICA Revolution'? Evidence from Uganda. Dev Econ 44(2):232–251
Kijima Y, Otsuka K, Sserukuuma D (2008) Assessing the impact of NERICA on income and poverty in Central and Western Uganda. Agric Econ 38(3):327–337
Ligon E, Sadoulet E (2007) Estimating the effects of aggregate agricultural growth on the distribution of expenditures, Background paper for the world development report 2008. World Bank, Washington, DC
Moser C, Barrett C (2006) The complex dynamics of smallholder technology adoption: the case of SRI in Madagascar. Agric Econ 35(3):373–388
Otsuka K, Estudillo JP, Sawada Y (2009) Rural poverty and income dynamics in Asia and Africa. Routledge, London
Schultz TW (1975) The value of the ability to deal with disequilibria. J Econ Lit 13(3):827–846
Sserunkuuma D (2008) Assessment of NERICA training impact. A study report prepared for the Japan International Cooperation Agency
Sunding D, Zilberman D (2007) The agricultural innovation process: research and technology adoption in a changing agricultural sector. In: Gardner B, Rausser G (eds) Handbook of agricultural economics, vol 1A. Elsevier, Amsterdam
World Bank (2008) World development report 2008: agriculture for development. World Bank, Washington, DC

Chapter 7
Impact of NERICA Adoption on Rice Yield: Evidence from West Africa

Aliou Diagne, Soul-Kifouly Gnonna Midingoyi, and Florent M. Kinkingninhoun-Medagbe

Abstract There is an urgent need to accelerate the growth in domestic rice production in West Africa to reduce its unsustainable and risky dependency on rice imports. Also important is resistance to drought and other climatic risks in rice farming in West Africa where precipitation is low and uncertain. The improved drought-resistant upland rice varieties, NERICAs, were introduced to rice farming system in Côte d'Ivoire, Guinea, Gambia and Benin from the late 1990s through participatory varietal selection trials. Farmers then started disseminating them through their informal channels. The objective of this chapter are to assess the characteristics of NERICA adopters and the potential contribution of the NERICA varieties to the improvement of land productivity in upland rice farming by applying the potential outcome framework to farm household survey data collected in the four West African countries.

Keywords NERICA • Upland rice • NERICA yield • West Africa • Area expansion • Land productivity • Participatory varietal selection

A. Diagne (✉) • F.M. Kinkingninhoun-Medagbe
Policy, Innovation Systems and Impact Assessment Program,
Africa Rice Center (AfricaRice), 01 B.P. 2031, Cotonou, Benin
e-mail: A.Diagne@cgiar.org; f.medagbe@cgiar.org; kinkinflo@yahoo.fr

S.-K.G. Midingoyi
Programme d'Analyse de la Politique Agricole (PAPA),
Institut National de Recherche Agronomique du Benin (INRAB),
B.P. 128, Porto-Novo, Benin
e-mail: zoulkm2000@yahoo.fr

K. Otsuka and D.F. Larson (eds.), *An African Green Revolution: Finding Ways to Boost Productivity on Small Farms*, DOI 10.1007/978-94-007-5760-8_7,

7.1 Introduction

Africa's rice production has not been able to match the growth in demand. Rapidly rising imports (8% growth per annum since 1997) have been filling the widening gap between regional supply and demand. Rice imports stood at 7.8 million metric tonnes (mt) in 2008 in West Africa, representing an estimated 2.42 billion US dollars in scarce foreign exchange flowing from the region (FAOSTAT 2009). With such high dependence on imports, Africa is highly exposed to international market shocks with grave consequences for its food security and political stability as attested by the events that took place during the 2008 food crisis in many West African countries. Despite the double digit increase in rice production in 2008 and 2009 in West Africa as a result of the policies implemented by many countries to respond to the food crisis, West Africa's rice import is estimated to have risen to close to 9 million mt amounting to a total import bill of 3.16 billion US dollars (FAOSTAT 2009). Hence, there is an urgent need to accelerate the growth in domestic rice production in West Africa to reduce its unsustainable and risky dependency on rice imports. So far, the growth in Africa's domestic rice production has been achieved almost exclusively through area expansion with a very low and stagnant land productivity compared to other regions of the world (Larson et al. 2010). The low productivity of rice farming in West Africa is due mostly to the predominance of rainfed rice farming in upland and lowland ecologies, which occupy up to 74% of the total rice planted area while accounting for only 55% of the rice produced in West Africa (Lancon et al. 2001).

There are three main rice ecologies in West Africa: upland (also called rainfed upland) rice, which depends on rainfall for water supply; lowland rice, which depends much less on rainfall because it is cultivated in plots that are close to the underground water table; and irrigated rice. Rainfed rice farming is characterized by traditional practices. It is mainly intended for meeting household consumption needs and, at times, expenses during religious or traditional ceremonies, or funerals. Rainfed rice farming is exposed to climatic hazards, notably drought, and is characterized by insufficient labor (restricted to the household). Also, rainfed rice farms are often small and intensification is seldom practiced due to the high cost of inputs (improved seeds, pesticides, fertilizers).

The interspecific hybridization breeding program of the Africa Rice Center (AfricaRice, ex WARDA- West Africa Rice Development Association), which was started in 1991, resulted in the availability of new plant types in 1996, and it won its creator Monty Jones the 2004 World Food Price. The new plant types, which are dubbed NERICA (New Rice for Africa), are the result of interspecific crosses between the *Oryza sativa* rice species from Asia and the locally adapted and multiple-stress resistant *Oryza glaberrima* African rice species. The glaberrima parents of the NERICAs appear to offer a rich source of genetic resistance to drought, weed competition, blast and virus diseases, soil iron toxicity and acidity (Jones et al. 1997; Dingkuhn et al. 1998, 1999Audebert et al. 1998; Johnson et al. 1998). In addition to these desirable characteristics, some NERICA progenies have higher protein content and good cooking and eating qualities (Watanabe 2001; Watanabe et al. 1999a, b). Hence, for all these reasons, the NERICA rice varieties promise to significantly increase the income and improve the food security and nutritional status of rainfed upland rice farmers in Africa.

The NERICAs were introduced to rice farmers in Côte d'Ivoire and Guinea from 1996 to 1997 respectively, and in Gambia and Benin from 1998 through participatory varietal selection (PVS) trials (WARDA 1999). Farmers then started disseminating them through their informal channels. Exploring the effect of NERICA adoption on rice productivity with respect to the gender concern, seems to be also relevant, given the place of women in rice production in Sub-Saharan Africa. The objective, of this chapter are to analyse the potential contribution of the NERICA varieties in raising land productivity in upland rice farming and to assess the impact of farmer adoption of the NERICA varieties on average rice yield by using farm household survey data collected in five West African countries (Benin, Cote d'Ivoire, The Gambia and Guinea).

The chapter is organized as follows. Section 7.2 describes the sources of the two sets of data used in our analysis: the agronomic trials data used in assessing the potential contribution of the NERICA varieties in raising land productivity in upland rice farming and the household level data sets used to assess the impact of NERICA adoption on average farm yield. Section 7.3 presents briefly the methodologies uses in our empirical analyses. Section 7.4 presents the results of the analyses and compares them with evidence from related studies and data. Section 7.5 concludes the chapter.

7.2 Materials and Methods

7.2.1 Data Source

The data used in our analysis came from two sources. The first source is data from experimental trials conducted in 1999 by the AfricaRice (then WARDA) breeding program at its headquarters in M'bé in Côte d'Ivoire and at two of its on-farm research sites (Man and Korhogo) also in Côte d'Ivoire.[1] The agronomic trials used a randomized complete block design with three replicates at each site. They involved NERICA progenies, their two *sativa* and *glaberrima* parents (WAB56 104 and CG14), and a select set of *sativa* checks which varied across the trials. The set of NERICA progenies are also different across the trials. We do not have access to the original raw data containing the recorded yield (or score for the stress tested) for

[1] AfricaRice had five on-farm research sites in Côte d'Ivoire, which are known as "key sites". The key sites were selected in the early 1990s when AfricaRice moved its headquarters to Côte d'Ivoire from Liberia. The location of the key sites was chosen so as to cover all the different rice ecologies and the main rice producing regions in the forest and savanna agro-ecological zones of Côte d'Ivoire.

each replication. The data files we have access to contain only the mean yield (or score) across replications for each line and each site.[2]

The second source of data used in the analyses of this study is from surveys of rice farm households and farmers conducted at different periods in time in four countries: Cote d'Ivoire (1,500 sample farm households in 2000), Guinea (1,467 sample farm households in 2001), Benin (360 sample farm households in 2004) and The Gambia (600 sample rice farmers in 2006).[3] In all the four countries, a multi-stage stratified random sampling method was used to select the sample rice farmers with the last two stages consisting of selecting the sample villages and farmers located in all the regions where NERICA has been introduced. The selection of sample villages (50 villages in Cote d'Ivoire, 79 villages in Guinea, 24 villages in Benin and 70 villages in the Gambia) was, however, not entirely random as it purposely included villages where AfricaRice has been conducting on-farm and participatory varietal selection (PVS) research activities. For sampling at village level, a list of all villages where NERICA seed were introduced (called NERICA villages) was first constituted and a NERICA villages sample was then randomly selected from that list. Then, for each sample NERICA village, a list of neighbouring villages within 5–10 km where NERICA was not introduced (called non-NERICA villages) was constituted and 2–3 sample villages were randomly selected from that list. Thus, in Cote d'Ivoire 25 NERICA villages and 25 non-NERICA villages were selected in the forest and savanna regions. In Benin, 12 NERICA villages and 12 non-NERICA villages were selected in the central region. In The Gambia, 35 NERICA villages and 35 non-NERICA villages were selected in all the four agricultural regions of the country. In Guinea, the villages were selected among four agro-ecological zones where NERICA dissemination activities were being conducted. In each zone a further stratification was done into two types of prefectures: NERICA prefecture (where NERICA varieties had been introduced) and non NERICA prefectures (where NERICA varieties were not yet introduced). Within NERICA prefectures, two NERICA villages were selected and 3–4 non-NERICA villages selected for each selected NERICA village for a total of 79 villages. In all four countries, selection of farmers within the sample villages was done entirely randomly among the village population of rice farmers with the sample size varying across countries: 30 per village in Cote d'Ivoire, 15 per village in Benin and 20 per village in Guinea. In Gambia, in some villages five farmers were selected and in others 10 farmers were selected.

In each country, the data was collected at both village and farmer levels through a structured questionnaire. At the village level, the data collected included the rice varieties known in the village (modern and traditional) and village infrastructures and community variables. At the farmer level, the data included the rice varieties

[2] The original raw data files and much of the documentation related to the trials were lost when the war erupted in Côte d'Ivoire in 2002 and AfricaRice was forced to evacuate from its headquarters in Bouaké.

[3] A more detailed description of the household survey methodologies and data can be found in Diagne (2006), Diagne (2010), Diagne et al. (2009), and Dibba et al. (2012a,b),

known and cultivated by the farmer and other socio-demographic data. Prior to administering the questionnaire for the farmers, a list of the known varieties in the village was constructed from the village-level survey and each sample farmer was asked his or her knowledge and cultivation of the varieties known in his or her own village.

7.3 Analytical Methodology

The methodology used to analyze the experimental agronomic trials data consists of simple one sample t-tests of equality of a mean with a given value, the set of NERICA progenies used in each trial being considered as a sample from the population of NERICA lines. Thus, the mean across progenies of the mean yields or scores of the sample NERICA lines was compared with the means of the *sativa* parent, the *glaberrima* parent, and the *sativa* checks under (1) high versus low input uses, (2) drought versus non-drought conditions, (3) acid versus non-acid soil conditions and (4) susceptibility to gall midge attack. These comparisons under more or less controlled settings should provide a general idea of the agronomic characteristics of the NERICA varieties compared to the other varieties and their potential contribution to raising average rice yield in upland ecologies. Indeed, the *sativa* parent of the upland NERICA lines (WAB56-104) is an improved *sativa* variety from the AfricaRice breeding program known to have relatively good yield potential while their glaberrima parent is known to have good tolerance/resistance to many stresses found in farmers' fields.[4] The comparisons also provide some of the background information that helps understand and interpret the results out of the analysis of the farm-level survey data.[5]

Our approach to assessing the impact of adoption on yield was based on the counterfactual or potential outcomes framework introduced by Rubin (1974) which has now become the standard framework for rigorous impact assessment (Imbens and Wooldridge 2009). Under this framework, the adoption of NERICA is the "treatment", rice yield is the outcome (with two potential outcomes for every farmer: yield with adoption and yield without adoption) and the mean causal effect or impact of NERICA adoption on the average rice yield obtained by farmer randomly selected from the population to adopt NERICA is given by the *average treatment effect* (ATE) parameter.

In the literature, data arising from household surveys are called *observational* data. The biases that can arise when estimating mean causal effects with observational

[4] The NERICA parents were selected out of 316 improved and 275 traditional *O. sativa* and 1130 *O. glaberrima* accessions evaluated for morphological and agronomic traits during 1991–1992 (WARDA 1995)

[5] It should be noted, however, that many of the NERICA lines in this group either did not make it to the PVS trials or were rarely selected by farmers during the PVS trials and therefore ended up not being released.

data are of two types (Rosenbaum 2001; Lee 2005): *overt* bias and *hidden* bias. Overt bias is the difference in the observed outcome not caused by the treatment but which is due to differences in *observed* characteristics. Hidden bias is the difference in the observed outcome not caused by the treatment but which is due to differences arising from *unobservable* characteristics. Third problem of another kind which arises with observational data is the problem of "non-compliance" (Imbens and Wooldridge 2009; Heckman and Vytlacil 2005; Imbens and Angrist 1994). The non-compliance problem arises in observational studies because the subjects of treatments are people who may or may not stick to their assigned treatments if treatment was to be assigned randomly. In that case, a difference in an individual person's potential outcomes may not be due to the treatment but rather to the unobserved factors that cause that person not to stick to his or her assigned treatment. As a consequence, the average treatment effect for the entire population is different from the mean treatment effect that would obtain if treatment was randomly assigned and every person in the population complied with their assignment (Imbens and Rubin 1997; Imbens and Angrist 1994). However, only the later mean treatment effect can be given a *causal* interpretation. Imbens and Angrist (1994) provide a solution to the non-compliance problem: they proposed the *local average treatment effect* (LATE) parameter which is the mean treatment effect for the subpopulation of compliers as a measure of impact that can be given a *causal* interpretation. Because the adoption of a variety is a farmer's choice, we are faced with the non-compliance problem discussed above. Therefore, the ATE estimate of the impact of adoption on yield does not have a causal interpretation. Thus, we will estimate the LATE estimate in order to have an estimate of the impact of NERICA adoption on rice yield with a causal interpretation. Moreover, in this adoption context and with the use of awareness of NERICA as instrument in our estimation of the LATE parameter, the subpopulation of compliers as described above is made of NERICA potential adopters. In the Appendix we provide a more detailed description of the methodology we have followed to estimate the ATE and LATE parameters.

7.4 Results

7.4.1 NERICA Performance Compared to Its Parents and Sativa Checks

Table 7.1 compares the performance of NERICAs progenies with that of their *sativa* and *glaberrima* parents and other *sativa* checks under high and low input conditions and under selected stresses in upland rice ecologies. The mean yield of the NERICA progenies under low input conditions is significantly higher than that of both the *sativa* and *glaberrima* parents while not statistically different (at the 1% level) from that of the *sativa* parent under high input conditions. The findings of trials conducted more recently in West Africa by Sokei (2010) found the same results. In fact,

Table 7.1 Performance of NERICA progenies and their parents under low- and high-input conditions and major stresses in upland rice ecosystems

	Yield under high- and low-input conditions		Yield under drought and non-drought prone sites		Yield under acid and non-acid soil sites		Score for susceptibility to Gall Midge attack
	Low input	High input	Drought	Non-drought	Acid	Non-acid	
NERICA progeny lines							
N	19	19	16	16	25	25	100
Mean (across replications and lines)	1.16	3.07	1.62	2.05	1.99	3.19	40.72
Std (across lines)	0.16	0.50	0.39	0.36	0.27	0.40	8.95
Min	0.91	1.70	1.08	1.49	1.57	2.15	8.76
Max	1.47	3.86	2.27	2.66	2.63	3.96	58.69
Sativa parent (WAB56-104)							
Mean value (across replications)	0.90	3.26	1.19	1.96	1.35	3.26	52.77
T-test for equality with mean of NERICA lines							
t-Statistics	7.18	−1.63	4.43	1.01	11.88	−0.93	−13.46
p-Value	0.00	0.12	0.00	0.33	0.00	0.36	0.00
Glaberrima parent (CG14)							
Mean value (across replications)	0.64	1.00	1.24	0.86	1.23	1.00	
T-test for equality with mean of NERICA lines							
t-Statistics	14.09	18.19	3.94	13.25	14.12	27.67	
p-Value	0.00	0.00	0.00	0.00	0.00	0.00	
Sativa tolerant check[a]							
Mean value (across replications)			1.51	1.96			12.07

(continued)

Table 7.1 (continued)

	Yield under high- and low-input conditions		Yield under drought and non-drought prone sites		Yield under acid and non-acid soil sites		Score for susceptibility to Gall Midge attack
	Low input	High input	Drought	Non-drought	Acid	Non-acid	
T-test for equality with mean of NERICA lines							
t-Statistics			1.19	1.00			32.01
p-Value			0.25	0.33			0.00
Sativa susceptible check[b]							
Mean value (across replications)			0.49	1.12			38.36
T-test for equality with mean of NERICA lines							
t-Statistics			11.54	10.39			2.64
p-Value			0.00	0.00			0.01

Source: Authors, based on data from experiment in 1999 at AfricaRice headquarters (in M'bé) and at two of its on-farm research sites (in Man and Korhogo). The trials used a randomized complete block design with three replicates at each site: the NERICA progenies, their two *sativa* and *glaberrima* parents (WAB56 104 and CG14), and a select set of *sativa* checks, which varied across the trials. The set of NERICA progenies also differed across trials.)

[a] *Moroberekan*, a local *sativa* variety commonly used as check for drought

[b] IR20, an improved *sativa* variety used as check for drought

[c] Yield data are mean values for two sites (M'bé and Man) from a 1999 randomized complete block design trial with three replicates at each site

the yield performances of NERICAs with or without fertilizers application are found to be always higher and sometimes more than twice higher comparatively to *glaberrima* parents. With respect to sativa parents performance, they are found to be a bit higher or almost equal to NERICAs in conditions of fertilization, however some *sativas* are found to less perform than NERICAs particularly in non-fertilization condition.

Similarly, the mean yield of the NERICA progenies is significantly higher than that of both parents under drought conditions and in acid soils and statistically not different from that of the *sativa* parent under non-drought and non-acid soil conditions. One also notes that the mean yield of the NERICAs is not statistically different from that of the *sativa* drought tolerant check in both drought and non-drought conditions. The NERICAs appears, however, to be susceptible to the African gall midge attack as judged by the mean sensitivity score, which is significantly higher than even the *sativa* susceptible check. One can see, however, from the table that the sativa parent has a significantly higher sensitivity score than the NERICAs.

7.4.2 Impact of NERICA Adoption on Average Rice Yield

Table 7.2 shows the average rice yield in Benin, Cote-d'Ivoire, Gambia and Guinea and compares the mean yield of NERICA adopting and non-adopting farmers within each country. The mean yield of women and men within each country are also compared. One can see from the table that the estimated average rice yield is highest in Benin (1.79 t/ha) followed by Cote-d'Ivoire (1.2 t/ha) with, Gambia and Guinea having average rice yield of less than 1 t/ha (0.947 and 0.839 t/ha respectively). When NERICA adopters and non-adopters are compared, Benin and Gambia samples shows positive differences in rice yields of 937 and 117 kg/ha respectively. On the other hand, Cote d'Ivoire and Guinea sample shows negative differences of –258 and –42 kg/ha of rice between NERICA adopters and non-adopters. These differences are statistically significant except for Guinea.[6] Hence, we can conclude from the evidence provided by the sample means that, compared to those who have not adopted, NERICA adopters obtained higher average yield in two out of the four countries (Benin and Gambia), practically the same average yield in the third country (Guinea) and lower average yield in the fourth country (Cote d'Ivoire).

The dominance of non-NERICA farmers in terms of average yield is not, however, uniform across gender. Indeed, Table 7.2 shows that when we differentiate between the men-headed households and women-headed households within each group, NERICA-adopting women have a slightly higher average yield than non-

[6] These numbers are statistically significantly different from zero at 1% level for Benin and 5% for Gambia and Cote-d'Ivoire and not statistically significant for the Guinea sample at the 5% level.

Table 7.2 Average yield and difference in mean yield between NERICA adopters and non-adopters

	Benin			Cote-d'Ivoire			Gambia			Guinea		
	All	W	M	All	W	M	All	W	M	All	W	M
Number of observations	304	184	120	1,284	419	865	571	532	39	1,202	70	1,132
Number of NERICA adopters	57	32	25	46	19	27	232	214	18	262	8	254
Mean yield of NERICA	2.553**	2.682**	2.388**	0.942**	1.299**	0.691**	1.016**	1**	1.195**	0.807**	1.146**	0.796**
adopters (t/ha)	0.077	0.109	0.097	0.104	0.19 6	0.08	0.026	0.026	0.11	0.032	0.314	0.032
Mean yield of NERICA	1.616**	1.586**	1.665**	1.200**	1.221**	1.19**	0.899**	0.894**	0.968**	0.848**	1.245**	0.820**
non-adopters (t/ha)	0.032	0.042	0.050	0.030	0.053	0.037	0.018	0.018	0.08	0.019	0.091	0.019
Difference in mean yields	0.937**	1.096**	0.724**	–0.258*	0.078	–0.498**	0.117*	0.106*	0.227**	–0.042	–0.100	–0.025
between NERICA adopters and non-adopters (t/ha)	0.083	0.117	0.109	0.108	0.203	0.088	0.031	0.032	0.137	0.038	0.327	0.037

Robust standard errors in parentheses
*Significant at 5%; **Significant at 1%

adopting ones for Benin (+308 kg/ha) and Cote-d'Ivoire where NERICA-adopting women-headed households (with 1.3 t/ha) obtained approximately twice of the mean yield of men-headed households (0.7t/ha). In Gambia, NERICA-adopting women-headed households also have higher mean yield than NERICA-adopting men -headed households by a difference of +350 kg/ha. This result may be explained by the appreciated implication of women in rice farming in these countries and for whom rice production and post-harvest activities are of high importance in terms of income generation. This contrast well with what is observed in the sample in Guinea where women-headed households are not well-represented in rice farming.

As discussed above, the differences in mean yield between NERICA-adopting and non-adopting farmers may be due to differences in observable and non-observable farmer socioeconomic and environmental characteristics. Table 7.3 shows the means and standard deviations of some observed socio-demographic characteristics of farmers in the two groups. The results of the *t*-test of difference of the group means and the Kolmogorov-Smirnov test for equality of two distributions also shown in the table indicate that compared to non-adopting ones, NERICA-adopting farmers are significantly more likely to live in PVS villages and participate in PVS trials in all of the four countries. In particular for Benin, NERICA-adopting farmers are significantly more likely to have more land available for rice production and higher household size. NERICA-adopting farmers in Cote-d'Ivoire are significantly more likely to be older, be of the Bete ethnic group, live in the forest zone and practice upland rice farming. On the other hand, they are significantly less likely to be of the Senoufo ethnic group and to practice lowland farming. NERICA-adopting farmers in Gambia are significantly more likely to be more experienced in upland rice cultivation. In Guinea, NERICA-adopting farmers are significantly more likely to have higher number of years of schooling, to belong to Soussou ethnic group, to have higher household size, to be female-gender and to be in contact with the institution Sassakawa Global 2000.

Table 7.4 displays the estimated ATE (the mean impact in the population) of the impact of NERICA adoption on rice yield based on the OLS with interaction estimation method (see Appendix).[7] Only Benin and Gambia have positive and significant ATE estimates. Hence, the ATE estimate of the impact of NERICA-adoption on rice yield is 968 kg/ha for Benin and 127 kg/ha for Gambia. The ATE estimates based on gender differentiation are also positive and significantly different from zero at 1% level for Benin and Gambia with women farmers having higher impact than men in Benin.

The LATE estimates of the impact of NERICA adoption on average rice yield based on the Abadie (2003) estimator are also shown in Table 7.4. The results are shown for all farmers and for women-headed households and men-headed households separately. The first row shows the estimates of the population shares of NERICA potential adopters, while the remaining rows of the table show the LATE estimates based on the OLS LARF model without interactions. The LATE estimates

[7] Almost, the same trend in the results is observed for almost all the different ATE methods we used and qualitatively one reach same conclusion with any of the model used.

Table 7.3 Socio-demographics of NERICA adopters and non-adopters in the four countries

	Benin (2005)		Cote-d'ivoire (2001)		Guinea (2002)		Gambia (2006)	
Variable	Adopter	Non-adopters	Adopter	Non-adopters	Adopter	Non-adopters	Adopter	Non-adopters
Size	57	247	46	1,238	262	940	262	940
Village PVS			0.93(0.25)	0.38(0.48)*[a]				
Participation in PVS trials before 2000			0.28(0.46)	0.09(0.28)*[a]				
NERICA village	0.75(0.43)	0.46(0.5)*[a]			0.6(0.49)	0.41(0.49)*[a]	0.68(0.47)	0.37(0.48)*[a]
Available land for rice production	0.43(0.38)	0.32(0.35)*[a]						
Age	43.14(10.91)	42.58(13.63)	44.96(14.26)	40.55(13.29)*[a]	47.9(13.49)	47.94(13.61)	45.06(13.42)	44.57(14.11)
Woman	0.56(0.5)	0.62(0.49)	0.41(0.5)	0.32(0.47)	0.03(0.17)	0.07(0.25)*	0.92(0.27)	0.94(0.24)
Number of years of schooling	2.39(3.41)	2.14(3.26)	3.04(3.69)	2.32(3.55)	4.83(3.52)	4.34(3.43)*	4.21(3.1)	3.82(3.01)
Household size	7.16(2.91)	5.61(2.74)*[a]	7.8(6.47)	6.92(4.01)	10.44(5.73)	10.06(5.54)*	16.36(12.04)	15.76(13.97)
Ethnic group1			0.39(0.49)	0.17(0.37)*[a]	0.06(0.25)	0.22(0.42)*[a]	0.65(0.48)	0.69(0.46)
Ethnic group2			0.15(0.36)	0.41(0.49)*[a]				
Was born in the village			0.74(0.44)	0.79(0.41)	0.94(0.25)	0.90(0.55)	1.46(0.5)	1.43(0.5)
Specific region			0.85(0.36)	0.51(0.5)*[a]				
Years experience in upland rice-farming					0.75(0.43)	0.72(0.45)	10.33(10.56)	7.24(9.93)*[a]
Extension			0.04(0.21)	0.05(0.21)	0.32(0.47)	0.22(0.41)*[a]	0.14(0.35)	0.15(0.36)
Years of experience in lowland rice cultivation							12.92(10.81)	14.26(9.8)[a]

Source: data from a farm survey conducted by AfricaRice in Benin 2005
Extension: CIDTGVC for Cote-d'ivoire and SG2000 for Guinea
Ethnic group1: Bete for Cote d'ivoire, Soussou for Guinea and Mandinka for Gambia
Ethnic group 2: Senoufo ethnic for Cote-d'ivoire
Specific region: forest zone for Cote-d'ivoire
In Gambia, there is no significant difference between adopters and non-adopters with the variables, contact with DAS and NARI, receiving raining, access to credit, and the dummies of the regions
In Benin, there is no significant difference between adopters and non-adopters with the variables alphabetization, having a secondary activity, receiving formal education, within rice farmers association and length of residence in the village
In Cote d'ivoire, there is no significant difference between adopters and non-adopters with the variables having a secondary activity, was born in the village
In Guinea, there is no significant difference between Adopters and Non-adopters with the variables native of the village, Contact with AfricaRice and access to credit
* t-test for equality of the two means: reject the null hypothesis of equality at the 5% significance level
[a] Kolmogorov-Smirnov test for equality of the two distributions: reject the nul hypothesis of equality at the 5% significance level

Table 7.4 ATE/LATE estimates of the impact of NERICA adoption on average rice yield

	Benin			Cote d'Ivoire			Gambia			Guinea		
	All	W	M	All	W	M	All	W	M	All	W	M
ATE Estimates based on OLS with adoption interacted with the covariates												
Number of observations	304	184	120	1,284	419	866	571	532	39	1,202	70	1,132
Number of NERICA adopters	57	32	25	46	19	28	232	214	18	262	8	254
Mean impact in the population (ATE)	0.968** 0.081	1.103** 0.107	0.764** 0.119	0.118 (0.418)	0.512 (0.453)	−0.267 (0.437)	0.127** 0.031	0.116** 0.032	0.281** 0.111	−0.022 0.040	0.054 0.067	−0.027 0.040 0.575**
LATE Estimates based on an OLS local average response function (LARF) without interaction												
Number of observations	293	177	116	1,179	386	793	571	532	39	1,202	70	1,132
Number of NERICA adopters	55	31	24	43	16	27	232	214	18	262	8	254
Number of NERICA exposed	167	103	64	119	43	76	265	246	19	457	15	442
Estimated population share of NERICA potential adopters	0.248** 0.088		0.256** 0.064	0.35** (0.014)		0.33** 0.020	0.875** 0.039	0.870** 0.039		0.573** 0.110		
Mean impact in the subpopulation of NERICA potential adopters (LATE)	1.272** 0.235	1.364** 0.253	1.143** 0.217	−0.438 0.802	−0.438 0.802	−0.438 0.802	0.134** 0.034	0.135** 0.035	0.135** 0.034	−0.106 0.072	−0.106 0.072	−0.106 0.072

Robust standard errors in parenthesis
* Significant at the 5% significance level
** Significant at the 1% significance level

of NERICA adoption on rice yield based on the LARF model without interaction are all statistically different from zero with positive signs for Benin and Gambia (for all the population and for women-headed households and men-headed households subpopulations separately). Hence, the impact of NERICA adoption on rice yield is 1.272 t/ha in Benin and 0.134 t/ha in Gambia. For Cote-d'Ivoire and Guinea, these estimates are found to be negative but are not statistically significantly different from zero. The gender-based estimates show that women-headed households and men-headed households have the same impact (0.135 t/ha) for Gambia, but in Benin women-headed households (with 1.364 t/ha) have higher impact than men-headed households (1.143 t/ha). The results of LATE estimates based on the OLS LARF model with interaction also show similar trends.[8]

7.4.3 Discussion

The results from the agronomic trials data show that in general under good conditions (high input use or no stress) a NERICA variety will give on average the same yield as the *sativa* parent and *sativa* checks. The data also show that on average the yield loss under stress is much lower for NERICA compared to its *sativa* parent and *sativa* checks. In all cases (whether under high or low input, acid or non-acid soils or drought or non-drought conditions), NERICA yield is significantly higher than its *glaberrima* parent. One instance where NERICA did poorly is for susceptibility to the African gall midge, which it has probably inherited from its *sativa* parent. Dalton and Guei (2003) reported that, in Cote d'Ivoire, the two traditional upland varieties, *Iguape cateto* and *Moroberekan,* are the most popular upland varieties and they are cultivated widely in both the savannah and the forest ecologies.[9] Assuming that the two varieties yield lower in general than the improved *sativas* like the NERICA parent, based on the agronomic trials results we should expect most of the farmers growing them to experience higher yield when they adopt NERICA. We should also expect a farmer who grows improved *sativas* and uses small quantities of inputs or has his or her rice plots in acid soils to experience increase in yield when he or she adopts NERICA. Farmers living in drought-prone areas should also gain by adopting NERICA but only in terms of lower yield loss when there is a drought. In other words, for such farmers his or her gain from adopting NERICA is in terms of an insurance against drought. On the other hand, the agronomic trials data tell us that farmers who are using the best improved *sativa* varieties in non-acid soils and non-drought-prone areas with sufficient inputs should not see on average a change in their rice yields when they adopt NERICA. In conclusion, the agronomic trials data tell us to expect some farmers to experience higher yield when they adopt NERICA but not others, depending on their environmental conditions, levels of

[8] Some of the LARF models did not generate estimates, which explains the blank space in the Table 7.4. This is due to many reasons including insufficiency of observations to generate these estimates.

[9] According to Dalton and Guei (2003), the two varieties are "purified" landraces that were introduced in Côte d'Ivoire in the 1960s.

input use and the varietal choices they have made before adopting NERICA. In other words, we should expect a heterogeneous impact of NERICA adoption in the farming population. What evidence have we found with the survey data in Benin, Cote-d'Ivoire, Gambia and Guinea? The comparison of average rice yield among the four countries have revealed notable differences among the four countries, which may be due to specific agro-ecological conditions and varying policy and institutional environment for rice production in each country. In particular, the relatively high yield obtained for Benin compared to the other three countries may be due to the availability and intensity of input-use as already found by Agboh-Noameshie et al. (2007). In fact, Benin is the third Sub-Saharan cotton production country after Mali and Burkina-Faso (Matthess et al. 2005) and the cotton company make fertilizer available to farmers through input-credit system every year. Many farmers use part of the fertilizer intended for cotton for their food crops. For Cote-d'Ivoire case, the estimated sample mean yield of 1.2 t/ha is within the range of average yield reported for upland rice farmers. This reflects the fact that the sample is predominantly made up of upland rice farmers although the average yield calculated here is across all ecologies. Dalton (2004) reports a lower average yield of 995 kg/ha for upland varieties from a sample of 50 farmers from some of the PVS villages in the forest zone included in this study.

However, all of the mean estimates for the four countries are lower than the average upland rice yield reported in a recent study of NERICA uptake in two states of Nigeria by Spencer et al. (2006) and which was slightly above 1.5 t/ha. They are also much lower than the average yield of 2.3 t/ha for NERICA varieties in Uganda estimated by Kijima et al. (2006) based on a sample of 254 NERICA farmers. However, their discussion regarding the soil fertility, crop rotation and fertilizer use conditions among the surveyed NERICA farmers suggest that the high average NERICA yield in Uganda may have been the results of exceptionally favourable conditions in terms of soil nutrients availability.

With respect to the difference in observed mean between NERICA-adopter and non-adopter, results show that Benin and Gambia NERICA-adopting farmers had an average rice yield statistically higher than that of non-NERICA adopting farmers. The gender- differentiated results for these two countries also show a similar pattern. This appears to show a superior yield performance of NERICA compared to other varieties grown in these countries. However, we don't see the same pattern in the results for Cote d'Ivoire where NERICA-adopting farmers are shown to have an average rice yield statistically significantly lower than that of non-adopters and for Guinea where there no differences between the two groups. This should be somewhat surprising for those who are familiar with the anecdotes in Guinea regarding the yield of NERICA varieties vis-à-vis other varieties.[10]

[10] The NERICA uptake study in Nigeria by Spencer et al. (2006) did not collect yield data; so we cannot compare. On the other hand, the Kijima et al. (2006) study in Uganda is about NERICA actual and potential yield. But the study is based on a sample exclusively made of NERICA farmers and on yield data made exclusively of NERICA varieties. This makes it impossible to compare in any meaningful way the results of that study with ours.

However, our results also show that NERICA-adopting and non-adopting farmers differ significantly with respect to several of their socio-demographics and the ecological location characteristics variables which we have observed. As argued in the methodological section, these differences can well explain the observed differences in yield between the two groups. Differences in the unobserved socio-demographics and environmental characteristics can also explain the observed differences in yield. In fact, in an analysis of technical efficiency of rice farmers in Côte d'Ivoire using survey data that contained detailed farm- and plot-specific information on the environmental conditions collected during 1993–1995 from essentially the same areas as in this study, Sherlund et al. (2002) found that inter-farm heterogeneity in environmental conditions (pest and weed infestation, plant disease, soil and plot location characteristics and rainfall) explained to a large extent the differences in rice output among farmers and their estimated technical inefficiencies. Moreover, when we differentiate between men-headed households and women-headed households within each group for Cote d'Ivoire and Guinea, we find no statistically significant differences between the average yield of female NERICA farmers and that of non-NERICA ones.

Further confirmation of the importance of the socio-demographics and environmental conditions in explaining the observed differences in mean yield between NERICA and non-NERICA farmers is provided by the ATE and LATE estimates of NERICA adoption on rice yield. Indeed the ATE and LATE parameters which measure the true causal effect of NERICA adoption were estimated using various alternative methods and they all gave more or less the same pattern of results: positive and significant impact in Benin (in the range of 1 t/ha) and Gambia (in the range of 130 kg/ha) and no impact in Cote d'Ivoire and Guinea. The gender-differentiated results also show a similar consistent patterns across the various alternative methods of ATE and LATE estimation: NERICA adoption is found to have positive significant impact for women-headed households (at least for Gambia and Benin) and with the impact generally higher than that of men-headed households.[11] This is a sign of some level heterogeneity in the impact of NERICA adoption in the population, which is consistent with the results of the analysis of the experimental agronomic trial data. The finding that the impact of NERICA adoption may be higher for women-headed households than for men-headed households can be explained by the fact that in the past male farmers tended to have easier access to improved varieties than women-headed households and that, in the absence of stresses, the NERICA varieties may not have a significant yield advantage over the existing improved varieties. This is certainly consistent with the evidence from then agronomic trials data.

[11] A similar result is found for Cote-d'Ivoire with the LATE estimates based on an exponential LARF with interaction with the estimate for the female potential adopters being statistically different from zero at the 5% significance level and relatively large (+741 kg/ha).

Considering all the results above, a fundamental question could be raised: How do we explain the significant number of actual and potential NERICA adopters who are not experiencing any increase in rice yield with NERICA adoption in Guinea and Cote d'Ivoire? The answer to this question is not evident in the evidence from the agronomic trials data unless one assumes that those farmers will stop cultivating NERICA once they realize that they are not having any yield increase. But this assumption is not consistent with the evidence presented in Diagne (2006a) who found a significant estimated repeat adoption rate and a long run population potential adoption rate of up to 76%. The answer to the second question lies elsewhere, in the non-yield varietal attributes that explain the behaviour of farmer adoption of varieties. Indeed, an important finding in the Dalton (2004) study (cited above) of the influence of rice varietal attributes on farmer varietal choice was that yield was not a significant determinant of farmer willingness to pay for seed of different varieties. Instead, two production attributes – cycle length and plant height – and three consumption traits – colour, swelling capacity and tenderness – were found to be significant determinants of farmer willingness to pay for seed of the different varieties. The varieties which included NERICA and other improved and traditional varieties were selected by farmers in the context of the AfricaRice PVS trials. Adesina and Baidu-Forson (1995) and Adesina and Zinnah (1993) found similar results in models of the determinants of adoption of sorghum and rice varieties in Burkina Faso, Guinea and Sierra Leone. These findings are consistent with the interpretation of our finding regarding the statistically not significant mean impact of NERICA adoption on rice yield in the general population to mean that a large number of farmers, especially those in the forest zones, adopt the NERICA because of its non-yield varietal attributes such as its short growth cycle, height, and consumption and grain qualities. In particular, NERICA is well known to have a much shorter growth cycle than most farmer varieties (up to 30 days shorter) and this attribute is almost always the first one cited by farmers when they are asked about what they like about NERICA. The good cooking and eating attributes of some of the NERICA varieties have also been documented by Watanabe et al. (1999b).

7.5 Conclusion

The main objective of the breeding work that led to the NERICA varieties was to combine the high yielding attribute of the *Oryza sativa* rice species with the resistance attributes of the indigenous *Oryza glaberrima* to the various biotic and abiotic stresses of the African environment. Another long-sought attribute for a good upland variety is the ability to give acceptable yield under the low-input use conditions typical of upland rice farming in Africa. Judging from the evidence from the experimental trials data analyzed in this study, one can conclude that these two objectives have been met to some extent. Indeed, the NERICA varieties essentially provide an assurance to upland rice farmers against loss under the stresses that penalize sativa varieties without having to rely on the low yielding glaberrima for that. Futhermore, the poor upland farmers who cannot afford fertilizer can still obtain yield that is relatively higher on average compared to the improved sativa varieties. On the other hand, the evidence from the analysis of the farm household survey data is mixed in the sense that significant positive impact of NERICA adoption was found only in two out the four countries examined. This result, which confirms the heterogeneity of the impact of NERICA adoption on yield as predicted by the agronomic trials data, does not in itself constitute a sign that the two objectives of the breeding program was not met. On the contrary, it may well be that in the two countries where the mean impact of NERICA adoption was insignificant the majority of rice farmers did not experienced any significant stress during the survey year. In that case, the lack of impact of NERICA adoption on rice yield is well consistent with the two objectives of the breeding are program met. In a more general context, as argued by Dalton (2004) and Pingali et al. (2001), we should question the almost exclusive focus on yield performance as the deciding criterion in most varietal evaluation and release programs. This excessive focus on yield may be partly explained by the difficulties that breeders and more generally non-producers and non-consumers of a particular crop have to perceive and appreciate the benefit of improvement in the non-yield varietal attributes of that crop. These difficulties and other methodological ones may also explain why valuing these benefits has rarely been attempted in impact assessment studies of crop genetic improvement. It is important that agricultural economists and other social scientists start devoting more effort to the studies of the demand for crop varietal attributes and to the assessment of the impact of their improvement in order to correctly account for the full impact of crop genetic improvement on farmers' livelihoods and welfare.

Appendix: ATE and LATE Estimation Methodologies

Several methods have been proposed in the literatures to remove (or at least minimize) the effects of overt and hidden biases and deal with the problem of non-compliance or endogenous treatment variable in the estimation of mean causal impact using observational data (see Imbens and Wooldridge 2009, for a review). The methods can be classified under two broad categories based on the types of assumptions they require to arrive at consistent estimators of causal effects. First, there are the methods designed to remove overt bias only and to estimate the ATE parameter. These methods that estimate the ATE parameter consistently are based on an assumption known under various names in the literature: as "ignorability", "unconfoundedness", "selection on observables" or conditional independence (CI) assumption (Imbens and Wooldridge 2009; Rubin 1974; Rosenbaum and Rubin 1983). The assumption postulates the existence of a set of observed covariates, which, when controlled for, renders the treatment independent of the two potential outcomes.[12]

Second, there are the instrumental variables (IV) methods of estimating mean causal effects that are designed to remove both overt and hidden biases (including the bias resulting from endogenous treatment) and to estimate the LATE parameter (Heckman and Vytlacil 2005; Imbens 2004; Abadie 2003; Imbens and Angrist 1994). The IV based methods assumes the existence of at least one variable **z** called instrumental variable that explains treatment status but is redundant in explaining the outcomes once treatment status is controlled for. In this chapter we have used farmer awareness of the existence of NERICA varieties as instrument. Indeed, the variable indicating awareness or not of the existence of NERICA is a "natural" instrument for NERICA adoption (the treatment variable). Indeed, firstly one cannot adopt a NERICA without being aware of its existence and we do observe some farmers adopting NERICA (i.e. awareness cause adoption). Second, it is natural to assume that NERICA awareness affects overall rice yield only through adoption (i.e. merely being aware of existence of NERICA without adoption does not affect the yield of a farmer). Hence, two of the three requirements for a valid instrument in classical IV models are satisfied by the NERICA awareness status variable. The third requirement for a valid instrument in classical IV models is the instrument not to be correlated with the unobserved determinants of the outcome (i.e. yield in this case). However, this third requirement of classical IV models, which in this context is essentially equivalent to assuming that that awareness of NERICA is random in the population, is not really necessary for the identification of the LATE parameter as several authors have noted (e.g. Abadie 2003; Imbens and Angrist 1994). Indeed, it suffices that instrument is independent of the unobserved determinants of the outcome *conditional* on some observed vector of covariates and Abadie (2003) has derived a LATE estimator based on this much weaker conditional independence assumption. Of course, the assumption that awareness of NERICA is random in the population is unrealistic in this context.[13] Therefore, we have used in our analysis Abadie's LATE estimator.

---

[12] These methods that estimate the ATE parameter consistently are either pure parametric regression-based methods, or they are based on non-parametric or semi parametric methods. The non-parametric methods include the nonparametric regression-based methods and various matching estimators. The simplest regression based method is an OLS procedure that consist of regressing the adoption dummy variable and a vector $\boldsymbol{x}$ of observed covariates on the observed yield variable y. The estimated coefficient of the adoption variable is then an estimate of the impact of adoption on yield. This simple OLS procedure implies that the impact of adoption is constant across the population. Also, for the OLS estimate to be consistent one must assume in addition to the conditional independence assumption that (1) the linear relationship between yield and adoption and the covariates is valid; and (2) farmers are not basing their adoption on the anticipated gain from adoption. The implication of a constant impact across the population can be avoided by interacting the adoption dummy variable with some of the covariates $\boldsymbol{x}$. Matching methods, which have become increasingly popular for removing overt bias, involve pairing treatment and comparisons units that are similar in some metrics in terms of their observables characteristics.

[13] See Diagne (2006) for discussion and evidence against this hypothesis.

In our estimation of LATE using the Abadie (2003) estimator, we have postulated an exponential conditional mean yield response function with and without interaction to guaranty both the positivity of predicted yield and heterogeneity of the treatment effect across the subpopulation of NERICA potential adopters.

References

Abadie A (2003) Semiparametric instrumental variable estimation of treatment response models. J Econom 113(2):231–263
Adesina AA, Baidu-Forson J (1995) Farmers' perceptions and adoption of new agricultural technology: evidence from analysis in Burkina Faso and Guinea, West Africa. Agric Econ 13(1):1–9
Adesina AA, Zinnah MM (1993) Technology characteristics, farmer perceptions and adoption decisions: a tobit model application in Sierra Leone. Agric Econ 9(4):297–311
Agboh-Noameshie RA, Florent K.-M, Aliou D, Franklin S, Biaou G (2009) Who benefit more from NERICA varieties? Gender impact on production and income in Benin. Paper presented at the conference on perspectives on impact evaluation, an international initiative for impact valuation, Cairo, Egypt, 29 Mar–02 Apr 2009
Audebert A, Dingkuhn M, Jones MP, Johnson DE (1998) Physiological mechanisms for vegetative vigor of interspecific upland rices – implications for weed competitiveness. Japn J Crop Sci 67(2):358–359
Dalton TJ (2004) A household hedonic model of rice traits: economic values from farmers in West Africa. Agric Econ 31(1):149–159
Dalton TJ, Guei RG (2003) Productivity gains from rice genetic enhancements in West Africa: countries and ecologies. World Dev 31(2):359–374
Diagne A (2006) The diffusion and adoption of NERICA rice varieties in Cote d'Ivoire. Dev Econ 44(2):208–231
Diagne A (2010) Technological change in smallholder agriculture: bridging the adoption gap by understanding its source. In: de Janvry A, Oluoch-Kosura W (eds) Special issue on agriculture for development in sub-Saharan Africa. Afr J Agric Resour Econ 5(1)
Diagne A, Midingoyi GS, Kinkingninhoun-Medagbe FM (2009) Impact of NERICA adoption on rice yield in West Africa. Paper presented at the symposium "Promoting a green revolution in Africa" organized at the XXVII international conference of agricultural economists, Beijing, 17–22 Aug 2009
Dibba L, Diagne A, Fialor SC, Nimoh F (2012a) Diffusion and adoption of new rice varieties for Africa (NERICA) in the Gambia. Afr Crop Sci J 20(1):141–153
Dibba L, Simon CF, Aliou D, Fred N (2012b) The impact of NERICA adoption on productivity and poverty of the small-scale rice farmer in the Gambia. Food Sec. doi 10.1007/s12571-012-0180-5
Dingkuhn M, Jones MP, Johnson DE, Sow A (1998) Growth and yield potential of *Oryza sativa* and *Oryza glaberrima* upland rice cultivars and their interspecific progenies. Field Crops Res 57(1):57–69
Dingkuhn M, Johnson DE, Sow A, Audebert AY (1999) Relationships between upland rice canopy characteristics and weed competitiveness. Field Crops Res 61(1):79–95
Food and Agriculture Organization of the United Nations (FAO) (2009) FAOSTAT. FAO, Rome
Heckman JJ, Vytlacil E (2005) Structural equations, treatment effects, and econometric policy evaluation. Econometrica 73(3):669–738
Imbens G (2004) Nonparametric estimation of average treatment effects under exogeneity: a review. Rev Econ Stat 86:4–29
Imbens GW, Angrist JD (1994) Identification and estimation of local average treatment effects. Econometrica 62(2):467–476

Imbens GW, Rubin DB (1997) Estimating outcome distributions for compliers in instrumental variable models. Rev Econ Stud 64(4):555–574

Imbens GW, Wooldridge JM (2009) Recent developments in the econometrics program evaluation. J Econ Lit 47(1):5–86

Johnson DE, Dingkuhn M, Jones MP, Mahamane MC (1998) The influence of rice plant type on the effect of weed competition on *Oryza sativa* and *Oryza glaberrima*. Weed Res 38(3):207–216

Jones MP, Dingkuhn M, Johnson DE, Fagade SO (1997) Interspecific hybridization: progress and prospect. Proceedings of the workshop: Africa/Asia joint research, interspecific hybridization between African and Asian rice species (*Oryza glaberrima* and *Oryza sativa*), West African Rice Development Association, Bouaké, Cote d'Ivoire

Kijima Y, Serunkuuma D, Otsuka K (2006) How revolutionary is the 'NERICA Revolution'? Evidence from Uganda. Dev Econ 44(2):232–251

Lancon F, Erenstein O, Touré AA (2001) Economics research of rice farming in West Africa at the West Africa Rice Development Association (WARDA). Paper presented at the workshop – economic analysis of agricultural technologies and rural institutions in West Africa: achievement, challenges and application to rice farming research, Japan International Research Center of Agricultural Sciences (JIRCAS), Tsukuba, Japan, 12–13 July 2001

Larson DF, Otsuka K, Kajisa K, Estudillo J, Diagne A (2010) Fostering a Green Revolution in rice: can Africa replicate Asia's experience? In: Pandey S, Byerlee D, Dawe D, Dobermann A, Mohanty S, Rozelle S, Hardy B (eds) Rice in the global economy: strategic research and policy issues for food security. International Rice Research Institute, Los Banos

Lee MJ (2005) Micro-econometrics for policy, program and treatment effects, Advanced texts in econometrics. Oxford University Press, Oxford

Matthess A, van den Akker E, Chougourou D, Midingoyijun S (2005) Le coton au Bénin: compétitivité et durabilité de cinq systèmes culturaux cotonniers dans lecadre de la filière. GTZ, Deutsche Gesellschaft für Technische Zusammenarbeit, Eschborn, Germany

Pingali PL, Rozelle S, Gerpacio R (2001) The farmer's voice in priority setting: a cross country experiment in eliciting technological preferences. Econ Dev Cult Change 49(3):591–609

Rosenbaum PR (2001) Observational studies, 2nd edn. Springer, New York

Rosenbaum PR, Rubin DR (1983) The central role of the propensity score in observational studies for causal effects. Biometrika 70(1):41–55

Rubin D (1974) Estimating causal effects of treatments in randomized and non-randomized studies. J Educ Psychol 66(5)):688–701

Sherlund SM, Barrett CB, Adesina AA (2002) Smallholder technical efficiency controlling for environmental production conditions. J Dev Econ 69(1):85–102

Sokei Y (2010) Complementary technology of upland NERICA varieties. 8th African rice initiative steering committee meeting, Cotonou, Benin, 19–21 Apr 2010

Spencer DSC, Dorward A, Abalu G, Philip D, Ogungbile D (2006) Evaluation of adoption of NERICA and other improved upland rice varieties following varietal promotion activities in Nigeria: a study for The Gatsby and Rockefeller Foundations. Final Report

WARDA (1999) Annual report 1999

Watanabe H (2001) Grain quality characteristics of *Oryza glaberrima* Steud and interspecific progenies, with a view to breed high protein rice. In: Takagi H, Itô O, Iwanaga M (eds) Guraberima-ine no tokusei hyôka to riyô no kanôsei [*Oryza glaberrima* genetic resources: evaluation and use], Working Report No. 22, Japan International Research Center for Agricultural Sciences, Tsukuba

Watanabe H, Futakuchi K, Jones MP, Sobambo B (1999a) Characteristics of protein content in *Glaberrima* and their interspecific progenies with *Sativa*. Jpn J Crop Sci 68(extra issue 1):206–207

Watanabe H, Futakuchi K, Jones MP, Teslim I, Sobambo B (1999b) Grain quality of *Glaberrima/Sativa* progenies in relation to their parents. Jpn J Crop Sci 68(extra issue 1):204–205

Chapter 8
Maize Revolutions in Sub-Saharan Africa

Melinda Smale, Derek Byerlee, and Thom Jayne

Abstract Maize remains crucial for food security in Sub-Saharan Africa. In some regions, the predominance of the crop in farming systems and diets implies that yield gains have the potential to jump-start a Green Revolution like those experienced in Asia for rice and wheat. However, despite episodes of success, the evidence compiled here suggests that very little progress has been made toward achieving this potential in recent years. Reversing this condition remains crucial to agricultural growth and food security in Africa.

Over the long term, large investments and sustained political commitment are needed to ensure strong plant breeding and seed systems to serve smallholders, predicated on improved crop management practices to protect soils and cope with unreliable rainfall, and access to appropriate labor-saving technologies. More innovative extension and advisory systems are also needed to facilitate farmer learning and adapt techniques and technologies to local environmental and social conditions. Better financial services, perhaps including new forms of insurance, are needed for smallholders.

Keywords Green Revolution • Maize revolution • Sub-Saharan Africa • Maize productivity • Smallholder • Structural adjustment

---

M. Smale (✉) • T. Jayne
Department of Agricultural, Food and Resource Economics, Michigan State University, 202 Agriculture Hall, East Lansing, MI 48824-1039, USA
e-mail: msmale@msu.edu; jayne@msu.edu

D. Byerlee
Independent Consultant (previously, World Bank), 3938 Georgetown CT NW, Washington, DC, USA
e-mail: dbyerlee@gmail.com

K. Otsuka and D.F. Larson (eds.), *An African Green Revolution: Finding Ways to Boost Productivity on Small Farms*, DOI 10.1007/978-94-007-5760-8_8,

8.1 Introduction

Over the past three decades, economists have described maize research and development in Sub-Saharan Africa as an "emerging maize revolution" (Byerlee and Eicher 1997), a "stop-and-go revolution" (Howard and Mungoma 1997), a "delayed green revolution" (Smale 1995), an "obscured revolution" (Gilbert et al. 1993), and a "failure" (Kydd 1989). Most often categorized as a qualified success (Eicher 1995), the maize productivity gains achieved through smallholder adoption of improved seed and fertilizer during the 1980s were driven in part by the appropriateness of the technologies themselves and in part by state policies that encouraged their use through supporting markets and prices. Although these policies successfully promoted maize production in many countries, they imposed massive costs on national treasuries and contributed to the fiscal crises that most African governments experienced during the 1980s and early 1990s (Jayne and Jones 1997; Smith et al. 1997).

The structural adjustment programs that followed were designed to shift state involvement in markets from direct operations to public goods expenditures, bolstering private investments. In many cases, fledgling private sectors were unable to fill the void left by the withdrawal of the state and public investment has declined. Programs had a mixed record, and were often construed as imposed by the World Bank and IMF against the wishes of politicians and farm lobbies that had benefited from state marketing systems.

Policy experiments during the past 15 years since structural adjustment have ranged between two extremes. Consistent with the tenets of structural adjustment, governments such as those of Mozambique and Uganda have relied primarily on markets and regulated trade in order to coordinate food production and marketing. By contrast, governments in Malawi and Zambia have revived the "development state" concepts of the 1970s in order to promote national food security (Kydd 2009).

This chapter updates Byerlee and Eicher's (1997) review of the performance of the maize supply chain in Sub-Saharan Africa. We take a circumspect view of maize technical change in the region. An immigrant crop, maize is today the most widely-grown staple food of Sub-Saharan Africa and an important wage good in many countries. Despite past successes, continued investment in maize productivity remains crucial to agricultural growth and food security. For example, investment in maize research is required to produce a new generation of improved varieties that are drought-tolerant, pest-resistant, and nutrient-efficient. In addition to appropriate seed, diversified maize farming systems and improved crop management practices will be essential for restoring soils in order to achieve productivity gains. To ensure adoption across the continent's heterogeneous production environments, farmers will need varied combinations of inputs and practices, diffused via pluralistic seed supply and advisory systems. Expanding markets in densely-populated areas with small-scale farms will require different approaches from areas with good potential, scattered populations and lower intensity of land use. Designing interventions to support market development will require persistent and careful monitoring of ongoing policy experiments.

8.2 Overview of Maize in Africa

8.2.1 Trends in Production[1]

Maize currently covers 25 Mha in Sub-Saharan Africa, largely in smallholder systems that produced 38 M in 2005–2008, primarily for food. From 2005 to 2008, maize represented an average of 27% of cereal area, 34% of cereal production and 8% of the value of all primary crop production (Table 8.1). This includes estimated area and production of green maize, which is highly valued as the harvest approaches at the end of the "hungry season". From 1961 to 2008, maize dropped slightly as a share of total area in primary crops, but not as a share of area of production of cereals, which has fluctuated between 32 and 45% over that time period.

The potential for expanding maize production in Sub-Saharan Africa is huge. Even after excluding protected and forested areas, an estimated 88 Mha of land that is not yet planted to maize is suited to the crop. Worldwide, this amount is equivalent to four times the area now planted to maize and over half of the additional land area that is suitable for maize (Deininger and Byerlee 2011). By far the largest proportion of this area is found in Sudan. Other areas with considerable potential for expansion are in Eastern and Southern Africa, including Mozambique, Angola, Zambia, Madagascar and Tanzania.

However, maize producers in these regions are often far from population centers with the markets and financial services that are conducive to technical change. Physical access to markets is far more restricted for farmers in Sub-Saharan Africa than for farmers in other regions of the developing rural world. Only a quarter of farmers in Sub-Saharan Africa are within 2 h of markets by motorized transport, as compared to nearly half of farmers in Asia and the Pacific, and 43% for the developing rural world. An estimated 75% of farmers are located more than 4 h to the nearest market, by motorized transport, as compared to 45% in Asia and the Pacific (Kate Sebastian, personal communication). Of course, most rural people in Sub-Saharan Africa have no access to motorized transport, so these figures understate the magnitude of the problem.

In sub-Saharan Africa, excluding South Africa, the highest growth in maize area, yields, and production from 1961 over the entire period has been in West Africa, and the least has been in Southern Africa where yields have stagnated at a little over 1 t/ha.[2] These differences are reflected in regional average yields, which are as high as 1.7 in West Africa and 1.5 in East Africa, but only 1.1 in Southern Africa. From 1961

[1] For consistency, despite well-known limitations, all figures reported in this section are calculated from FAOSTAT data available at http://faostat.fao.org. Regional names are those used by FAO, although countries included by region differ. Country lists are compared in the Appendix.

[2] An additional 2.8 million ha is grown in South Africa, mostly on large-scale commercial farms (averaging about 380 ha each), much of it yellow maize for animal feed. Owing to its apartheid legacy, smallholder maize contributes less than 15% of national production, and accounts for a only a minor fraction of household income of black rural families. Maize marketing and pricing policy issues focus primarily on keeping food prices at tolerable levels for urban consumers, and ensuring the continued viability of the large commercial farm sector, with very little attention to smallholder maize production or marketing. National yields have steadily improved to reach about 5 t/ha while area has declined. Except in drought years, South Africa produces a modest maize

Table 8.1 Maize area, production, yield and consumption in regions of Sub-Saharan Africa, 1961–2008

	Western Africa	Central Africa	Eastern Africa	Southern Africa[a]	Sub-Saharan Africa[a]	South Africa
Maize area (million ha, 2005–2008)	7.75	2.31	7.79	6.99	24.84	2.46
Maize production (million tons, 2005–2008)	12.86	2.42	11.62	7.62	38.21	8.55
Maize yield (2005–2008)	1.66	1.05	1.49	1.09	1.39	3.45
Growth in maize area (%/year, 1961–2008)	3.09	1.92	1.84	1.30	2.03	–0.89
Growth in maize production (%/year, 1961–2008)	4.80	2.90	3.02	1.30	2.99	0.98
Growth in maize yields (%/year, 1961–2008)	1.71	0.98	1.18	0.00	0.95	1.87
Average kg/cap/year (2003–2005)	24.4	24.9	26.9	81.8	39.6	104.2
Average percent of calories/cap/year (2003–2005)	8.6	12.4	19.3	36.1	19.1	30.8

Source: FAOSTAT. See the Appendix for country classification used in this table
[a] Excludes South Africa

to 2008, area growth accounted for two thirds of the overall 3% annual production growth in Sub-Saharan Africa; yield growth has averaged only 1% annually.

Growth rates vary considerably in each region of the continent and by decade (depending on endpoints chosen and the incidence of droughts), sometimes appearing negative but also much higher during episodes of success. In some cases, such as Zimbabwe and Zambia, trends in maize production have changed abruptly with policy shifts, and in other countries such as Angola and Mozambique, prolonged civil wars depress trends. The yield gap between countries in Sub-Saharan Africa and those with comparable production conditions is large, although it narrows if only rainfed areas are considered. It is important to recognize, however, that yield variability is much greater in Sub-Saharan Africa than elsewhere on a world scale (Box 8.1).

Box 8.1 How Do Yields in Sub-Saharan Africa Compare to Those of Other Tropical Regions?

Average yields and yield growth rates in other countries in tropical rainfed environments provide points of contrast. From 2005 to 2008, average maize yields were estimated at 3.8 t/ha in Brazil, 3.1 t/ha in Mexico, 2.5 t/ha in the Philippines, and 3.9 t/ha in Thailand, compared to 1.4 t/ha in Sub-Saharan Africa. Annual yield growth from 1961 to 2008 averaged 2.4, 1.8, 2.8, and 1.6% respectively in Brazil, India, Philippines and Thailand, on average about double the 1% growth in Sub-Saharan Africa. However, a careful analysis of sub-national data, suggests that netting out irrigated area which is important in all but Brazil, the gap between yields in Africa and rainfed areas elsewhere is smaller although still sizeable. For example, rainfed maize yields in Mexico are just over 2 t/ha and have been rising at about 1.9% per annum since the 1970s. Much of this yield gap would be due to higher and more widely adopted fertilizer use on Mexican maize.

Maize yield variability is extremely high in sub-Saharan Africa. Even among developing countries that have approximately the same mean yields, the variability of yields is nearly always higher in countries of Sub-Saharan Africa (Byerlee and Heisey 1997). Countries in Southern Africa have the highest coefficients of variation (Table 8.2). Zimbabwe's coefficient of variation in maize production from 1991 to 2007 was 41%, as compared to 33% in Malawi, 31% in Zambia, and only 11% in Kenya. Climatic factors are responsible for much of yield variability, which as discussed later also aggravates price variability. By contrast, in countries where rice is major food staple in Asia, coefficients of variation in production are in the single digits.

(continued)

surplus for export. Yield increases partly reflect deregulation of the industry and the reduction of maize area where it is no longer competitive because of lower yields and higher risks. Commercial farmers have also invested substantially to improve maize production. About one quarter of the area is irrigated. In Northern Cape Province, where all maize is irrigated, yields are around 10 t/ha. Farmers also use advanced maize hybrids, including genetically modified seed, and apply about 75 kg/ha of fertilizer nutrients, much higher than elsewhere in Africa. Given its uniqueness, we have chosen to not to include South Africa in analysis of regional data.

Box 8.1 (continued)

Table 8.2 Variability in maize and rice yields and prices around trend in major maize producing countries compared to rice in Asia, 1991–2007

Country and region	CV of yield (%)	CV of producer price (%)
Africa—maize		
Ethiopia	14.5	23.2
Ghana	7.2	37.6
Kenya	11.1	19.5
Malawi	32.9	39.3
Mozambique	23.8	22.0
Nigeria	6.5	20.6
South Africa	20.3	28.6
Tanzania	11.2	na
Uganda	8.2	na
Zambia	30.6	na
Zimbabwe	40.9	na
SE Asia—rice		
Cambodia	7.2	24.8
Indonesia	2.1	24.2
Laos	4.1	15.4
Malaysia	3.8	9.0
Myanmar	3.0	na
Philippines	5.5	7.0
Thailand	3.0	15.7
Global		
Maize	1.3	13.4
Rice	3.3	19.9

Source: Authors

Note: Computed from the standard error around a linear time trend, divided by the mean for the period

8.2.2 Trends in Maize Consumption and Trade

In high income countries, an estimated 70% of maize is destined for feed, only 3% is consumed directly by humans, and the remainder is used for biofuels, industrial products and seed. In Sub-Saharan Africa outside of South Africa, 77% of maize is used as food and only 12% serves as feed.

Maize, predominantly white, is consumed in Sub-Saharan Africa boiled or cooked. The two types of white maize (dent and flint) are largely associated with different food products (FAO 1997). Dent maize is soft and floury and used for porridges, while flint maize has a hard, vitreous endosperm and is used primarily for gruel or couscous.

In parts of Sub-Saharan Africa such as Malawi, flint maize has been preferred to dent because of smaller losses incurred in traditional storage and processing practices.

Maize has accounted for 22–25% of starchy staple consumption in Africa from 1980, representing the largest single source of calories, followed closely by cassava.[3] The significance of maize as a staple varies across the continent. The highest amounts of maize consumed are found in Southern Africa at 85 kg/capita/year as compared to 27 in East Africa and 25 in West and Central Africa. In Lesotho, Malawi, South Africa, Zambia and Zimbabwe, average consumption is over 100 kg/capita/year. These amounts represent more than 50% of total calories in Lesotho, Malawi and Zambia, 43% in Zimbabwe, and 31% in South Africa.

As a point of comparison, the 2007 average of rice consumption in Southeast Asia as a whole is 131 kg/capita/year and rice represents 55% of total calories. Thus, for some countries in Sub-Saharan Africa, maize is important enough in farm production, incomes and diets that yield gains could have impacts on producer and consumer welfare similar to those that occurred with improved rice in Southeast Asia. However, food staples are much more diversified in many areas of Sub-Saharan Africa than they are in Asia (Larson et al. 2010).

8.2.3 Urbanization and Trade

Net maize imports to Sub-Saharan Africa average around 1.5 Mt or less than 5% of total consumption. Maize imports are generally small in West and Central Africa, but play an important role in Eastern and Southern Africa. Trade within the region, both formal and informal, is also significant in some years. However, many countries resort to discretionary and unpredictable trade policy controls such as import and export bans, as well as direct state trading operations, which have curtailed the potential of regional trade to reduce price instability (Chapoto and Jayne 2009).

Few countries in Sub-Saharan Africa are competitive in global markets for exports, largely because of high transport and logistics costs; for the same reasons, most countries are competitive for import substitution. Given both greater productivity and improved infrastructure, the expansion of regional markets could eventually provide the basis for competition in export markets (World Bank 2009).

Within the next few decades, the majority of Sub-Saharan Africa's people will be living in urban centers and will depend on a diminishing minority of the population to produce food. Ongoing demographic change means that countries in eastern and

[3] By region, no trend is apparent in per capita maize consumption over the past five decades, although a slight increase is visible in Central Africa. However, in Ethiopia, maize as a percentage of daily energy has nearly doubled from 10% in 1961–1963 to 19% in 2003–2005.

Table 8.3 Staple food budget shares, urban centers in Kenya, Mozambique, and Zambia

Urban center	Year	Percent of food group in total value of consumption of main staples[a]				Percent of the four main staples in total food consumption
		Maize	Wheat	Rice	Cassava	
Nairobi, Kenya	1995	42.4	35.3	22.4	0.0	–
	2003	36.3	39.0	24.7	0.0	28.4
Urban Maputo Province	1996	2.6	50.7	35.0	11.7	42.8
	2002	8.9	57.4	28.9	4.8	27.0
Urban Northern Mozambique (includes Nampula city)[b]	2002	32.6	8.2	14.7	44.4	47.5
Lusaka, Zambia[c]	2007/2008	39.0	49.4	10.7	0.9	19.5
Kitwe, Zambia[c]	2007/2008	42.5	45.3	10.3	2.0	23.2
Mansa, Zambia[c]	2007/2008	45.8	28.2	10.0	16.0	23.8

Source: Mason et al. (2011)

[a] Main staples refers to maize, wheat, rice, and cassava. Budget shares of these four staple foods sum to 100% +/–0.1%. Shares for Nairobi and northern Mozambique are the percentage of total food purchases

[b] Cassava category also includes potatoes for urban northern Mozambique (separate figures for cassava only not available)

[c] Excludes foods purchased and consumed away from home – information not available

southern Africa regions are increasingly dependent on imports of staple foods. Net maize exports in East and Southern Africa as regions display downward trends, with substantial variability, over the past few decades (FAOSTAT). Net exports are negligible in West and Central Africa and show no trend. All countries in Southern Africa have a negative trend in net maize exports. In East Africa, there is a negative trend in net maize exports for two of six countries (Kenya and Rwanda). Kenya and Zimbabwe, net exporters of maize in the 1970s and 1980s, are now chronic importers. Malawi has also been a net maize importer in four of the past 6 years. The reduction of maize production subsidies in South Africa has also reduced the exportable surplus in that country, although it remains a reliable exporter.

Urban populations are growing at over 4% per year in Sub-Saharan Africa compared to less than 1% among rural populations. To feed urban populations, especially in coastal cities, maize imports would have been much larger but for rising imports of wheat and rice. For example, in the urban areas of East and southern Africa shown in Table 8.3, wheat and rice (much of which is imported) are more important than maize in consumption. The consumption shares of wheat and rice in urban areas are growing rapidly even in areas where maize has long been the primary staple crop, reflecting the overall decline in maize self-sufficiency in these countries as well as a shift in urban preferences toward "convenience staples" such as bread and rice (Jayne et al. 2010).

8.3 Technological Change

8.3.1 Special Challenges of Maize R&D

Morris (1998) provides an in-depth characterization of maize that distinguishes it from rice or wheat, the world's two other major cereals. Because maize is a cross-pollinating crop, a field of maize that is harvested and replanted will result in maize plants that differ from the preceding generation and from each other. Improved, open-pollinated varieties quickly lose their identity unless seed is frequently replaced. At the same time, cross-pollination enables breeders to exploit "hybrid vigor." The most rapid genetic improvements in cereals have been realized with hybrids in temperate maize (Fischer et al. 2009). Provided that farmers replace the seed each season, yield advantages of hybrids can be substantial. Whether farmers grow improved open-pollinated varieties (OPVs) or hybrids, however, they are reliant on a commercial seed industry to a much greater extent than growers of improved rice or wheat.

Maize is also more photosensitive, yet is grown over a wider range of altitudes and latitudes than any other food crop, under temperatures ranging from cool to very hot, on wet to semi-arid lands, and in many different types of soil. Environmental heterogeneity leads to continual interaction of genotype with environment and the formation of new maize types in farmers' fields through natural outcrossing and farmer selection. Well-adapted germplasm is highly specific to location. Thus, noteworthy advances in temperate maize among industrialized countries cannot be transferred to tropical environments of developing countries, and progress achieved in one tropical environment cannot be easily replicated in another. The preferences of Sub-Saharan Africans for white maize have also constrained progress in breeding.

Heisey and Edmeades (1999) report that compared to wheat and rice, maize is more likely to be grown in areas that are regarded as "marginal" from a physical or economic standpoint. They argue that crop management interventions may have greater potential for significant impact on maize production in these environments than genetic solutions, though may be costlier to develop and diffuse across the continent's heterogeneous landscapes and may ultimately reach fewer farmers than stress-tolerant germplasm.

8.3.2 Development of Maize R&D Systems

Episodes of successful maize breeding and adoption in Eastern and Southern Africa have been reviewed in detail by Smale and Jayne (2003) and Lynam et al. (2010). The products of early scientific efforts, initiated on the eve of independence in Kenya (1963), Zambia (1964) and Malawi (1964), and several decades before the independence of Zimbabwe (1980), were promising. These served as the basis for generations of new maize hybrids and other improved varieties that spread rapidly among smallholders in newly formed African states.

Just after independence in 1963, Kenya's research program in Kitale, located in the part of the highlands populated by European settlers, released a varietal hybrid (Hybrid 611) that was a cross between an improved, open-pollinated maize variety developed from local stock and landrace stock from the Americas (Ecuador 573). H611 diffused among large- and small-scale farmers in the high potential areas of Kenya as rapidly as the hybrids that swept across the U.S. Corn Belt in the 1930s and 1940s (Gerhart 1975). H611 has since served as the basis of many of the hybrids released by the national maize program (Hassan 1998).

Similarly in Zimbabwe, just after independence in 1980, smallholder Africans rapidly adopted the R-200 series of maize hybrids. Originally bred for European settlers, they were suitable for cultivation on sandy soils in low-rainfall areas and performed relatively well for smallholders (Rohrbach 1988). Following independence in 1964, Zambia's maize breeders also introduced an impressive array of both hybrids and improved open-pollinated varieties (Howard 1994).

The lack of a large farm sector prior to independence and local consumption preferences delayed Malawi's maize revolution. Malawi's smallholder farmers preferred flint maize types that processed and stored well on their farms. At that time, regional breeding efforts were focused on dent maize types suited to large-scale milling, and flint breeding materials from outside Malawi were not easy to find. Malawi's breakthrough hybrids (MH17 and MH18), not accomplished until 1990, resulted from an innovative top-cross of Malawian lines, including SR-52, with flint maize populations. The earlier-maturing of the two, MH18 often escaped dry spells, processed and stored well on-farm, and yielded more than local maize even when both were unfertilized (Heisey and Smale 1995). According to surveys undertaken by the National Statistical Office in 2006, MH17 and MH18 were still planted to over half of the area of improved maize in this densely-populated, maize-dependent nation (Sauer and Tchale 2009).

The International Maize and Wheat Improvement Center (CIMMYT) played a key role in the development of the successful hybrids in Malawi. Although CIMMYT had a regional presence in maize breeding from the late 1970s, it only seriously invested from 1985 when it established a research station at Harare in Zimbabwe. It has since maintained a strong presence in the region and many of the more recently released hybrids and OPVs contain CIMMYT parentage.

In West Africa, where there were no settler farmers like those found in Eastern and Southern Africa, maize hybrids were not developed. The scientific breakthrough in West Africa came with the release of the open pollinated varieties, TZB and TZPB, developed by IITA, during the 1970s. These varieties combined high yields with resistance to rust and blight (TZPB) and drought tolerance (TZB), spearheading the Nigerian maize revolution in the 1980s (Smith et al. 1997). They have been also widely adopted elsewhere in West Africa. Later varieties focused on streak virus resistance and are the basis for currently grown varieties (Alene et al. 2009). Overall, the number of varieties released in the region jumped from less than one annually in the 1970s to five in the 1980s, to 12 in the late 1990s.

Strong national programs are critical to successful R&D systems. Investment in national R&D programs increased rapidly from the 1970s but then stagnated in the 1990s. Spending on R&D fell in about half of the countries of Sub-Saharan Africa during the 1990s (Beintema and Stads 2006). Over the whole period staffing increased faster than funding, so that funding per scientist fell to less than half of the levels of 1971. Research capacity has also been affected by staffing discontinuities in the national maize breeding programs, and shifting emphasis between efforts to breed hybrids as compared to improved, open-pollinated varieties. Two of the three maize breeding programs recently reviewed by Lynam et al. (2010), in Ghana and Malawi, have lost all the senior maize breeders that were instrumental in earlier successful maize varietal releases. Lynam et al. (2010) identified only Kenya, with six maize breeding programs and six PhDs in maize breeding, as having substantial capacity in maize breeding. The share of maize in the national research budget is as high as12% in Kenya, which is similar to the share of rice in Asian research systems (Beintema and Stads 2006).

Investments in the crop improvement programs at international research centers also declined during the 1990s. IITA's budget for maize research fell from $10m in 1988 to $5m in 2000 (Alene et al. 2009). However, re-investment in center breeding programs by the Bill and Melinda Gates Foundation and growing emphasis on regional breeding programs has in part substituted for the centralized breeding programs of the International Agricultural Research Centers (Lynam et al. 2010).

Despite the fluctuating fortunes of maize breeding programs, research impacts in the region have been demonstrated by Manyong et al. (2003), Alene et al. (2009) and Morris et al. (2003). For example, Alene et al. (2009) estimated rates of return to research exceeding 40% in West Africa from 1971 to 2005.

Partly offsetting weaknesses in public research systems has been a sharp rise in private sector interest in plant breeding and the seed sector. In a review of varietal releases in 13 countries (excluding South Africa), Setimela et al. (2009) found that 250 varieties and hybrids had been released during the period, 2002–2006 (or nearly four per country per year). Over 60% were private hybrids. Most activity was evident in Kenya, Malawi, Zambia and Zimbabwe. Many of these hybrids were probably based on inbreds produced by CIMMYT, IITA or national public sector maize programs.

8.3.3 Adoption of Improved Maize

The most recent estimates place the adoption of improved open-pollinated varieties and hybrids at 44% of maize area in Eastern and Southern Africa in 2006–2007 excluding South Africa, and 60% in West and Central Africa in 2005 (Tables 8.4 and 8.5). These data suggest a substantial increase in adoption over the past decade, primarily in West and Central Africa. Morris et al. (2003) estimated that in the late 1990s, excluding South Africa, only 36% of maize area was planted to modern maize (mostly hybrids). Manyong et al. (2003) estimated that 37% of maize area in West and Central Africa was planted to modern maize in the late 1990s (mostly to improved open-pollinated varieties).

However, adoption figures for individual countries in Eastern and Southern Africa are erratic, depending in part on the estimation method. Data assembled from a range of sources indicate that adoption was as high in 1990 as it is now in that region, dipping in the mid-1990s (Table 8.4). Adoption of modern maize in Kenya

Table 8.4 Adoption of improved maize varieties (% of maize area) in Eastern and Southern Africa, 1990, 1996 and 2006

	2006				1996			1990		
	Improved OPV	Hybrid	Modern maize	Adjusted for saved seed	Improved OPV	Hybrid	Modern maize	Improved OPV	Hybrid	Modern maize
Eastern Africa	7	26	33	37	6	26	32	15	25	40
Ethiopia	5	14	19	21	3	5	8	16	5	21
Kenya	4	68	72	74	9	62	71	8	62	70
Tanzania	6	12	18	22	2	2	4	12	6	18
Uganda	21	14	35	54	4	4	9	50	10	60
Southern Africa	9	29	38	52	4	22	26	6	42	48
Angola	4	1	5	10	11	0	11	–	–	–
Malawi	15	7	22	50	1	13	14 (37)	3	11	14
Mozambique	10	1	11	22	9	0	9	17	1	18
Zambia	4	69	73	81	1	22	23	5	72	77
Zimbabwe	6	74	80	93	0	82	82	0	96	96
Eastern and Southern Africa	10	25	35	44	4	23	28	10	33	43

Source: Langyintuo et al. (2008), Hassan et al. (2001), and Lopez-Pereira and Morris (1994)

Note: Langyintuo et al. (2008) adjust the actual seed sales in 2006/2007 by improved OPV sales in previous two seasons for adjusted adoption rate

For both sources, improved OPV and hybrid rates are calculated as percent of seed sales. Hassan et al. include Tanzania in Southern Africa, and also include Lesotho (71%) and Swaziland (78%), but exclude South Africa (96%). Including South Africa, they calculate that the adoption rate for Southern Africa in 1996 is 47 and 43% for Eastern and Southern Africa. No data are reported for Angola in 1990. The second figure for Malawi is estimated by Smale et al. (1998) based on official area estimates. Seed sales estimates are lower

Table 8.5 Adoption of improved maize varieties (% of maize area) in Western and Central Africa, 2005, 1998, and 1981

	2005	1998	1981
West and Central Africa	60	37	4
Benin	41	25	3
Burkina Faso	75	46	3
Cameroon	44	28	8
Cote d'Ivoire	52	–	4
Ghana	89	53	1
Mali	38	23	3
Nigeria	61	40	6
Senegal	95	89	4

Source: Alene et al. (2009) and Manyong et al. (2003)
Note: Authors estimate that over 95% of modern maize planted in West and Central Africa is improved OPV, based on breeder surveys in each year, similar to those conducted by CIMMYT sources in Table 8.5
No data are reported for Cote d'Ivoire in 1998
Manyong et al. (2003) include Togo (1.3%), Chad (70%)
DR of Congo (31%) and Guinea (23%) in the regional adoption rate

appears to have leveled at 70–75% of maize area. In Zimbabwe, adoption rates reached 96% as early as 1990 (Loìpez-Pereira and Morris 1994).

Slow turnover of maize hybrids on farms may explain stagnating yields in Eastern and Southern Africa. For example, in 1996–1998, the estimated average age of varieties grown in Ethiopia was 14 years and one variety 20 years old constituted a third of maize seed sales (CIMMYT, personal communication). In 1992, the average age of all maize varieties and hybrids grown by farmers in Kenya was 14.5 years (Hassan 1998). H614D, derived from H164 released in 1986, was planted on 42% of maize area in 1992 and continued to occupy 51% of maize area in 1998 and 48% in 2010 (Hassan 1998; F. M. Ndambuki, Kenya Seed Company, personal communication, May 11). Outside of the high potential areas where H614 has superior adaptation, new adoption patterns are indeed emerging, but the range of hybrids on farms in Kenya still does not reflect the large number now registered for sale.

Despite a later start, adoption of improved maize is now higher in West and Central Africa. A mere 4% of maize area in West and Central Africa was planted to improved maize varieties in 1981 (Alene et al. 2009). West Africa appears to have also experienced a more robust rise in the adoption of improved varieties since 1990, although it is especially difficult to reliably estimate areas under improved open-pollinated varieties.[4] Ghana and Nigeria were the prominent success stories of

[4] Because of the difficulty in measuring areas planted to improved OPVs, in particular, estimates should be considered with caution. Almost all of the maize area in West African countries, with the exception of Nigeria, is planted to improved open-pollinated varieties. Given that the private seed system has not been active, it is likely that farmers practice seed saving for much more than the recommended number of years and because of cross-pollination, it may be difficult to differentiate improved from unimproved materials.

the 1980s. Impressive rates of adoption also occurred in Mali, where maize is grown in cotton-based systems, although the area is relatively small.

Despite abundant evidence of dynamic change, the data reported in Tables 8.5 and 8.6 indicate that roughly half of Sub-Saharan Africa's maize area continues to be planted to farmers' varieties, though through cross-pollination and farmer selection, breeders often suggest that many of these have been influenced by proximity to improved maize.

8.3.4 Fertilizer Use

As shown above, even in countries where improved maize covers much of maize area, only modest yield gains seem to have been achieved. Although the use of improved maize can be a catalyst for increasing farmers' use of other inputs, and especially fertilizer, such broad-based change has only occurred in some parts of Sub-Saharan Africa. Most farmers do not adopt the additional production practices needed to sustain yield improvement. This is particularly noticeable with respect to practices for maintaining and enhancing soil fertility, even though the shortening of the bush fallow rotation as a consequence of population pressure has made poor soil fertility the major constraint to raising productivity in many areas.

For all of Sub-Saharan Africa about 40% of fertilizer is used on maize, implying that the average dose is only about 17 kg/ha of nutrients compared to the developing country average of 100 and the industrialized country average of 270 kg/ha on the same crop (Morris et al. 2007; Heisey and Norton 2007). While it is incorrect to surmise that modern maize "depends" on fertilizer, modern maize does generally trace a steeper response curve for fertilizer than do traditional farmers' varieties. Maize is a heavy consumer of fertilizer, leading fertilizer demand in industrialized countries among major cereals, and the second most heavily fertilized crop on a global scale, after potatoes (Heisey and Norton 2007).

Farmers grow improved varieties without fertilizer in many areas of Africa, especially in marginal areas, such as the drier zones of Kenya and Zimbabwe, but also some relatively favored areas, such as Ghana. Higher adoption rates for improved seed than fertilizer reflect the high costs of fertilizer in Africa, lack of input availability, and farmers' cash constraints.

Even where fertilizer is used, it is often used inefficiently. A single recommendation is provided for wide areas, which does not account for the diversity of smallholder situations and the acute cash constraints under which they operate. Mistiaen (2006) analysis, which employs a benchmark productivity measure computed by matching farm household survey data to optimal fertilizer response functions for maize based on agronomic field experiments in Kenya, indicates that achieving technical efficiency could improve average yields by about 60%.

Low agronomic efficiency results from poor soil and moisture conditions, which can be remedied by adding organic sources of nitrogen. Fertilizers cannot profitably increase crop yields if soils are severely degraded. Recent research in Kenya has confirmed that removing fertilizer supply constraints will encourage use by wealthier farmers who cultivate better soils but have no impact on use by poorer farmers who grow maize on degraded land where fertilizer response is not enough to make its use profitable (Marenya and Barrett 2009a, b). Survey data commonly indicate that the contribution of fertilizer to food grain yields varies tremendously across farms even within the same villages. Households in Xu et al.'s (2009) Zambia study are characterized by a great variation in the marginal productivity of nitrogen, even in the same agro-ecological and soil conditions, which most likely reflects differences in farmers' management ability, knowledge about appropriate application rates, and whether they are able to acquire fertilizer in a timely manner. Simply bringing fertilizer response rates among farmers in the bottom half of the distribution up to the mean would contribute substantially to household and national food security (Nyoro et al. 2004).

Experts recommend greater emphasis on integrating organic matter, such as manure from livestock or post-harvest crop waste, to raise soil carbon levels and make nutrients from fertilizers more available to plants. In Malawi, Sauer and Tchale (2009) found that controlling for other factors, maize yield response to fertilizer was higher with integrated soil fertility management. Similarly, a decade of experimentation in Malawi by Snapp et al. (2010) provides evidence that integration of semi-perennial legumes (such as pigeon pea, which produces grain) can provide a foundation for sustainable crop management. Modest application of nitrogenous fertilizer to monoculture maize was effective at doubling yield, but a rotation system with semi-perennial legumes reduced the variability of yields, produced grain with 45–70% higher protein, and improved nitrogen recycling. Across sites, profitability and farmer preferences, expressed by spontaneous adoption, were in accord with these findings.

8.3.5 *Other Crop Management Practices*[5]

Extension efforts increasingly emphasize the use of more legumes, intercropping, organic manure, reduced tillage, herbicides and agroforestry, and there are some indications that farmers are adopting such practices (Holden and Lunduka 2010a). Intercropping may also be rising in some maize-based systems. Based on panel data collected by Tegemeo Institute of Egerton University, Ariga and Jayne (2010) found a rising trend in the proportion of maize area planted in more complex intercropped patterns from the mid 1990s.

Experience from many African countries has shown that seasonal labor availability is an important constraint on the acceptance of improved management practices such as plant spacing and weeding that are relatively labor intensive. If these

[5] Technologies and management practices to reduce post harvest losses should be added to the list of opportunities for improving the efficiency of the maize supply chain. Various estimates put post harvest losses for maize grown by smallholders in the humid tropics of Africa at 15–20%.

are recommended as a package with fertilizer and seed, the profitability of other components is also affected. Even where land is in short supply, seasonal labor shortages often decisively influence farmers' choice of technology for several reasons: hand-hoe agriculture demands a great deal of labor, off-farm work is important in many areas, and a pool of landless rural laborers is not available when demand for labor is greatest (Low 1988). It is therefore critical to evaluate recommended management practices in terms of their effect on the returns to labor.

Reflecting this labor constraint, farmers in the savanna of western Africa and much of southern Africa and Ethiopia have adopted animal traction in maize-based systems. However, a seasonal draft power constraint often emerges because animals are in short supply or in poor condition during the peak demands for land preparation (Collinson 1982). This has led to efforts to develop conservation tillage practices that eliminate tillage, retain crop residues and integrate legumes. While early experience has sometimes been positive (e.g., in Zambia, Haggblade et al. 2010), adoption is still limited and considerable research is needed to adapt conservation agriculture practices to locally-specific biophysical and socioeconomic conditions (Giller et al. 2009).

There is little doubt that research on crop and resource management to overcome seasonal labor constraints, and maximize returns to cash inputs, while conserving the soil base and enhancing soil fertility over the longer run, will go a long way toward increasing productivity and sustainability of maize-based systems. Research on these constraints has increased sharply in the past decade, but success, measured in terms of adoption, has not been impressive. Practices must be adapted to locally-specific situations in order to account for agroclimatic circumstances, population pressure, labor availability, and the stage of infrastructural and institutional development. Special efforts are also needed to transfer and adapt them once developed, given that many are knowledge-intensive, highlighting the importance of extension.

8.3.6 Agricultural Extension

Without doubt, maize farmers have been major beneficiaries of the expansion of national extension systems. Extension was a driving force behind the diffusion of improved maize technology to smallholders in all the countries that have experienced wide uptake of improved maize technologies (Kenya, Nigeria, Ghana, Zimbabwe, Zambia, Mali). Despite these successes, management problems arose as the number of extension staff increased and operating budgets for travel and farm visits decreased due to fiscal constraints. In Kenya, De Groote et al. (2006) found a striking decline in access to extension services from 58% of maize growers in 1992 to only 30% in 2002, even as access to credit grew from 8 to 26%.

General disenchantment with extension has led to many efforts to 'fix' public extension. One of the most influential of such efforts was the training and visit (T&V) model of organizing extension, supported by the World Bank from 1975 to 1995 in 27 countries of Africa. The T&V approach aimed to improve performance of extension systems by strengthening their management and formulating specific and regular extension messages (Anderson et al. 2006). T&V projects helped extension agencies reach greater numbers of farmers and sometimes spearheaded rapid adoption of maize technologies (Cleaver 1993; Balcet and Candler 1981). However, where rigorous evaluations of impacts of T&V extension on productivity have been conducted as in Kenya, the results were disappointing (Gautam 2000). In addition, the T&V system exacerbated fiscal sustainability and lacked real accountability to farmers (Anderson et al. 2006). By the early 1990s, a World Bank evaluation found that at least half of the extension projects in Africa were rated as "unsatisfactory" due to the use of a top down rigid model with insufficient attention to heterogeneous production conditions and circumstances of farmers in rain-fed areas (World Bank 1994).

Another approach was initiated by Sasakawa-Global 2000 (SG2000), an NGO, to demonstrate available yield-enhancing technology to farmers and policy makers in Ghana in 1986. SG 2000 has assisted public extension workers to conduct thousands of large (0.5 ha) demonstrations on farmers' fields to show the potential of a new technological package of seed and fertilizer in 14 countries in Sub-Saharan Africa (http://www.saa-tokyo.org/english/country/). Maize has been by far the major crop included in the SG2000 programs.

The SG 2000 project in Ghana claimed the most success. The extensive coverage of on-farm demonstrations was undoubtedly a major factor in the wide adoption by Ghanaian farmers of maize seed-fertilizer technology. An even larger program in Ethiopia, initiated in the early 1990s under the Participatory Demonstration and Training Extension System, integrated extension with provision of seed, fertilizer and credit. Once scaled up, the program reached about 40% of the roughly 10 million farm households in Ethiopia over a 10-year period (3.6 million demonstrations in 1999 alone) and demonstrated that the adoption of seed–fertilizer technologies could more than double maize yields. Despite these efforts, adoption of maize technologies in Ethiopia is still low and a viable private sector input distribution system has yet to emerge (Spielman et al. 2010).

The SG 2000 country projects have demonstrated that maize yields can reach 4–5 t/ha from average national yields of 1–1.5 t/ha, serving as a reminder that rapid adoption of new technologies is possible in medium-to high potential areas when relevant technology is combined with input delivery systems and market opportunities. When programs withdrew, the realities of overcoming input supply discontinuities, extending supply chains into remote rural areas, and forging solvent local agro-enterprises persisted.

Since the 1990s, a spectrum of other extension innovations have been introduced in Sub-Saharan Africa, with many systems moving to pluralistic approaches with different models often being used within a country (Davis 2008). Extension is still largely publicly funded, but funds often flow through local governments, NGOs and farmer organizations that have a controlling interest in fund allocation. To provide more accountability, many governments moved away from centralized systems and transferred to local governments the responsibility for delivering extension and, in some cases, financing it, in line with wider efforts to decentralize government closer to its local constituents. Although these are good reasons to decentralize extension, various challenges, including political capture by local elites, have often compromised progress in delivering more effective advisory services.

Uganda's National Agricultural Advisory Services empowered farmer organizations by providing them matching grants to contract NGOs and private providers to deliver specific advisory services. This program significantly increased gross farm revenues from 2004 to 2007 but impacts have differed by region, and have been greater for high-value enterprises and male farmers, but also for poor farmers (Benin et al. 2011).

One extension model is the Farmer Field School, originally designed as a way to introduce integrated pest management in Asia. The schools have been introduced, mostly on a pilot basis, in several African countries, and their scope has been broadened to other practices and technologies (van den Berg and Jiggins 2007). Evidence of impacts, although still limited, suggests that the approach can significantly enhance farmers' knowledge of new options. In the pilot districts where the approach has been used in Kenya, Tanzania and Uganda, incomes rose by some 61% on average, and women farmers and farmers without formal schooling gained most (Davis et al. 2010). Critical reviews of the evidence, most related to use of integrated pest management, suggest that Farmer Field Schools have not generated changes beyond local communities (Davis 2008), tending to favor more privileged farmers within those communities (Tripp et al. 2005). Tripp, Wijeratne and Piyadasa, as well as van den Berg and Jiggins 2007 express concern that the assessment of FFS has been narrow, potentially biased, and focused on the short-term. In an econometric analysis based on comparison of changes between control and treatment groups, Feder, Murgai and Quizon found that the training had no statistically significant impact on the yields or the pesticide use among the participants or others in the same communities, raising questions concerning the high costs per participant and the financial sustainability of the approach.

Evaluation of extension experiments is limited to date (Anderson and Feder 2003). Still, a range of options are now available for improving the performance of extension systems. The challenge now is to scale up successful innovations and close out ineffective systems.

8.4 Emerging Policy Environments

The experience of maize technical change in Sub-Saharan Africa underscores the role of policy as a determinant of development, adoption and impact. This section discusses recent policy experience with respect for seed, fertilizer, input subsidies, and maize markets, highlighting Ethiopia, Kenya, Malawi, and Zambia. The case of Ethiopia represents strong state intervention in input markets (a seed-fertilizer "technology push") with a liberalized grain market. Kenya's government has retained some control over maize grain markets, but has largely liberalized fertilizer markets and to some extent seed markets. Zambia's and Malawi's governments exert strong control over both input and maize grain markets.

8.4.1 Seed Policies

The burgeoning demand for maize grain in Sub-Saharan Africa would suggest a healthy farmer demand for certified seed, but even though maize seed markets may be better developed than are markets for most other crops, the practice of farmer seed-saving remains common. For example, although hybrids have been widely adopted in Zambia, survey data suggest that during the 2007/2008 maize growing season, 59% of maize growers use non-traded or recycled seed (also see Chap. 9 for Kenya and Uganda).

The central role of the seed industry has been repeatedly emphasized in policy analyses of Eastern and Southern Africa, and much progress has been made in developing the private seed sector in this region based on hybrid seed. In 2009, seed companies, including a few public companies, accounted for 98% of the market. More than half were private national companies, and about one fifth each were multinational and regional private companies (Langyintuo et al. 2008).[6] Langyintuo et al. (2008) concluded that the major bottlenecks in the seed industry of Eastern and Southern Africa were lack of awareness of the availability and value of existing varieties, the high investment costs to set up seed companies, outdated and rigid seed policies, and lack of credit and skilled human resources. Seed policies known to impede the development of the seed supply chain include lengthy variety release and seed certification requirements, which delay product lead times, and import-export restrictions on seed, and taxation policies. In Kenya with its elaborate regulatory framework, new varieties take the longest to reach farmers' fields. Efforts to harmonize seed laws and regulations within the region have been underway for many years in both Eastern and Southern Africa, and in West Africa in order to speed varietal release across the region by allowing approval of a variety throughout a region once one country has approved it. However, implementation progress has been very slow.

In Kenya, despite liberalization and the entry of numerous new seed companies, Kenya Seed Company (KSC), the parastatal organization, still accounted for 86% of maize seed sales in 2004, reflecting its exclusive access to hybrids produced by KARI. Nonetheless, the distances traveled by farmers to the nearest hybrid seed retailer shortened between 1997 and 2007 (Ariga and Jayne 2010). According to Ariga and Jayne, greater progress has been made in the lowland and mid-altitude zones, with the release of improved varieties by KARI and by private seed companies. De Groote et al. (2006) report rising use of improved maize seed in the lowlands, reflecting the efforts of KARI to develop new varieties, and particularly hybrids, for that zone.

In contrast to the case of Kenya, Ethiopia's seed system remains state-based and top down, integrating extension, seed, fertilizer and credit into fixed packages. Improved seed production and multiplication is carried out by the Ethiopian Seed

[6] Nongovernmental organizations and national research organizations accounted for a scant 4% of all seed marketed in the region.

Enterprise (ESE), a fully state- owned company that is the only formal source of seed for most crops. After the market reforms of the 1990s, seed production and distribution was opened to the private sector, but by 2004, there were only eight active firms, most of them involved in hybrid maize seed as subcontractors to ESE. In 2004, approximately 70% of maize seed, mostly hybrid, was still produced by ESE (Alemu et al. 2007). An even smaller level of private sector activity is seen in the distribution and retail side of the market—Pioneer Hybrid is the next largest player in the industry, producing 16% of the seed, but relying on the public sector to distribute about half of it to farmers. Not surprisingly, purchased seed in 2007–2008 accounted for just 20% of the area under maize cultivation.

8.4.2 Fertilizer Policies

Both supply and demand constraints have hindered the emergence of viable fertilizer markets in Sub-Saharan Africa (Heisey and Norton 2007; Morris et al. 2007). Since nearly all fertilizer is imported, the cost of fertilizer is dependent on transport costs, and landlocked countries are particularly disadvantaged with respect to this bulky input. Transport and logistics costs in African have been found to be three to four times higher than they are in the US, explaining the fact that in general farmers in Sub-Saharan Africa pay at least double the price for fertilizer relative to farmers in Asia and the US (Heisey and Norton 2007: Morris et al. 2007). The high seasonality of demand for fertilizer in rainfed systems and the bulkiness of the product lead to relatively slow stock turnover, considerable storage requirements, and high finance charges, resulting in risk for distributors and dealers.

On the demand side, high cost, combined with low agronomic efficiency, makes the use of inorganic fertilizers unprofitable for many farmers in Sub-Saharan Africa. Against this background, it is not surprising that most maize producing countries in Sub-Saharan Africa followed trends in Asia and chose to subsidize fertilizer sales up until the mid-1980s or even later, when fiscal crises curtailed or ended them (Heisey and Norton 2007). Extensive subsidies were fiscally unsustainable, and coupled with a parastatal input marketing system led to highly inefficient and inequitable fertilizer distribution.

Liberalization of fertilizer markets has been implemented to varying degrees across countries and with very mixed success. The liberalization of Kenya's fertilizer markets is considered to have been most successful (detailed in Ariga and Jayne 2010). After the elimination of fertilizer price and import controls in the early 1990s national fertilizer consumption doubled by 2007. Survey data collected from 1997 to 2007 by Tegemeo Institute indicate that smallholder fertilizer use per hectare of maize cultivated grew by 34%. The distance traveled by farmers to the nearest fertilizer retailer declined dramatically, reflecting increased investment in fertilizer retailing by private dealers. Inflation-adjusted fertilizer marketing margins between Mombasa and inland markets have narrowed, and nutrient-to-grain price ratios at the farm gate have become more favorable. Despite these gains, there is considerable potential for further efficiency gains through improving soil and moisture management to enhance yield response to fertilizer on the demand side, and reducing distribution costs through investments in eroded rail, road, and port infrastructure on the supply side.

In contrast to Kenya, Ethiopia continues the state-led, package-based approach today. The Government of Ethiopia liberalized the fertilizer sector after the end of the Derg and by 1996 several private firms were importing fertilizer, and 67 private wholesalers and 2,300 retailers had taken over a significant share of the domestic market (Spielman et al. 2010). However, the private sector rapidly exited within a few years of its entry, and was at first replaced by "private" holding companies with strong ties to the ruling party and then by cooperative unions. The parastatal, Agricultural Input Supply Enterprise, continues to be a major importer and distributor of fertilizer. In addition, since 1994, about 90% of fertilizer has been delivered on credit at below-market interest rates and guaranteed by regional governments, displacing sales from the private sector, including a substantial share sold on a cash basis.

Fertilizer consumption per ha has increased only marginally over the past decade, and there is evidence that many farmers have dis-adopted seed-fertilizer technology packages over time. A study of Ethiopian smallholders found that half of farmers surveyed reported that fertilizer arrived after planting, and one-third reported underweight bags (Bonger et al. 2004). Loan recovery, using extension agents and local officials, was generally successful until the collapse of maize prices in 2002 forced rescheduling that incurred significant fiscal costs. Spielman et al. (2010) conclude that although state-led policies have generated some positive impacts in Ethiopia, they also reduce the quality and timeliness of inputs services, limit farmers' options, incur hidden costs, and entail the risk of large fiscal outlays.

Xu et al.'s (2009) study in Zambia illustrates that, in the more remote areas, where farmers faced nitrogen-maize price ratios that were 20% higher than elsewhere, fertilizer use was profitable only for a minority of farmers. At the same time, fertilizer use was profitable for farmers in the more accessible areas only when its delivery was timely. Subsidized fertilizer under government programs in Zambia has often been distributed late. The authors report that government programs have also caused private traders to wait and see where subsidized fertilizer is being distributed before deciding where to distribute commercial fertilizer, exacerbating the problem of late delivery even for commercial fertilizer.

8.4.3 Smart Subsidies

The urgency of arresting soil nutrient mining combined with rising fertilizer prices in recent years have stimulated interest in ways to raise fertilizer use through a new generation of so-called smart input subsidies (Morris et al. 2007; World Bank 2008; Minde et al. 2008; Dorward et al. 2008). Input subsidies are "smart" if;

- the crop productivity and food security benefits outweigh what might have been achieved through alternative investments (not only direct but also considering the opportunity costs of resources used)
- they stimulate investment in input distribution by private suppliers and agro-dealers and the development of a robust input distribution system
- they target farmers who would not otherwise use purchased inputs in areas were economic yield response to fertilizer can be achieved, and
- they have a clear exit strategy.

Input vouchers redeemable at private input dealers and targeted to farmers who use little fertilizer have been the main vehicle for implementing smart subsidies. Malawi's is one of the most studied cases of subsidy and voucher programs. During the 1980s, the provision of subsidized seed, fertilizer and credit was tied to purchases by a parastatal marketing board. However, in 1995, prices of all inputs and crops except maize were fully liberalized and the extension service began promoting other crops and activities.

In 1996–1997, in response to a crisis situation, the Starter Pack Initiative was introduced to "jump-start" maize production by providing enough seed and fertilizer for 0.1 ha of maize, and seed of other crops, for all smallholders. After several seasons of exceptionally good harvests, donors began to complain about the welfare nature of the scheme, urging its replacement with the Targeted Inputs Program (TIP). The TIP scaled down the number of beneficiaries and replaced hybrids with improved OPVs, which were viewed as more suitable for smallholders. Delayed deliveries, poor weather, and late maize imports led to high prices and increasing food scarcity during the 2001–2002 season. A similar scenario occurred in 2005–2006.

In response to the 2005/2006 crisis, the government initiated the Agricultural Input Subsidy Programme (AISP). AISP provides about 50% of farm households with vouchers for 100 kg of fertilizer and small quantities of maize (and lately legume) seed, with mainly privately imported fertilizers delivered principally, and in some years exclusively, by two parastatal input suppliers. During the 2005/2006 season over one million input coupons were distributed for a fiscal cost of US$32 million. Since then the program has been scaled up each year to reach US$242 million in 2008/2009, largely paid by the Government of Malawi. Corresponding to rising fertilizer prices, the subsidy paid 91% of fertilizer costs in 2008/2009. The program has been perceived as a test case for potential implementation elsewhere in Africa.

Since the policy motivation for governments to subsidize fertilizer is to enable smallholders to attain higher maize yields, establishing positive impacts on productivity is fundamental. In an analysis of 3 years of plot-level data collected from 450 households in Central and Southern Malawi, Holden and Lunduka (2010a) found that access to subsidized fertilizer had a significant positive effect on maize yields. However, Dorward et al. (2010) concluded that the benefits of the program are difficult to assess due to controversies about national statistics on maize production, which are likely to be overestimated. With reasonable assumptions about maize yield response to fertilizer, the authors do find that the program has generated a positive, though modest benefit cost ratio in 3 of the 4 years since the subsidy program was initiated.

Despite the reported increase in maize production, it is not clear that the program has enhanced food security. Domestic maize prices have been high in 3 out of 4 years of the program, incurring a major hardship for poor people, including the 60% of farmers who are net maize buyers. There is a tendency for the program to focus on production objectives and producer welfare, but to ignore consumers, and thus the conditions necessary for overall food security. Also, based on farm panel data over a 6 year period, Ricker-Gilbert and Jayne (2009) find that the receipt of the subsidy in multiple prior years had little enduring effect on recipient households' incomes or asset wealth after they stopped receiving the subsidy.

Targeting has posed continuous difficulties. Household surveys suggest that the 2006/2007 program was highly variable across locations in terms of targeting criteria, but that there was a tendency to reach households which were productive full-time farmers. Female-headed and poorer households were less likely to receive coupons (Holden and Lunduka 2010b). Holden and Lunduka (2010b) also report the presence of secondary markets for coupons—not from households that initially received the coupons, but from other leakages in the distribution system. The secondary market for fertilizer coupons also favored wealthier households. The authors ask whether targeting is more effective at reaching poor and vulnerable people than would be a general subsidy.

Given the type of household reached it is not surprising that the voucher program displaces commercial sales. If the voucher for subsidized fertilizer is received by a farmer who would otherwise have bought fertilizer at a commercial price, then the voucher program may shift the composition of retailer's profits from commercial fertilizer to subsidized fertilizer, with uncertain effects on the total quantity of fertilizer applied to the farmer's field. Ricker-Gilbert et al. (2011) found that the displacement rate is considerably lower among the poorest farmers. They report an overall displacement rate of commercial fertilizer by subsidized fertilizer of 0.29, meaning that each additional kilogram of subsidized fertilizer distributed under the government program contributes an additional 0.71 kg to total fertilizer use.

Some "crowding out" of commercial suppliers by government subsidy programs has also been demonstrated in Zambia, where an additional kilogram of fertilizer distributed under the subsidy program added 0.92 kg to the amount of fertilizer used by farmers (Xu et al. 2009). Where the private sector was already active, this leverage was only 0.12, suggesting that the subsidy program led to the withdrawal of some private retailers. By contrast, where fertilizer was targeted to areas where the private sector was inactive, and to poorer households, the leverage was as much as 1.7 kg per household, suggesting the potential for "crowding in" in such areas.

As can be expected given the history of fertilizer subsidies in Sub-Saharan Africa, program sustainability continues as a major issue. The costs of the Malawi program have exceeded the planned budget and represented 72% of the total budget of the Ministry of Agriculture, and 16% of the national budget, in 2009. Mann's (2007) concludes that the AISP, similar to the subsidies of the 1980s, is too large to be sustained, and three times as costly as the earlier Starter Pack Program, which had achieved considerable success (but was rejected by donors as too expensive). Malawi and Zambia have implemented nearly continuous fertilizer subsidy programs each year for the past several decades and no feasible exit strategy is apparent.

8.4.4 Stabilizing Maize Markets

The price spikes in global grain markets during 2008 focused public attention on the vulnerability of the rural and urban poor to volatility in food and fertilizer prices, although these issues are by no means new. A compilation and review of empirical research in a conference sponsored by the World Bank (World Bank 2005; Byerlee et al. 2006) led to several general conclusions regarding maize grain markets in Sub-Saharan Africa. First, poor producers and consumers in Africa, which include many smallholder farmers, are more exposed to sharp movements in the price of maize relative to those who depend on rice in Southeast Asia (Table 8.3). Second, landlocked countries in southern Africa that depend on maize are most exposed to domestic sources of shocks, as are other landlocked African countries, such as Ethiopia. In these countries, food production is highly variable, and national capacity to operate on world markets to smooth supply variability is limited by high transport costs and foreign exchange constraints. For example, maize prices in Ethiopia can fluctuate widely between import parity of $250 or more and export parity prices that may be as low as $50. Consistent net importers of maize with better infrastructure, such as Kenya, can smooth prices through trade, although they risk exposure to sharp spikes in world prices, as occurred in 2008.

Not surprisingly, the high level of price instability for a staple crop such as maize has invited efforts to stabilize prices, even during the post-structural adjustment period. Yet discretionary interventions in grain markets often reduce participation by the private sector in countries where reform from parastatal to market-led approaches remains incomplete. Maize markets are more volatile in Malawi than in other countries of southern Africa, despite the fertilizer subsidy and recorded production gains. Continued suspicion with respect to the capabilities and intentions of the private sector has led to greater involvement of the nation's parastatal, Agricultural Development and Marketing Corporation (ADMARC), in maize marketing. Tschirley and Jayne (2010) conclude that market shortages and stock-outs at ADMARC have sometimes led to huge price surges. However, there is growing private sector entry in maize marketing and encouraging evidence on the number of traders to whom farmers can sell maize and proximity to point of sales.

Over the past few decades in Kenya, synergies between the liberalization of the input and maize markets and public investments led to tangible investments by the private sector in not only seed and fertilizer retailing but also maize marketing (Ariga and Jayne 2010). Maize marketing margins have also contracted, as well as the distance traveled to the point of maize sale. However, maize sales remain highly concentrated among farmers. The Tegemeo Institute panel data confirm that less than 2% of the farms account for 50% of the overall marketed maize surplus from the smallholder sector. Most smallholders, which account for 96% of all the farm households in Kenya, were consistently buyers of maize in the three seasons for which data were collected (which included one good production year and two average years).

Kenya has pursued a policy of high food prices with import tariffs in the range of 25–50% and until 2005, restrictions on maize inflows from neighboring countries.

The operations of the maize marketing board (NCPB) have raised the level of maize prices in the country by offering support prices well above market levels (Jayne et al. 2008). Grain price supports and/or stabilization policies that raise mean price levels over time will have income distributional effects that run counter to stated goals of reducing poverty. Mean-neutral forms of price stabilization would most likely avoid these adverse distributional effects, and by reducing risks, would also help to promote diversification toward higher-valued crops by maize-purchasing households (Fafchamps 1992). Thus, the question for state maize price stabilization or price support is not whether these policies can generate positive benefits for surplus-producing farmers, but whether such benefits could reasonably be expected to exceed the costs of higher food prices for the majority of the population.

Over the long term there is a need to encourage the transition to market-based food systems and build capacity in private markets. Generalized measures to support market efficiency, such as investments in transport, storage, information systems and market regulations will serve to reduce the volatility of maize prices in Sub-Saharan Africa. To create space for private markets to operate, governments need a predictable, well-defined food security strategy that is implemented sequentially. For example, blanket subsidies and restrictions on grain trade, such as pan-territorial and pan-seasonal prices, would need to be removed for private traders to have an incentive to store and move grain from surplus to deficit areas.

Risk management instruments, such as warehouse receipts and futures and options markets offer another option. Futures and options markets are expanding rapidly in the developing world. South Africa has a well established exchange that other countries in the region can and sometimes do tap (Dana et al. 2006). Variable tariffs and small strategic grain reserves continue to receive some support as short-run, transition policies. Such market-oriented interventions should be backed by safety nets to deal with consequences of extreme prices on vulnerable populations.

The promotion of regional trade is one of the most effective "quick-wins" for reducing food price volatility in smaller countries (World Bank 2005). Regional production varies less than production in individual countries, and despite large and positive correlations in maize production among countries, there is generally scope for intra-regional trade in all but the worst years. Govereh et al. (2008) demonstrate that natural "marketsheds" span borders throughout Sub-Saharan Africa. However, for regional markets to function, countries need to agree to ban export restrictions in times of high prices and use other means to protect the vulnerable population.

8.5 Conclusions

Maize remains crucial for food security in Sub-Saharan Africa. In some regions, the predominance of the crop in farming systems and diets implies that yield gains have the potential to jump-start a Green Revolution like those experienced in Asia for rice and wheat. However, despite episodes of success, the evidence compiled here suggests that very little progress has been made toward achieving this potential since Byerlee and Eicher's (1997) review. Moreover, while maize remains the most important food security crop for millions of rural households, chronic food insecurity persists even where progress in maize production has been achieved, as in Malawi and Ethiopia.

The fact that domestic maize production cannot keep up with the food requirements of expanding urban populations is reflected in the growing consumption of rice and wheat in cities and towns, most of which is imported. African smallholders are generally competitive in maize production, at least with imports, and import substitution and integrated regional markets provide ready markets for greater maize production. Demand for maize to feed livestock is expected to grow rapidly, further taxing food supplies.

Green Revolution-style intensification is expected to succeed best in the densely populated and relatively high potential areas such as the East African highlands, Malawi, and parts of Nigeria where maize is the dominant staple. Yet even in these areas, yield growth has been slow, and although the adoption of improved maize varieties has increased in many areas, it has often fluctuated as a consequence of policy shifts. In areas where improved maize varieties have been widely adopted, genetic yield gains are dampened by the use of old varieties. Use of fertilizers and other crop management practices remains limited. Combined with soil nutrient mining and degradation, this poses fundamental challenges in sub-Saharan Africa's rainfed production systems.

In many areas, too, access to land has become so constrained that surplus maize production is unattainable for many smallholders even with successful adoption of seed fertilizer technologies. A strategy to diversify maize production systems could provide higher returns to scarce land and improve food security, provided that retail maize markets are dependable. In semi-arid and more marginal environments, where the risk of drought is high, such a strategy will include suitable higher-value crops and livestock products.

Sub-Saharan Africa also has large areas of low population density that are suitable for expanding maize production and where it is not surprising that intensification technologies have not yet been adopted, given relative land abundance. In these areas, such as in much of the savanna and miambo woodlands, adoption of labor-saving technologies together with sustainable soil management practices will be the key to expanding the area under maize (World Bank 2009). Many of these areas are relatively remote and appropriate public investments in infrastructure and technology, combined with private investment in commercial farming, offer the opportunity for Africa to be a major exporter of maize in the future.

Over the long term, large investments and sustained political commitment are needed to ensure strong plant breeding and seed systems to serve smallholders, predicated on improved crop management practices to protect soils and cope with unreliable rainfall, and access to appropriate labor-saving technologies. More innovative extension and advisory systems are also needed to facilitate farmer learning and adapt techniques and technologies to local environmental and social conditions. Better financial services, perhaps including new forms of insurance, are needed for smallholders.

Harder questions concern how these investments should be sequenced, and how they should be tailored to the highly heterogeneous, maize-based farming systems of Sub-Saharan Africa. This review has highlighted the importance in maize technical change of establishing and maintaining conducive policies. These are equally, if not more, important for agricultural transformation than seed, fertilizer and management practices. Although pockets of success are visible, policy reform has generally been incomplete and policy interventions, including donor priorities, have often been *ad hoc* and unpredictable. The new initiatives of this decade, founded on 'market smart' approaches, have strayed quickly from their original path, and are not likely to be sustainable. There is now a risk of repeating the mistakes of the 1970s and 1980s by focusing on silver bullets such as large-scale input voucher programs, rather than investing in a broad-based strategy for long run productivity growth.

Appendix: Country Classification Used to Analyze FAOSTAT Data

The country classification used in this chapter differs from that used by FAOSTAT, with the exception Western Africa. Data was loaded for each country and summarized according to the following classifications:

Eastern Africa	Southern Africa	Western Africa	Central Africa
Burundi	Angola	FAO	FAO without
Comoros	Botswana		Angola
Eritrea	Lesotho		
Ethiopia	Malawi		
Kenya	Madagascar		
Mauritius	Mozambique		
Reunion	South Africa		
Rwanda	Swaziland		
Somalia	Zambia		
Tanzania	Zimbabwe		
Uganda			

References

Alemu D, Mwangi W, Nigussie M, Spielman DJ (2007) An analysis of maize seed production and distribution systems in Ethiopia's rift valley, vol 72, Research report. Ethiopian Institute of Agricultural Research, Addis Ababa

Alene AD, Menkir A, Ajala SO, Badu-Apraku B, Alanrewaju AS, Manyong VM, Ndiaye A (2009) The economic and poverty impacts of maize research in West and Central Africa. Agric Econ 40(5):535–550

Anderson JR, Feder G (2003) Rural extension services, vol 2976, World Bank policy research working paper. World Bank, Washington, DC

Anderson JR, Feder G, Ganguly SS (2006) The rise and fall of training and visit extension: an Asian mini-drama with an African epilogue. World Bank, Development Research Group, Rural Development Team, Washington, DC

Ariga J, Jayne TS (2010) Unlocking the market: fertilizer and maize in Kenya. In: Spielman D, Pandya-Lorch R (eds) Millions fed: proven successes in agricultural development, Chapter 14. International Food Policy Research Institute, Washington, DC

Balcet J-CL, Candler W (1981) Farm technology adoption in Northern Nigeria. World Bank, Washington, DC

Beintema NM, Stads GJ (2006) Agricultural research and development in Sub-Saharan Africa: an era of stagnation, agricultural science and technology indicators. International Food Policy Research Institute, Washington, DC

Benin S, Nkonya E, Okecho G, Randriamamonjy J, Kato E, Lubade G, Kyotalimye M (2011) Returns to spending on agricultural extension: the case of the National Agricultural Advisory Services (NAADS) program of Uganda. Agric Econ 42:249–267

Bonger T, Ayele G, Kumsa T (2004) Agricultural extension, adoption and diffusion in Ethiopia, vol 1, Research report. Ethiopian Development Research Institute, Addis Ababa

Byerlee D, Eicher CK (1997) Africa's emerging maize revolution. Lynne Rienner, Boulder

Byerlee D, Heisey PW (1997) Evolution of the African maize economy. In: Byerlee D, Eicher CK (eds) Africa's emerging maize revolution. Lynne Rienner, Boulder

Byerlee D, Jayne TS, Myers RJ (2006) Managing food price risks and instability in a liberalizing market environment: overview and policy options. Food Policy 31(4):275–287

Chapoto A, Jayne TS (2009) Effects of maize marketing and trade policy on price unpredictability in Zambia, vol 38, Food security country working paper. Michigan State University, Lusaka

Cleaver K (1993) Making agricultural extension work in Africa. In: Russell NC, Dowswell CR (eds) Policy options for agricultural development in Sub-Saharan Africa. Sasakawa Global 2000, Mexico

Collinson P (1982) Farming systems research in Eastern Africa: the experience of CIMMYT and some national agricultural research services, 1976–81, vol 3, MSU international development papers. Department of Agricultural Economics, Michigan State University, East Lansing

Dana J, Gilbert CL, Shim E (2006) Hedging grain price risk in the SADC: case studies of Malawi and Zambia. Food Policy 31(4):357–371

Davis K (2008) Extension in Sub-Saharan Africa: overview and assessment of past and current models, and future prospects. J Int Agri Ext Edu 15(1):15–28

Davis K, Nkonya E, Kato E, Mekonnen DA, Odendo M, Miiro R, Nkuba J, Okoth J (2010) Impact of farmer field schools on agricultural productivity and poverty in East Africa, Discussion paper. International Food Policy Research Institute, Washington, DC

De Groote H, Kimeju S, Owuor G, Wanyama J (2006) Market liberalization and agricultural intensification in Kenya. Contributed paper presented at 26th conference of the International Association of Agricultural Economics, Gold Coast, Australia, August

Deininger K, Byerlee D (2011) Rising global interests in farmland: can it yield sustainable and equitable benefits. World Bank, Washington, DC

Dorward A, Chirwa E, Boughton D, Crawford E, Jayne TS, Slater R, Kelly V, Tsoka M (2008) Towards "smart" subsidies in agriculture? Lessons from recent experience in Malawi. Overseas Development Institute, London

Dorward A, Chirwa E, Jayne T (2010) The Malawi agricultural inputs subsidy programme, 2005/6 to 2008/9. Report prepared for the "Yes Africa can: success stories from a dynamic continent" project, World Bank Group, Africa Region

Eicher CK (1995) Zimbabwe's maize-based Green Revolution: preconditions for replication. World Dev 23(5):805–818

Fafchamps M (1992) Cash crop production, food price volatility and rural market integration in the third world. Am J Agric Econ 74(1):90–99

Fischer RA, Byerlee D, Edmeades GO (2009) Can technology deliver on the yield challenge to 2050? Paper presented at the expert meeting on how to feed the world in 2050. Food and Agriculture Organization of the United Nations, Rome

Food and Agriculture Organization of the United Nations (FAO), International Maize and Wheat Improvement Center (1997) White maize: a traditional food grain in developing countries. FAO/CIMMYT, Rome/Mexico

Gautam M (2000) Agricultural extension: the Kenyan experience. World Bank, Washington, DC

Gerhart J (1975) The diffusion of hybrid maize in western Kenya. International Maize and Wheat Improvement Center, Mexico

Gilbert E, Phillips LC, Roberts W, Sarch MT, Smale M, Stroud A (1993) Maize research impact in Africa: the obscured revolution. Division of Food, Agriculture and Resources Analysis, Office of Analysis, Research and Technical Support, Bureau for Africa, United States Agency for International Development, Washington, DC

Gilbert RA, Bezner-Kerr R, Kanyama-Phiri GY (2010) Biodiversity supports ecosystem services for a greener revolution in Africa. Proc Natl Acad Sci 07(48, November):20840–20845

Giller KE, Tittonell P, Witter E, Corbeels M (2009) Conservation agriculture and smallholder farming in Africa: the heretics' view. Field Crops Res 114(1):23–34

Govereh J, Haggblade S, Nielson H, Tschirley D (2008) Maize marketsheds in Eastern and Southern Africa. Report prepared by Michigan State University for the World Bank under contract No. 7144132, June

Haggblade S, Tembo G, Kaboré D, Reij C, Ajayi OC, Franzel S, Mafongoya P, Place F (2010) Sustainable soil fertility management systems. In: Hand S, Hazell P (eds) Successes in African agriculture. Johns Hopkins University Press, Baltimore

Hassan RM (1998) Maize technology development and transfer. CAB International, CIMMYT, KARI, Wallingford, Oxford

Hassan RM, Mekuria M, Mwangi W (2001) Maize breeding research in Eastern and Southern Africa: current status and impacts of past investments made by public and private sectors, 1966–1997. CIMMYT, Mexico

Heisey PW, Edmeades GO (1999) Maize production in drought-stressed environments: technical options and research resource allocation. CIMMYT 1997/1998 world maize facts and trends. International Maize and Wheat Improvement Center, Mexico

Heisey PW, Norton G (2007) Fertilizers and other farm chemicals. In: Evenson RE, Pingali PL (eds) Handbook of agricultural economics, vol 3. Elsevier, Amsterdam

Heisey PW, Smale M (1995) Maize technology in Malawi: a Green Revolution in the making? International Maize and Wheat Improvement Center, Mexico

Holden S, Lunduka R (2010a) Too poor to be efficient? Impacts of the targeted fertilizer subsidy program in Malawi on farm plot level input use, crop choice and land productivity. Report to Norwegian Agency for Development Cooperation. Department of Economics and Resource Management, Norwegian University of Life Sciences

Holden S, Lunduka R (2010b) NORAD-project: impacts of the fertilizer subsidy program in Malawi: targeting, household perceptions and preferences. Mimeo, Norwegian University of Life Sciences

Howard JA (1994) The economic impact of improved maize varieties in Zambia. Ph.D. thesis, Department of Agricultural Economics, Michigan State University, East Lansing

Howard J, Mungoma C (1997) Zambia's stop-and-go maize revolution. In: Byerlee D, Eicher CK (eds) Africa's emerging maize revolution. Lynne Rienner, Boulder

Jayne TS, Jones S (1997) Food marketing and pricing policy in Eastern and Southern Africa: a survey. World Dev 25(9):1505–1527

Jayne TS, Myers RJ, Nyoro J (2008) The effects of government maize marketing policies on maize prices in Kenya. Agric Econ 38(3):313–325

Jayne TS, Mason N, Myers R, Ferris J, Mather D, Beaver M, Lenski N, Chapoto A, Boughton D (2010) Patterns and trends in food staples markets in Eastern and Southern Africa: toward the identification of priority investments and strategies for developing markets and promoting smallholder productivity growth, vol 104, International development working paper. Michigan State University, East Lansing

Kydd J (1989) Maize research in Malawi: lessons from failure. J Int Dev 11(1):112–144

Kydd J (2009) A new institutional economic analysis of the state and agriculture in Sub-Saharan Africa. In: Kirsten JF, Dorward AR, Poulton C, Vink N (eds) Institutional economics perspectives on African agricultural development. International Food Policy Research Institute, Washington, DC

Langyintuo AS, Mwangi W, Diallo AO, MacRobert J, Dixon J, Banziger M (2008) An analysis of the bottlenecks affecting the production and deployment of maize seed in Eastern and Southern Africa. International Maize and Wheat Improvement Center, Harare

Larson DF, Otsuka K, Kajisa K, Estudillo J, Diagne A (2010) Fostering a Green Revolution in rice: can Africa replicate Asia's experience? In: Pandey S, Byerlee D, Dawe D, Dobermann A, Mohanty S, Rozelle S, Hardy B (eds) Rice in the global economy: strategic research and policy issues for food security. International Rice Research Institute, Los Banos

Loìpez-Pereira MA, Morris ML (1994) Impacts of international maize breeding research in the developing world, 1966–1990. International Maize and Wheat Improvement Center, Mexico

Low ARC (1988) Farm household economics and the design and impact of biological research in Southern Africa. Agric Adm Ext 29(1):23–34

John L, Gilbert E, Elliot H, Blis F (2010) Evolving a plant breeding and seed system in Sub-Saharan Africa in an era of donor dependence. A report for the global initiative on plant breeding capacity building, Food and Agriculture Organization, Rome

Mann CK (2007) Malawi's starter pack: a smarter subsidy to empower the poor with science-based agriculture. Africa Policy J 4 (Spring/Summer, 2008), Harvard Kennedy School of Government

Manyong VM, Kling JG, Makinde KO, Ajala SO, Menkir A (2003) Impact of IITA germplasm improvement on maize production in West and Central Africa. In: Evenson RE, Gollin D (eds) Crop variety improvement and its effect on productivity. CAB International, Wallingford, Oxford

Marenya PP, Barrett CB (2009a) Soil quality and fertilizer use rates among smallholder farmers in Western Kenya. Agric Econ 40(5):561–572

Marenya PP, Barrett CB (2009b) State-conditional fertilizer yield response on Western Kenyan Farms. Am J Agric Econ 91(4):991–1006

Mason N, Jayne TS, Chapoto A, Donovan C (2011) Putting the 2007/2008 global food crisis in longer-term perspective: trends in staple food affordability in urban Zambia and Kenya. Food Policy 36(3):350–367

Minde I, Jayne T, Ariga J, Govereh J, Crawford E (2008) Fertilizer subsidies and sustainable agricultural growth in Africa: current issues and empirical evidence from Malawi, Zambia, and Kenya. Report for regional strategic agricultural knowledge support systems for Southern Africa, based at International Water Management Institute, Pretoria, South Africa

Mistiaen J (2006) Poor and inefficient? Scientists, farmers and the maize yield gap in Kenya, Draft paper. World Bank, Washington, DC

Morris ML (1998) Maize in the developing world: waiting for a Green Revolution. In: Morris ML (ed) Maize seed industries in developing countries. Lynne Rienner, Boulder

Morris ML, Mekuria M, Gerpacio R (2003) Impacts of CIMMYT maize breeding research. In: Evenson RE, Gollin D (eds) Crop variety improvement and its effect on productivity. CAB International, Wallingford, Oxford

Morris ML, Kelly VA, Kopicki RJ, Byerlee D (2007) Promoting increased fertilizer use in Africa: lessons learned and good practice guidelines. World Bank, Washington, DC

Nyoro J, Kirimi L, Jayne TS (2004) Competitiveness of Kenyan and Ugandan maize production: challenges for the future, vol 10, Working paper. Egerton University/Tegemeo Institute, Nairobi

Ricker-Gilbert J, Jayne TS (2009) Do fertilizer subsidies affect the demand for commercial fertilizer? An example from Malawi. Contributed paper presented at the international association of agricultural economists conference, Beijing, China, 16–22 Aug

Ricker-Gilbert J, Jayne TS, Chirwa E (2011) Subsidies and crowding out: a double-hurdle model of fertilizer demand in Malawi. Am J Agric Econ 93(1):26–42
Rohrbach DD (1988) The economics of smallholder maize production in Zimbabwe: implications for food security, vol 11, MSU international development paper. Michigan State University, East Lansing
Sauer J, Tchale H (2009) The economics of soil fertility management in Malawi. Rev Agric Econ 31(3):535–560
Setimela PS, Badu-Apraku B, Mwangi W (2009) Variety testing and release approaches in DTMA project countries in sub-Saharan Africa. CIMMYT, Harare
Smale M (1995) Maize is life: Malawi's delayed green revolution. World Dev 23(5):819–831
Smale M, Jayne TS (2003) Maize in Eastern and Southern Africa: 'seeds' of success in retrospect, vol 97, EPTD discussion paper. International Food Policy Research Institute, Washington, DC
Smale M, Phiri A, with contributions from Chikafa GA, Heisey P, Mahatta F, Msowoya MNS, Mwanyongo EBK, Sagawa HG, Selemani HAC (1998) Institutional change and discontinuities in farmers. Use of hybrid maize and fertilizer in Malawi: findings from the 1996–7 CIMMYT/MOA survey. Economics Working Paper 98-01, CIMMYT
Smith J, Weber G, Manyong MV, Fakorede MAV (1997) Fostering sustainable increases in maize productivity in Nigeria. In: Byerlee D, Eicher CK (eds) Africa's emerging maize revolution. Lynne Rienner, Boulder
Spielman DJ, Byerlee D, Alemu D, Kelemework D (2010) Policies to promote cereal intensification in Ethiopia: the search for appropriate public and private roles. Food Policy 35(3):185–194
Tripp R, Wijeratne M, Piyadasa VH (2005) What should we expect from farmer field schools? A Sri Lanka case study. World Dev 33(10):1705–1720
Tschirley D, Jayne TS (2010) Exploring the logic of Southern Africa's food crises. World Dev 38(1):76–87
van den Berg H, Jiggins J (2007) Investing in farmers: the impacts of farmer field schools in relation to integrated pest management. World Dev 35(4):663–686
World Bank (1994) Agricultural extension: lessons from completed projects. World Bank, Washington, DC
World Bank (2005) Managing food price risks and instability in an environment of market liberalization. Agriculture and Rural Development Department, World Bank, Washington, DC
World Bank (2008) World development report 2008: agriculture for development. World Bank, Washington, DC
World Bank (2009) Awakening Africa's sleeping giant prospects for commercial agriculture in the Guinea savannah zone and beyond. World Bank, Washington, DC
Xu Z, Guan Z, Jayne TS, Black R (2009) Factors influencing the profitability of fertilizer use on maize in Zambia. Agric Econ 40(4):437–446

[illegible]

Chapter 9
Maize, Soil Fertility, and the Green Revolution in East Africa

Tomoya Matsumoto and Takashi Yamano

Abstract The low application of inorganic fertilizer in Sub-Saharan Africa is one of the major constraints to achieving a Green Revolution in this region. In this study, we estimate the direct impact of the soil fertility on the maize yield and examine if the soil fertility increases the returns to inorganic and organic fertilizer based on comparative study of Kenya and Uganda. The results of the analyses indicate that the Kenyan maize farmers have applied the inorganic fertilizer roughly at the optimal level in one out of the two survey years on both the purchased high-yielding varieties and local/recycled maize varieties. In Uganda, even the low application of inorganic fertilizer is not profitable because of the high relative price. Regarding the returns to inorganic fertilizers on degraded soils, we do not find any increasing marginal returns of such fertilizers to the soil fertility.

Keywords Inorganic fertilizer • Organic fertilizer • Green Revolution • High yielding varieties • Land degradation • Soil nutrients • Soil carbon content • Soil fertility • East Africa

9.1 Introduction

The low application of inorganic fertilizer in Sub-Saharan Africa (hereafter Africa) is one of the major constraints to achieving a Green Revolution in this region (IFDC 2006). Although there have been many studies to explore the reasons behind

T. Matsumoto (✉)
Division of Policy Research, National Graduate Institute for Policy Studies,
7-22-1 Roppongi, Minato-ku, Tokyo 106-8677, Japan
e-mail: tmatsumo@grips.ac.jp

T. Yamano
Division of Social Science and Economics, International Rice Research Institute,
CG Block, NASC Complex, DPS Marg, Pusa, New Delhi, India
e-mail: t.yamano@cgiar.org

K. Otsuka and D.F. Larson (eds.), *An African Green Revolution: Finding Ways to Boost Productivity on Small Farms*, DOI 10.1007/978-94-007-5760-8_9,

the low application of the inorganic fertilizer, some competing hypotheses remain (Morris et al. 2007; Kelly 2006). Market-based hypotheses suggest that farmers are responding to the high fertilizer price, which has been the result of high transportation and marketing costs in Africa (Jayne et al. 2003; Gregory and Bumb 2006). Non-market based hypotheses emphasize farmers' lack of knowledge on inorganic fertilizer and high yielding varieties, as well as financial constraints (surveyed in Morris et al. 2007).

One of the non-market constraints is the land degradation which could lower the returns to the fertilizer application (Adesina 1996; Marenya and Barrett 2009). Degraded soils have low capacity to hold water and external soil nutrients, and, thus, external fertilizers have low returns on degraded soils. The low returns on the degraded soils would force farmers to reduce the already low inorganic fertilizer application, which in turn may contribute to further land degradation (Smaling et al. 1997; Henao and Baanante 2006; IFDC 2006).[1] A recent study by Marenya and Barrett (2009) advances the research on this issue by employing plot level soil carbon content data in western Kenya and finds that the marginal return to the inorganic fertilizer is low and not profitable in maize production when the soil carbon content is low. Their results suggest that conventional policies to encourage farmers to use inorganic fertilizer would be ineffective on depleted soils. Their analyses, however, are based on cross sectional data and do not control for the possible endogeneity of the input use or the selection of high yielding variety (HYV) seeds.

In this study, we follow the approach taken by Marenya and Barrett (2009) but use 2-year panel data of farm households in Kenya and Uganda where we have maize production data on 6,329 plots, of which we have soil fertility data for more than 70% of the plots. Kenya and Uganda provide an interesting comparison because Kenya has one of the highest productivities for maize in Africa, while Uganda has one of the lowest (Smale and Jayne 2003; Sserunkuuma 2005). We control for the HYV selection by using the household fixed effects semiparametric endogenous switching model, developed by Kyriazidou (1997). In particular, we estimate the direct impact of the soil fertility on the maize yield and examine if the soil fertility increases the returns to inorganic and organic fertilizer. The results of the analyses indicate that the Kenyan maize farmers have applied the inorganic fertilizer roughly at the optimal level in one out of the two survey years on both the purchased HYV and local/recycled HYV maize.[2] In Uganda, even the low application of inorganic fertilizer is not profitable because of the high relative price. Regarding the returns to external fertilizers on degraded soils, we do not find any increasing marginal returns of external fertilizers to the soil fertility.

The chapter is organized as follows. Section 9.2 explains the production model that captures the interactions between the soil fertility and the fertilizer inputs and

[1] Although there is a debate over the possible overestimations of the previous estimates of soil losses, many experts agree that the land degradation is a critical constraint to African agriculture (Koning and Smaling 2005; Pender et al. 2006).

[2] Many farmers in Kenya and Uganda recycle purchased HYV maize after harvesting. We group the recycled HYV maize with the local maize as we explain later in Sect. 9.2.

describes the semi-parametric endogenous switching model used in this chapter. Section 9.3 explains the household panel data and soil fertility data. The surveys in both countries were conducted by the same research project which employs comparable questionnaires across countries and time. Section 9.4 presents the estimation results on maize production in both countries. Finally, we discuss the policy implications based on the results in Sect. 9.5.

9.2 Model on Soil Fertility and Returns to Fertilizer on Maize

9.2.1 Soil Fertility and Crop Production

There are several pathways though which soil fertility contributes to crop production. Directly, soil provides nutrients to crops, and, indirectly, soil affects how easily external inputs are absorbed by the crops (Tiessen et al. 1994; Palm et al. 2001; Bationo and Mokwunye 1991). As a proxy for soil fertility, Marenya and Barrett (2009) use the carbon content. The soil carbon content is also a proxy for soil organic matter (SOM), which consists of the decayed tissues of plants and animals taken from animal excreta and is increasingly taken as a strong indicator of soil fertility and land degradation because SOM tightly controls many soil properties and major biogeochemical cycles (Ngugi et al. 1990; Manlay et al. 2007).

Some soil characteristics are not fixed over the long run. Organic fertilizer, for instance, can directly alter soil characteristics. Thus, the impacts of organic fertilizer application have a long-term impact on crop production through changing the soil characteristics. Thus, the current soil characteristics reflect the past applications of organic fertilizer to some extent. In the following analysis, therefore, we consider the organic fertilizer application as a flow variable and the soil carbon content as a stock variable.

9.2.2 Production Function

Regarding the production function, we consider a soil nutrient indicator which is a function of three factors. Let us denote N_{pit} as the soil nutrient indicator of plot p of household i at time t:

$$N_{pit} = N(E_i, C_{pit}, O_{pit}), \tag{9.1}$$

where E_i is the basic soil condition, such as the soil carbon content, assumed to be time-invariant for a short time period; C_{pit} is the quantity of the inorganic fertilizer application (kg/ha), and O_{pit} is the quantity of the organic fertilizer application (t/ha). We assume that the basic soil condition, E_i, is common across the maize plots within a household and fixed over time. As we discussed in the previous sub-section, we use the soil carbon content as a single indicator of the soil condition in the following analyses by following Marenya and Barrett (2009).[3]

For the production function, we consider a simple yield function of the Cobb-Douglas form. The yield, kilograms per hectare, denoted by Y_{pit}, is given as follows:

$$Y_{pit} = A_i L_{pit}^{\beta_L} S_{pit}^{\beta_S} N_{pit}^{\beta_N} e^{\omega_{ipt}}, \tag{9.2}$$

where L_{pit} is the plot size (ha), S_{pit} is the seed quantity planted (kg/ha), A is the Hicks neutral technology parameter or the total factor productivity.[4] We assume that A is time-invariant at least for the short time period of 2–3 years. ω is assumed to capture a productivity shock affected by weather conditions or other idiosyncratic factors. By taking logs of the yield function (9.2), we have

$$y_{ipt} = a_i + \beta_L l_{ipt} + \beta_S s_{ipt} + \beta_N n_{ipt} + \omega_{ipt}, \tag{9.3}$$

where the lowercase variables represent the logs of their corresponding uppercase variables. The functional form of the nutrient indicator given by Eq. (9.1) is unknown but we take a second-order approximation of the log of N so that it captures the interaction effects of the external inputs and the soil carbon content, which is given by,

$$n_{pit} = \ln N(E_i, C_{pit}, O_{pit}) = \gamma_0 + \sum_x \gamma_x x + \sum_x \sum_{x'} \gamma_{xx'} x x' \tag{9.4}$$

for $x,\ x' \in \{E, C, O\}$. We expect that the interaction terms between the soil carbon content and the external inputs have positive impacts on the crop production. By plugging Eq. (9.4) into Eq. (9.3), we have

$$y_{pit} = a_i + \beta_L l_{pit} + \beta_S s_{pit} + \sum_x \delta_x x_{pit} + \sum_x \sum_{x'} \delta_{xx'} x_{pit} x'_{pit} + \omega_{pit}, \tag{9.5}$$

where the coefficients δs are the product of β_N and γ_s; that is, $\delta_v = \beta_N \gamma_v$, $v \in \{E, C, O, EC, EO, CO, EE, CC, OO\}$.

[3] As we discuss in Sect. 9.3, we have only one soil observation per household. Thus, we assume that the soil carbon content is fixed across maize plots within a household and over time. Although it is not clear how long the soil carbon content is stable over time, it seems that the soil carbon content is more stable than other soil nutrients, such as nitrogen content.

[4] We do not include family labor in the model because family labor information was not sought in the second round of the surveys in both Kenya and Uganda. The family labor module was removed from the questionnaire in the second round because the quality of the family labor information was considered poor in the first round of the surveys. We implicitly assume that family labor input is adjusted optimally when the other input levels change. In the regression models, we estimate the household fixed effect models. Thus, as long as the family labor input remains at the same level, the omission of the family labor may not cause a serious bias.

Simple OLS regression of y on the observables with pooled samples, however, may provide biased estimates. First, the unobservable total factor productivity, a_i, could be correlated with the inputs. Fortunately for us, we have panel data. Thus, by estimating the fixed effects model, we can at least remove the time-invariant unobserved factors, although we cannot identify the coefficients of the time-invariant independent variables, such as E_i and E_i^2, in the fixed effects (FE) model. We rewrite the estimation Eq. (9.5) as:

$$y_{ipt} = \alpha_i + \beta_L l_{ipt} + \beta_S s_{ipt} + \sum_{x \neq E} \delta_x x_{ipt} + \sum_{x \neq E} \sum_{x'} \delta_{xx'} x_{ipt} x'_{ipt} + \omega_{ipt}, \tag{9.6}$$

where $\alpha_i = a_i + \delta_E E_i + \delta_{EE} E_i^2$. The fixed effects factors, collected in α_i, will be dropped from the FE model.

Another issue to be considered is the correlation of the productivity shock with the input variables. Specifically, rainfall would be correlated with the input applications and yield simultaneously because the agriculture production in our survey regions is predominantly rain-fed, and farmers determine the level of input use according to the level of rainfall. In this chapter, this issue is dealt with by introducing time-region dummies as covariates. With this treatment, we can control for other region level time-variant factors as well. For notational simplicity, we denote z as a vector of the independent variables, including the time-region dummies and φ as a vector of parameters corresponding to z. Subsequently, we describe the model simply as follows:

$$y_{ipt} = \alpha_i + z'_{ipt} \varphi + \varepsilon_{ipt}. \tag{9.7}$$

9.2.3 Endogenous Switching Model

We also need to consider the effect of HYV seed adoption in the yield function. As shown by previous studies on maize in Kenya and Uganda, HYV and local seeds have different yields (Hassan and Karanja 1997; Nyoro et al. 2004; Sserunkuuma 2005). Farmers in Kenya and Uganda also recycle HYV seeds for many seasons, as in other countries in SSA (Chap. 8). After one season, the newly purchased HYV seeds lose their high responsiveness to inorganic fertilizer. Indeed, we find that the recycled HYV seeds have a yield distribution which is more similar to the local seeds than the newly purchased HYV seeds (Appendix Fig. 9.A.1). Thus, in this chapter, we group the recycled HYV maize seeds with the local maize seeds and label them as "local/recycled HYV" maize seeds and we label the newly purchased HYV maize seeds as "purchased HYV" seeds. To consider the differences in yield and returns to fertilizer use according to seed type, the extended model is given by

$$y_{pit} = d_{pit} y^1_{pit} + (1 - d_{pit}) y^0_{pit}, \tag{9.8}$$

where d_{pit} is a binary indicator taking 1 if the purchased HYV seeds are planted on plot p of household i at time t and 0 otherwise: y^j_{ipt} ($j \in \{0,1\}$) is the potential outcome when the HYV adoption status j is exogenously given. Then the yield function is expressed as follows:

$$y^j_{ipt} = \alpha^j_i + z'_{ipt}\phi^j + \varepsilon^j_{ipt},\ j \in \{0,1\}. \tag{9.9}$$

We observe only either one of the two potential outcomes. Plugging $y^j s$ into Eq. (9.8), we obtain

$$y_{ipt} = \alpha^0_i + d_{ipt}(\alpha^1_i - \alpha^0_i) + z'_{ipt}\phi^0 + d_{ipt}z'_{ipt}(\phi^1 - \phi^0) + \varepsilon^0_{ipt} + d_{ipt}(\varepsilon^0_{ipt} - \varepsilon^1_{ipt}). \tag{9.10}$$

There are two possible problems: the presence of the unobserved effects α^j_i and the potential endogeneity of the independent variables specifically on the HYV adoption and its interaction terms. To obtain consistent estimates, we apply the two-step estimation method for the panel data sample selection model, developed by Kyriazidou (1997). To apply her model, we need to obtain consistent estimates of the selection equation with individual fixed effects. In our case, the selection equation will be given by $d_{ipt} = 1\{\eta_i + w'_{ipt}\xi + v_{ipt}\}$, where η_i is a time-invariant household-specific effect, w is a vector of independent variables, and v is unobserved disturbance. Specifically, we use Logit estimation to obtain the consistent estimates, ξ. In the second step, using ξ, the parameters of the yield equation are estimated by

$$\varphi^j = \left[\sum_{i=1}^{n}\frac{1}{M_i(M_i-1)}\sum_{m\neq m'}\varphi_{imm'}(z_{im}-z_{im'})(z_{im}-z_{im'})' d^j_{im}d^j_{im'}\right]^{-1} \times\left[\sum_{i=1}^{n}\frac{1}{M_i(M_i-1)}\sum_{m\neq m'}\varphi_{imm'}(z_{im}-z_{im'})(y_{im}-y_{im'})' d^j_{im}d^j_{im'}\right], \tag{9.11}$$

where M_i is the number of observations of household i, the subscript m is a substitute for the subscript pt solely for notational simplicity, $d^1 = d$ and $d^0 = 1 - d$, and $\varphi_{imm'}$ is a kernel weight that becomes large when $w'_{im}\hat{\xi}$ and $w'_{im'}\hat{\xi}$ are close. Namely, $\varphi_{imm'} \equiv \frac{1}{h}K\left(\frac{(w_{im}-w_{im'})'\hat{\xi}}{h}\right)$, where $K(\cdot)$ is a kernel function and h is a bandwidth. The intuition behind this estimation method is that by taking the difference in the yield function between two observations within a household when their predicted single indexes $w'\hat{\xi}$ obtained in the first stage regression for the selection model take the same value, not only do time-invariant household fixed-effects, α_i, disappear but so do the selection biases. Using the variables transformed by taking the difference in the above manner, we may be able to apply OLS regression and obtain consistent estimators. In practice, there may not be two observations taking exactly the same value of the predicted single indexes within an individual. To handle this issue, the Kyriazidou estimation applies a weighted regression in which heavier weights are assigned to the differences of the two samples with closer values on their predicted single indexes.

9.3 Data and Descriptive Analysis

9.3.1 Data

The data used in this chapter come from household-level panel surveys in Kenya and Uganda, collected as part of the Research on Poverty and Environment and Agricultural Technology (R*e*PEAT) Project. All surveys employ comparable questionnaires across countries and time. In addition, soil samples were collected from maize fields when the first rounds of the surveys were conducted. The surveys in Kenya were conducted in 2004 and 2007. The first round of the surveys covered 899 randomly selected households located in 100 sub-locations scattered in central and western regions of Kenya.[5] In the second round, seven sub-locations in Eastern province were dropped because of the scale reduction of the survey project. Thus, in this chapter, we drop the samples from Eastern province in Kenya for the analysis below since we apply statistical methods relying on the longitudinal features of the data. In addition, attrition also reduced the number of households interviewed. As a result, out of the 777 targeted households, 725 households were revisited for the survey, resulting in an attrition rate of 6.7%.[6]

The surveys in Uganda cover 94 rural Local Council 1 (LC1)s that are located across most regions in Uganda, except the North where security problems exist.[7] From each rural LC1, ten households are randomly selected, resulting in a total of 940 small farm households. The second round was conducted in 2005, and 895 households out of the 940 original households visited in the first round were interviewed. Thus, the attrition rate was low at 4.8%.[8]

Along with the first rounds of the surveys in Kenya and Uganda, we collected soil samples from the largest maize plot or one of the other cereal plots if maize was not cultivated at each sample household. If no cereal crops were cultivated by a household, no soil samples were taken. The soil samples were collected at a depth of 0–20 cm from five different positions within each plot of a sample household and mixed. Later, the samples from Kenya and Uganda were sent to the soil laboratory at the World Agroforestry Center (ICRAF) in Nairobi and were tested by a new method called near-infrared reflectance spectroscopy (NIRS), following protocols developed by Shepherd and Walsh (2002) and Cozzolino and Moron (2003).

[5] These two waves of surveys in Kenya were conducted by Tegemeo Institute, with financial and technical help from National Graduate Institute for Policy Studies (GRIPS).

[6] We estimated the determinants of the attrition from the surveys and found that none of the independent variables is significant at the 5% level (Appendix Table 9.A.1). Thus, we think that the attrition mostly occurred randomly and do not expect serious attrition biases.

[7] The surveys in Uganda were conducted jointly by Makarere University, Foundation for Advanced Studies on International Development (FASID), and National Graduate Institute for Policy Studies (GRIPS).

[8] The attrition rate is less than 5%. None of the independent variables in the determinants of the attrition model is significant even at the 10% level (Appendix Table 9.A.1). Thus, we do not think the attrition biases serious.

We have matched the soil information to 77% of the maize plots in Kenya and 67% of the maize plots in Uganda. The major reason for not having the soil information on some of the maize plots is simply because some soil samples were either lost or spoiled before being tested in the laboratory. Because the soil samples were collected at the time of the first survey, we do not have soil information on the maize plots of households who did not produce maize or any other cereals in the first round of the surveys. The Probit regression models for the soil sample attrition indicate that most of the household variables are not correlated with the attrition (Appendix Table 9.A.1). The major determinants of the soil sample attrition are the region dummies which represent the soil sample losses and spoilages. Thus, we do not think that the soil sample attrition is systematically correlated with the household characteristics to create attrition biases. In addition, because we estimate the household fixed effects models, we think that if any attrition biases exist, they would be small.

In Table 9.1, we compare the maize production and input applications between the purchased HYV seeds and the local/recycled HYV seeds. The adoption of the newly purchased HYV is about 59% in Kenya, while it is 21% in Uganda. As expected, the maize yield is higher for the purchased HYV seeds than the local/recycled HYV seeds, and it is higher in Kenya than in Uganda. In Kenya, the yield of the purchased HYV maize is about 2.2 t/ha, which is 0.5 t higher than the yield of the local/recycled HYV maize. The difference between the two groups is statistically significant and partly driven by the differences in the quantities of the input applications. For instance, 86% of the purchased HYV maize plots receive at least some inorganic fertilizer, while only 58% of the local/recycled HYV maize plots do so. In terms of the quantity, the average amount of inorganic fertilizer applied on the purchased HYV maize plots is about 119 kg/ha, which is about twice as much as the amount applied on the local/recycled HYV maize plots.

In contrast, in Uganda, the maize yield is low for the two maize seed groups, and the difference between the two groups is small, at about 0.2 t/ha. The small difference between the two maize seed groups may be due to the low applications of external fertilizer on both seed groups in Uganda. For instance, only 3 and 6% of the maize plots receive inorganic and organic fertilizer, respectively. Although the purchased HYV maize plots receive more inorganic fertilizer than the local/recycled HYV maize plots, the average quantity of the inorganic fertilizer application on the HYV maize plots is only 9 kg/ha.

Among the maize plots with soil data, we find that the average carbon content is 2.5% in Kenya and 2.4% in Uganda (Table 9.1). Thus, the average carbon content is about the same in the two countries. In Kenya, the purchased HYV seeds are cultivated in better soils than the local/recycled HYV seeds, while in Uganda the purchased HYV seeds are cultivated on poorer soils than the local/recycled HYV seeds in Uganda. The average carbon content is not so different across seed types within country and across countries. To examine the relationship between the soil fertility and the maize production further, we divide the samples based on the soil carbon content next.

Table 9.1 Summary statistics of maize production in Kenya and Uganda

		Seed type		
	All (1)	Purchased HYV (3)	Local/Recycled HYV (2)	Difference (2)–(3) (4)
Kenya				
Number of plots	3,131	1,848	1,283	
Maize yield (kg/ha)	1,986	2,172	1,718	454*
Maize plot size (ha)	0.33	0.34	0.30	0.05**
Seed planted (kg/ha)	28.4	26.5	31.0	–4.5⁺
Proportion of chemical fertilizer used	0.74	0.86	0.58	0.28**
Chemical fertilizer use (kg/ha)	94.7	119.4	59.2	60.2**
Nitrogen chemical fertilizer (kg/ha)	18.41	23.17	11.56	11.62**
Proportion of organic fertilizer Used	0.50	0.50	0.50	0.00
Organic fertilizer use (kg/ha)	1,935	2,258	1,471	787**
Proportion of samples with soil data	0.77	0.74	0.80	–0.06
Carbon content (%)	2.48	2.59	2.33	0.26**
pH	6.15	6.08	6.25	–0.18**
Nitrogen-maize price ratio in 2004	13.4			
Nitrogen-maize price ratio in 2007	16.0			
Uganda				
# Plots	3,198	680	2,518	
Maize yield (kg/ha)	1,561	1,719	1,518	202
Maize plot size (ha)	0.31	0.37	0.29	0.08**
Seed planted (kg/ha)	24.7	22.9	25.2	–2.3*
Proportion of chemical fertilizer used	0.03	0.12	0.01	0.11**
Chemical fertilizer use (kg/ha)	2.4	9.1	0.6	8.5**
Nitrogen chemical fertilizer (kg/ha)	0.76	2.95	0.17	2.78**
Proportion of organic fertilizer used	0.06	0.07	0.06	0.01
Organic fertilizer use (kg/ha)	86	142	71	71
Proportion of samples with soil data	0.67	0.66	0.68	–0.02
Carbon content (%)	2.35	2.15	2.40	–0.25**
pH	6.64	6.69	6.63	0.05*
Nitrogen-maize price ratio in 2003[a]	22.3			
Nitrogen-maize price ratio in 2006[a]	33.7			

Note: The recycled HYV seeds are grouped together with the local seeds. The yield distribution of the recycled HYV seeds has a similar distribution to the local seeds rather than the purchased HYV seeds (Appendix Fig. 9.A.1)

[a] The nitrogen-maize price ratios are obtained from eastern Uganda, where farmers apply inorganic fertilizer

⁺ Significant at 10%; * Significant at 5%; ** Significant at 10%

9.3.2 Soil Fertility and Maize Inputs and Outputs

In Table 9.2, we divide the samples into four quartiles based on the carbon content. Note that we only include those samples with soil carbon content information in Table 9.2. In Kenya, the soil carbon content increases from 1.3% in the lowest quartile to 4.0% in the highest quartile. The proportion of the purchased HYV adoption increases from 46 to 68% from the lowest to the highest carbon content quartiles, respectively. Thus, the Kenyan farmers plant the purchased HYV maize seeds on fertile plots. The maize yields of both the purchased HYV and the local/recycled HYV seeds are highest in the highest quartile, and the maize yields remain about the same level among the lowest three quartiles. In particular, the maize yield of the local/recycled HYV seeds is very high at 3.6 t/ha in the highest quartile, while it remains around 1.3 t/ha among the lowest three quartiles. We have checked if the high yield in the highest quartile is due to outliers but find the results robust. Although the information is not reported in Table 9.2, we find that the median yield of the local/recycled HYV seeds is 1.4 t/ha in the highest quartile, while it is about 0.8 t/ha in the other three quartiles. Thus, it seems that the local/recycled HYV maize seeds are responsive to the soil carbon content at the high soil carbon content level, even though we need to be careful not to link the high maize yield directly to the soil carbon content only. The quantity of the organic fertilizer application, for instance, is about 3.4 t/ha in the highest quartile, while it is about 1.3 t/ha in the lowest quartile.

In contrast, the maize yield and input applications have no clear correlation with the soil carbon content in Uganda. The average maize yield is around 1.5 t/ha, regardless of the seed types and the soil carbon content quartiles. Inputs also do not have any clear relationships with the soil carbon quartiles. The geological distribution of the maize production in Uganda may explain why no clear relationships exist between the soil carbon content quartiles and the maize inputs and outputs. In the eastern region of Uganda, the soil is poor, but the maize technology is much more advanced than the maize production in the Central and Western regions because it is closer to the Kenyan border. In the western region, where banana is the most important staple crop, the soil is good, but the maize production technology is not advanced. To control for the geographical differences and other observed characteristics of the maize production in Kenya and Uganda, we rely on regression analyses.

9.4 Regression Results

9.4.1 Adoption of Purchased HYV

First, we present the regression results of the (purchased) HYV adoption in Table 9.3, separately for Kenya and Uganda. For each country, we present the results from the random-effects (RE) Logit estimation and the household level fixed-effects (FE)

Table 9.2 Input and output level in maize production by soil carbon content in Kenya and Uganda

	All	Quartile of soil carbon content			
		Lowest	2nd	3rd	Highest
	(1)	(2)	(3)	(4)	(5)
Kenya					
Number of plots	2,403	545	533	647	678
Carbon content (%)	2.48	1.28	1.82	2.42	4.01
Ratio of purchased HYV	0.57	0.46	0.46	0.65	0.68
Yield (kg/ha): Purchased HYV	2,109	2,125	1,841	1,962	2,374
Yield (kg/ha): Local/ Recycled HYV	1,765	1,216	1,266	1,326	3,634
Seed use (kg/ha)	26.8	23.1	28.8	26.8	28.2
Ratio of chemical fertilizer used	0.70	0.64	0.62	0.74	0.79
Chemical fertilizer use (kg/ha)	87.0	96.1	76.0	84.5	90.7
Ratio of organic fertilizer used	0.56	0.53	0.58	0.54	0.57
Organic fertilizer use (kg/ha)	2,287	1,293	2,016	2,144	3,436
Uganda					
Number of plots	2,151	595	606	491	459
Carbon content (%)	2.35	1.33	1.82	2.43	4.27
Ratio of purchased HYV adopted	0.21	0.25	0.18	0.20	0.20
Yield (kg/ha): Purchased HYV	1,532	1,786	1,536	1,255	1,413
Yield (kg/ha): Local/ Recycled HYV	1,579	1,377	1,954	1,380	1,531
Seed use (kg/ha)	25.3	25.7	26.2	27.3	21.2
Ratio of chemical fertilizer used	0.03	0.03	0.02	0.02	0.03
Chemical fertilizer use (kg/ha)	1.4	0.8	2.6	0.7	1.3
Ratio of organic fertilizer used	0.04	0.03	0.03	0.05	0.07
Organic fertilizer use (kg/ha)	42	23	62	35	48

Note: In this table, we only include maize plots that are matched with the soil samples

Logit estimation. As the basic explanatory variables, we include household, plot, and community level variables. In addition, to control for the region-specific time-variant effects, such as climate and market conditions, we include seven region dummies, a season dummy, a second survey round dummy, and the interaction terms of these dummies. Because the RE Logit model allows us to estimate the coefficients of time-invariant variables, we add some time-invariant household and

Table 9.3 Determinants of the newly purchased HYV seed adoption in Kenya and Uganda

	Kenya		Uganda	
	RE logit	FE logit[a]	RE logit	FE logit[a]
	(1)	(2)	(3)	(4)
Plot characteristics				
Carbon content	0.1216		0.1100	
	(1.30)		(0.70)	
1{The maize plot is rented}	0.0476	0.2849	0.8961	1.0079
	(0.19)	(1.17)	(2.62)**	(2.86)**
Walking time to the plot (min)	0.0103	0.0093	0.0045	–0.0035
	(1.72)	(1.74)	(1.23)	(0.81)
Household characteristics				
ln (Total size of owned land in ha)	0.3199	0.5724	0.0357	0.1516
	(1.38)	(1.35)	(0.15)	(0.48)
ln (Value of physical assets in USD)	0.1824	0.0333	0.4568	0.2580
	(1.73)	(0.28)	(2.52)*	(1.19)
1{Female household head}	–0.3745		0.4160	
	(1.36)		(0.82)	
Years of schooling of male adult	0.0551		0.1349	
	(1.94)		(2.64)**	
Years of schooling of female adult	0.0209		–0.1455	
	(1.02)		(2.80)**	
Number of adult males	0.0228		0.2644	
	(0.25)		(2.39)*	
Number of adult females	0.1947		0.3680	
	(2.22)*		(3.29)**	
Community characteristics				
1{DAP price info. NOT available}	–0.8827	–1.5551	–4.0430	–2.8287
	(0.61)	(0.86)	(4.46)**	(2.97)**
DAP price/maize price	0.0339	–0.5149	–0.5429	–0.4402
	(0.06)	(0.79)	(3.69)**	(2.94)**
Male hourly wage/maize price	–0.6268	–0.0315	0.1469	0.2025
	(1.76)	(0.06)	(0.60)	(0.85)
1{HYV seed price info.NOT available}	–55.3951	–15.9666	–28.4899	–30.5371
	(0.00)	(0.01)	(0.00)	(0.03)
HYV seed price/maize price	0.2163	0.0775	0.0768	0.0211
	(4.13)**	(1.27)	(2.87)**	(0.69)
Constant	–3.7599		–2.0889	
	(2.59)**		(1.79)	
Region * Season * Year dummies	Included	Included	Included	Included
Hausman's test for FE vs. RE on coefficients of common covariates	$÷^{2}(11)=35.37$**		$÷^{2}(11)=48.90$**	
Observations	2,156	1,295	2,015	978
Number of households	591	286	486	199

Note: Absolute value of z statistics in parentheses

* Significant at 5%; **Significant at 1%

[a] In the fixed-effects Logit estimation, the households who do not alter the state of the HYV seed adoption across plots or seasons are dropped. In addition, the (almost) time-invariant explanatory variables are excluded

soil characteristics into the model. Although the RE model has the advantage of providing the estimation results on time-invariant variables, the RE model estimates could be biased because of omitted variables. Indeed, the Hausman test, presented at the bottom of Table 9.3, indicates that the RE model estimates are not consistent with the FE estimates. Thus, on the time-varying independent variables, we interpret the results from the FE model.

The RE model results indicate that the soil carbon content has no relationship with HYV adoption in both Kenya and Uganda. Farmers do not appear to consider the soil quality when they choose to apply the purchased HYV. The other results from the RE model are consistent with the common observations. The education level of men in the household has a positive association with the HYV adoption in both countries, and the numbers of men and women in the household increase the HYV adoption in general in both countries. The education level of women in the household in Uganda has a negative association with the HYV adoption. This is the only unexpected result, and the reason for this finding is not clear. The results from the RE model indicate that the asset value has a positive association with the HYV adoption in both countries. In the FE model, the estimated coefficient of the asset value becomes smaller and not significantly different from zero. The land size, which is another wealth indicator, has no significant impacts on the HYV adoption in both countries. In Uganda, the results of both the RE and FE models indicate that the farmers adopt the HYV maize more frequently on the rented-in plots than the owned plots, which may reflect the possible actions taken by the tenant farmers who want to maximize the immediate returns from the rented-in plots.

Regarding the community level price variables, we find that the relative price variables do not have significant impacts on HYV maize adoption in Kenya. This could be because the input market is well developed in Kenya and the relative prices are nearly constant across regions. Thus, the regional and time dummies may absorb the impacts of the relative prices. In Uganda, on the other hand, the input market is not well developed. Thus, in the central and western regions of Uganda, we do not even have information on the relative price of DAP simply because it is not available. DAP is the most commonly used fertilizer type in Kenya and Uganda, according to our panel surveys. Indeed, the results indicate that the HYV adoption rate is significantly lower in areas where the DAP price information is missing. The relative price of DAP over the maize output price also has a negative impact on the HYV maize adoption. Because HYVs use fertilizer intensive technology and require a certain amount of inorganic fertilizer application, a high relative price of DAP is likely discourage the farmers from adopting the purchased HYV seeds more than the price of the HYV seeds itself.

9.4.2 Maize Yield Function

Next, we present the results from the yield model, separately for Kenya and Uganda in Tables 9.4 and 9.5, respectively. In each table, we present the results from the three models: the pooled OLS, the household fixed effects model, and the household

Table 9.4 Determinants of log of maize yield (kg/ha) in Kenya

	Pooled OLS model		FE model[a]		FE endogenous selection model[a]	
	Purchased HYV plots (1)	Local/Recycled HYV plots (2)	Purchased HYV plots (3)	Local/Recycled HYV plots (4)	Purchased HYV plots (5)	Local/Recycled HYV plots (6)
ln (Maize plot size in ha)	−0.2578	−0.2980	−0.3440	−0.4068	−0.3512	−0.4218
	(8.60)**	(8.16)**	(7.21)**	(7.32)**	(6.87)**	(6.29)**
ln (Seed kg/ha planted)	0.3803	0.3554	0.3571	0.3749	0.4092	0.4321
	(7.75)**	(7.49)**	(5.70)**	(6.03)**	(4.79)**	(5.49)**
ln (Carbon content)	0.4568	1.0130				
	(2.30)*	(5.60)**				
$\ln^2$ (Carbon content)	−0.0696	−0.2901				
	(0.78)	(3.08)**				
Nitrogen content of chemical	1.3971	1.6396	1.3891	1.3417	0.5926	2.0820
Fertilizer input (100 kg/ha)	(4.33)**	(4.01)**	(2.67)**	(2.21)*	(0.95)	(3.44)**
Nitrogen^2	−0.5735	−0.3037	−0.5508	0.0336	−0.0921	−0.7703
	(3.09)**	(1.10)	(2.21)*	(0.10)	(0.24)	(1.45)
Organic fertilizer (t/ha)	−0.0066	0.0657	0.0230	0.0676	0.0383	0.0751
	(0.48)	(3.05)**	(1.30)	(2.67)**	(1.88)	(2.24)*
Organic^2	−0.0004	−0.0019	−0.0007	−0.0020	−0.0005	−0.0015
	(1.56)	(3.22)**	(2.21)*	(3.17)**	(1.12)	(1.25)
Nitrogen × ln (Carbon content)	−0.4130	−0.5494	−0.2562	−0.2366	0.3497	−0.9179
	(1.65)	(1.53)	(0.66)	(0.47)	(0.84)	(1.39)
Organic × ln (Carbon content)	0.0268	0.0214	0.0154	0.0254	−0.0004	0.0047
	(2.32)*	(1.21)	(1.03)	(1.25)	(0.03)	(0.17)
Nitrogen × Organic	0.0047	−0.0930	−0.0034	−0.1306	−0.0116	−0.0590
	(0.46)	(3.27)**	(0.27)	(3.96)**	(0.63)	(1.06)
Constant	4.2312	3.5159	5.0117	4.4756		
	(24.52)**	(21.95)**	(28.25)**	(25.98)**		

Region × Season × Year dummies	Included		Included	Included	Included	Included
Observations	2,371		1,165	773	1,165	773
Number of households			356	220	356	220
E[Maize yield (kg/ha)| HYV/ non-HYV]	2065.0	1391.1	2080.2	1387.5	2080.2	1387.5
E[∂ lnY/∂ ln Carbon| HYV/ non-HYV][b]	0.32**	0.57**				
E[∂ lnY/∂ Nitrogen| HYV/ non-HYV][b]	0.81**	1.03**	0.92**	0.964**	0.82**	1.13**
E[∂ lnY/∂ Organic| HYV/ non-HYV][b]	0.019*	0.025**	0.028*	0.011**	0.022*	0.041*
Hausman's test vs. FE model					$\chi^2(9)=-$	$\chi^2(9)=15.6$

Absolute value of t statistics in parentheses

* Significant at 5%; **Significant at 1%

[a] Households with less than two observations in the same state of HYV seed adoption are excluded from the regression

[b] The significance level attached to the mean value corresponds to the test statistic for the joint test of the related coefficients being equal to zero simultaneously

Table 9.5 Determinants of log of maize yield (kg/ha) in Uganda

	Pooled OLS model		FE model[a]		FE endogenous selection model[a]	
	Purchased HYV plots	Local/Recycled HYV plots	Purchased HYV plots	Local/Recycled HYV plots	Newly purchased HYV plots	Local/Recycled HYV plots
	(1)	(2)	(3)	(4)	(5)	(6)
ln (Maize plot size in ha)	−0.1700	−0.1550	−0.1696	−0.3004	−0.2886	−0.2996
	(2.97)**	(5.56)**	(2.14)*	(8.51)**	(2.84)**	(5.72)**
ln(Seed kg/ha planted)	0.6203	0.6071	0.6151	0.5772	0.5891	0.4848
	(10.76)**	(23.09)**	(7.58)**	(16.71)**	(5.14)**	(9.23)**
ln (Carbon content)	−0.0723	0.1445				
	(0.21)	(1.02)				
$\ln^2$ (Carbon content)	−0.0726	0.0392				
	(0.35)	(0.58)				
Nitrogen content of chemical Fertilizer input (100 kg/ha)	0.1469	−7.5816	3.8045	−13.5939	3.0635	−19.9570
	(0.03)	(0.82)	(0.73)	(1.50)	(0.67)	(0.01)
Nitrogen²	3.7317	0.8901	−3.8466	1.8568	−0.5241	2.3248
	(0.80)	(0.36)	(0.73)	(0.90)	(0.02)	(0.00)
Organic fertilizer (t/ha)	0.9868	−0.1206	−0.6800	0.2552	1.5585	−0.0612
	(1.28)	(0.45)	(0.18)	(0.88)	(0.16)	(0.18)
Organic²	−0.0468	0.0077	0.0727	−0.0304	−1.1731	0.0323
	(0.43)	(0.30)	(0.05)	(0.97)	(0.15)	(0.54)
Nitrogen × ln (Carbon content)	1.2016	11.2553	−1.2713	18.9528	−1.7808	28.0156
	(0.30)	(1.00)	(0.27)	(1.66)	(0.44)	(0.02)
Organic × ln (Carbon content)	−0.9244	0.1687	0.5763	−0.0097	−0.4270	0.0934
	(1.78)	(0.68)	(0.21)	(0.04)	(0.04)	(0.33)
Nitrogen × Organic	−17.7050	0.0000	−14.8094		−20.3941	
	(0.73)	(.)	(0.61)		(0.01)	

Constant	5.0469 (24.48)**	4.6093 (43.12)**	5.0797 (24.35)**	4.4380 (45.97)**		
Region × Season × Year dummies	Included		Included	Included	Included	Included
Observations	2,084		356	1,461	356	1,461
Number of households			112	366	112	366
R-squared	0.98		0.34	0.40	0.34	0.40
E[Maize yield (kg/ha)| HYV/ non-HYV]	1532.0	1338.2	1572.0	1403.0	1572.0	1403.0
E[∂ lnY/∂ lnCarbon| HYV/ non-HYV][b]	–0.20	0.23**				
E[∂ lnY/∂ Nitrogen| HYV/ non-HYV][b]	0.20**	0.92	2.62	0.46	1.35	0.80
E[∂ lnY/∂ Organic| HYV/ non-HYV][b]	–0.41	0.007	–0.97	0.25	0.36	0.010
Hausman's test vs. FE model					$\chi^2(9)=0.23$	$\chi^2(8)=6.12$

Absolute value of t statistics in parentheses

* Significant at 5%; **Significant at 1%

[a] Households with less than two observations are excluded from the regression

[b] The significance level attached to the mean value corresponds to the test statistic for the joint test of the related coefficients being equal to zero simultaneously

fixed semiparametric endogenous switching model. In general, all three estimation models provide robust estimates, with a few exceptions. In the pooled OLS model, all the independent variables are interacted with the purchased HYV maize dummy, and we present the total impacts, not the differential impacts, of the interaction terms with the HYV dummy to make comparisons between the pooled OLS and the other models possible. The switching model controls for the household fixed effects as well as the selection between the purchased HYV and the local/recycled HYV maize, as we explained in Sect. 9.3.

Because the soil carbon content (measured in the natural log), the nitrogen content of the inorganic fertilizer (100 kg/ha), and the organic fertilizer application (t/ha) are all interacted with each other and have the squared terms, interpretations of the results could be complicated. Thus, at the bottom of the tables, we present the partial derivative of one input evaluated at the means. We also indicate if the partial derivates are jointly significant.

Regarding the soil carbon content, the results from the pooled OLS model suggest that the soil carbon content has a positive impact on the maize yield with a decreasing return on both of the seed types in Kenya. The elasticity evaluated at the means is about 0.32 for the purchased HYV maize, while it is about 0.57 for the local/recycled HYV maize. Because the average carbon content levels are about the same for the two seed types, according to the HYV adoption model in Table 9.3, the results suggest that the local/recycled HYV maize has a greater physical responsiveness to the soil carbon content than the purchased HYV maize. The impact of the organic fertilizer is also greater on the local/recycled HYV maize than on the purchased HYV maize. According to the endogenous switching model, the average impact of an additional 1 t of organic fertilizer application per ha increases the maize yield by 4.1% for the local/recycled HYV maize and 2.2% for the purchased HYV maize. The estimated coefficients are robust across the estimation models. Thus, it seems that the local/recycled HYV maize is more physically responsive to the organic matter, i.e., the soil carbon content and the organic fertilizer, than the purchased HYV maize.

The sizes of the estimated impacts of organic fertilizer may seem small. But note that the estimated coefficients of the organic fertilizer could be biased toward zero because of the possible attenuation biases created by the measurement errors in the organic fertilizer variables. First, it is difficult to measure the quantity of the applied organic fertilizer. Farmers may not remember clearly how much of the organic fertilizer they applied. Second, the quality of the organic fertilizer varies from one farmer to another. The quality of the organic fertilizer depends on the contents and how it is prepared. Thus, we should treat the estimated average impact of the organic fertilizer as a conservative estimate.

As expected, we find a large impact of the inorganic fertilizer application on the maize yield. The evaluated average impacts of the nitrogen content of the inorganic fertilizer, measured in 100 kg/ha, is 0.82 for the purchased HYV maize and 1.13 for the local/recycled HYV maize, according to the results of the endogenous switching model. Because of the decreasing return to the inorganic fertilizer, the smaller average impact on the purchased HYV maize than the local/recycled HYV maize could be explained partially by the larger quantity of the nitrogen application on the purchased HYV maize than on the local/recycled HYV maize. The average

Table 9.6 Relative prices and marginal returns of nitrogen application in Kenya and Uganda

	Marginal physical product (MPP)[a] (1)	Average relative price (RP) (2)	Test statistics if MPP = RP (3)
Kenya – Wave 1			
Purchased HYV maize	14.10** (0.60)	13.4	t=1.17
Local/Recycled HYV maize	11.05** (0.91)	13.4	t=−2.59**
Kenya – Wave 2			
Purchased HYV maize	19.89** (0.67)	16.0	t=5.77**
Local/Recycled HYV maize	16.13** (0.87)	16.0	t=0.15
Uganda – Wave 1			
Purchased HYV maize	23.44 (1.68)	22.3	t=0.68
Local/Recycled HYV maize	20.78 (10.80)	22.3	t=−1.88
Uganda – Wave 2			
Purchased HYV maize	24.96 (4.70)	33.7	t=−0.14
Local/Recycled HYV maize	25.23 (9.37)	33.7	t=−0.90

* Significant at 5%; **Significant at 1%. The * and ** in column (1) indicate that the estimated coefficients for the evaluated MPPs are jointly significant in Tables 9.4 and 9.5. The * and ** in column (3) indicate that the MPP and RP are statistically different

[a] $MPP = E[Y^* \partial \ln Y / \partial \text{Nitrogen} | \text{HYV/non-HYV}]$, where Y is maize yield per ha

nitrogen application on the purchased HYV maize is about 17.9 kg/ha, while the average nitrogen application on the local/recycled HYV maize is about 9.8 kg/ha. To investigate the different application rates, we need to calculate the Marginal Physical Product (MPP) and the profitability of the nitrogen application. We do that in Table 9.6, together for Kenya and Uganda.

Unlike the results from Marenya and Barrett (2009), the yield effect of external fertilizers does not differ depending on the soil fertility. The interaction terms between the carbon content and the inorganic and organic fertilizer applications are generally not significant. Although the interaction term between the organic fertilizer application and the soil carbon content is positive and significant in the pooled OLS model, it becomes insignificant once we control for the household fixed effects and the seed selection. Similarly, the interaction term between the nitrogen application and the organic fertilizer application loses its significance once the seed selection is controlled for.

In Table 9.5, we present the results from Uganda. Regarding the soil carbon content, we find that the elasticity of the carbon content, evaluated at the means, is 0.23 on the local/recycled HYV maize. We do not find, however, any significant impacts of the soil carbon content on the yield of purchased HYV maize. These results are consistent with those in Kenya that the soil carbon content has a larger impact on the local/recycled HYV maize than on the purchased HYV maize. We do not find any significant impacts, either individually or jointly, of the organic and inorganic fertilizer applications on the maize yield in Uganda. This is not surprising because of the very low applications of both fertilizers. As we show in Table 9.1, only 9.3 and 9.4% of the maize plots in Uganda received the inorganic or organic fertilizer, respectively.

9.4.3 Optimality of Nitrogen Fertilizer Application

Is the nitrogen fertilizer application at the optimal level? We answer this question by testing if the MPP of the nitrogen application is equal to the nitrogen-maize relative price. Thus, for each year and a given maize seed type, we calculate the MPP by multiplying the average maize yield with the partial derivative of the nitrogen application evaluated at the means:

$$MPP(kg/kg)_t^j = \overline{Y}_t^j \times E\left[\frac{\partial \ln(Y_t^j)}{\partial\, Nitrogen(100kg)_t^j}\right]/100, \qquad (9.12)$$

and conduct a test to see if the MPP is equal to the relative price.[9]

By reviewing numerous technical studies by agricultural scientists, Yanggen et al. (1998) report that the typical yield response rate, which is the additional output obtained in kg divided by the additional nitrogen applied, is 17 in East and Southern Africa. In Kenya, Mbata (1997) reports response rates of 12–18 in Central and Western Kenya. Marenya and Barrett (2009) report the average MPP at 22 in Western Kenya, although they find considerable heterogeneity.

In Table 9.6, we find that the MPP varies from 11 to 20 in Kenya and 21 to 25 in Uganda. Compared with the previous estimates, these estimates are within a reasonable range. The MPP is 14 for the purchased HYV during the first wave of the panel surveys in Kenya. The nitrogen-maize relative price is 13 during this period. The t-test indicates that the MPP is not different from the relative price, suggesting that the nitrogen application is roughly at the optimal level for the purchased maize during this period in Kenya. For the local/recycled maize, the MPP is lower than the relative price, suggesting a slightly over-application of the nitrogen. During the next survey period, the results are the opposite. We find an almost optimal application on the local/recycled HYV maize but somewhat an under-application for the purchased HYV maize. Because of unexpected events, both agro-ecologically and economically, it is not surprising that Kenyan farmers miss the optimal application levels occasionally. It is more important to point out that the MPPs move in the same direction as the relative-price over time. From the first to the second wave, the

[9] To obtain the nitrogen price, we have divided the DAP price by 0.18 because 100 kg of DAP contain 18 kg of nitrogen.

relative price increased from 13 to 16 and the MPPs of the purchased and local/recycled HYV maize also increased from 14 to 20 for the purchased HYV maize and from 11 to 16 for the local/recycled HYV maize. Thus, the results indicate that the Kenyan farmers are responding to the change in the relative price and successfully achieving the near optimal application level in 1 of the 2 years for both the purchased and local/recycled HYV maize.

The relative price is much higher in Uganda than in Kenya: it is 22 and 34 in the first and second wave, respectively. Because of the low use of the nitrogen fertilizer in Uganda, the MPPs are not precisely estimated. Despite the low precision, we find that the MPP on the purchased HYV maize during the first wave is 23, which is close to the relative price at 22. During the second wave in Uganda, the MPP is around 25 for both the purchased HYV and local/recycled HYV maize, when the relative price is 33. Thus, assuming a decreasing marginal return, even the low application of the nitrogen fertilizer is over-application and not profitable. The relative price in Uganda is simply too high to apply the inorganic fertilizer. The high relative price in Uganda is mostly because of the low maize price in Uganda, which is about 60% of the Kenyan price (see Appendix Table 9.A.2). Because it would cost more to send the inorganic fertilizer from eastern Uganda to central and western Uganda, the potential relative price would be higher in the central and western Uganda. Thus, to decrease the relative price, the maize price has to increase. Otherwise, the relative price remains too high for any farmers to apply the inorganic fertilizer.

9.5 Conclusions

To dramatically improve maize productivity in Sub-Saharan Africa, the current level of external fertilizer application is considered to be too low. Thus, we estimate the maize yield function in Kenya and Uganda to investigate the reasons for the low external fertilizer application on maize. Kenya has one of the highest productivities for maize in Sub-Saharan Africa, while Uganda has one of the lowest. Thus, a comparison between the two countries provides valuable lessons for other African countries. By comparing the marginal physical product (MPP) of the nitrogen application on the maize yield and the nitrogen-maize relative price, we find that Kenyan farmers have successfully achieved the optimal nitrogen application level in 1 of the 2 survey years on both the purchased and local/recycled HYV maize. We also find that they have responded to the relative price change over time. Thus, the results suggest that a market-based approach, such as reducing the inorganic fertilizer price or increasing the maize price or both, would be effective in encouraging farmers to use more inorganic fertilizer in Kenya. In Uganda, the application levels of external fertilizers are too low to identify precise estimates. Nonetheless, we find that the low inorganic fertilizer application is already over-application in Uganda because of the very high relative price. In both Kenya and Uganda, the potential success of a non-market approach, such as credit or extension provision, may be limited as long as the relative price remains at the present level. Note that as will be demonstrated in Chap. 12, there is a possibility that the majority of maize farmers in Uganda may not have enough knowledge about the impacts on yield of HYV maize as well as in organic fertilizer. In this case, the extension system have to be strengthened in this country.

Another major contribution of this chapter is the determination of whether the returns to external fertilizers differ depending on the soil fertility. According to the results in this chapter, we do not find any significant differences in the returns to the organic and inorganic fertilizer application depending on the soil fertility. Thus, the results suggest that policies that encourage inorganic fertilizer application would be effective even on degraded soils where maize farmers in our samples cultivate maize. This does not suggest, however, that the soil fertility is not important. We also find that the soil carbon content directly increases the maize yield both in Kenya and Uganda. Especially, we find larger impacts on the local/recycled HYV maize than on the purchased HYV maize. Thus, improving the soil fertility has a direct impact on the maize production. In this chapter, we are not able to identify the costs of improving the soil fertility. Because of the high relative price of the inorganic fertilizer, it is worth estimating the relative costs of improving the soil fertility. This remains to be investigated in the future.

9.6 Appendix

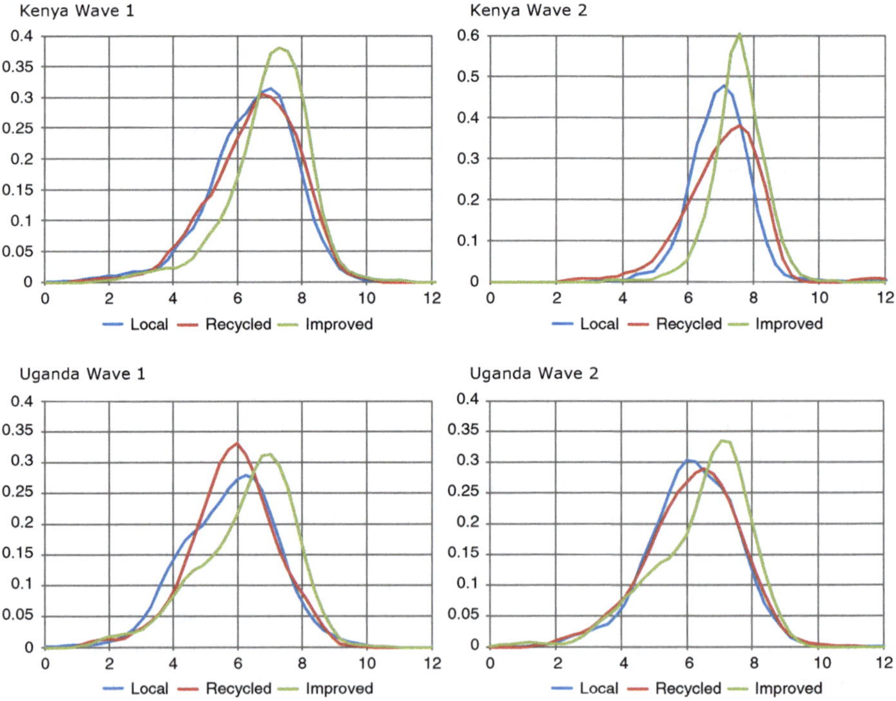

Fig. 9.A.1 Log of maize yield (kg/ha) by seed type

Table 9.A.1 Determinants of sample attrition in Kenya and Uganda (probit)

	Kenya		Uganda	
	Soil sample not available	Not interviewed in wave 2	Soil sample not available	Not interviewed in wave 2[a]
Dependent variable	(1)	(2)	(3)	(4)
Household characteristics in the initial survey year				
Log of land size (ha)	0.0039	−0.0239	−0.0240	0.0008
	(0.13)	(1.02)	(0.88)	(0.44)
Log of asset holdings (USD)	−0.0257	0.0074	0.0167	0.0002
	(1.80)[+]	(0.67)	(0.95)	(0.14)
1{female headed}	−0.0071	0.0265	0.0425	
	(0.19)	(0.94)	(0.80)	
Years of schooling of male adult	−0.0051	−0.0056	0.0000	0.0005
	(1.13)	(1.71)[+]	(0.00)	(0.87)
Years of schooling of female adult	0.0103	−0.0020	−0.0140	−0.0008
	(2.13)*	(0.56)	(2.39)*	(1.26)
Number of adult males	0.0074	−0.0199	−0.0109	0.0015
	(0.51)	(1.71)[+]	(0.74)	(1.43)
Number of adult females	−0.0297	−0.0073	−0.0176	0.0007
	(1.98)*	(0.64)	(1.17)	(0.85)
Kenya region dummies (reference region: Nyanza)				
Western	−0.2663	−0.0450		
	(6.19)**	(1.22)		
Rift valley	−0.1036	−0.0604		
	(2.53)*	(1.85)[+]		
Central	−0.1177	−0.0109		
	(3.00)**	(0.35)		
Uganda region dummies (reference region: East)				
Central			0.2190	−0.0005
			(5.54)**	(0.15)
West/South western			0.1545	
			(3.67)**	
E[y]	0.26	0.13	0.41	0.01
Number of households	825	825	938	621

Note: Reported coefficients are the change in probability for an infinitesimal change in each independent, continuous variable and the discrete change in the probability for dummy variables. Absolute value of z statistics in parentheses

[+]Significant at 10%; *Significant at 5%; **Significant at 1%

[a]The sample households living in West/South western are dropped from the regression since there is no variation in the dependent variable within the region

Table 9.A.2 Input and output prices on maize production in Kenya and Uganda by region

Region	Maize price Wave 1 (1)	Maize price Wave 2 (2)	DAP price Wave 1 (3)	DAP price Wave 2 (4)	Nitrogen price Wave 1 (5)	Nitrogen price Wave 2 (6)	HYV seed price Wave 1 (7)	HYV seed price Wave 2 (8)
	USD/100 kg		USD/100 kg [DAP/Maize]		USD/100 kg [Nitrogen/Maize]		USD/100 kg [HYV/Maize]	
Kenya								
Central	17.4	17.7	42.1	50.7	233.9	281.7	178.9	183.6
			[2.4]	[2.9]	[13.4]	[15.9]	[10.3]	[10.4]
Rift valley	14.3	16.3	34.8	50.9	193.3	282.8	170.8	174.4
			[2.4]	[3.1]	[13.5]	[17.3]	[11.9]	[10.7]
Western	15.8	18.4	40.8	51.9	226.7	288.3	163.6	178.8
			[2.6]	[2.8]	[14.3]	[15.7]	[10.4]	[9.7]
Nyanza	19.2	19.8	42.2	53.7	234.4	298.3	155.5	193.3
			[2.2]	[2.7]	[12.2]	[15.1]	[8.1]	[9.8]
All	16.7	17.8	40.3	51.2	223.9	284.4	167.1	182.3
			[2.4]	[2.9]	[13.4]	[16.0]	[10.0]	[10.2]
Uganda								
East	9.4	8.8	38.2	59.4	212.2	330.0	73.0	91.9
			[4.1]	[6.7]	[22.6]	[37.5]	[7.8]	[10.4]
Central	9.9	10.4	n.a.	n.a.	n.a.	n.a.	49.7	64.0
							[5.0]	[6.1]
West/SW	8.6	10.8	n.a.	n.a.	n.a.	n.a.	35.8	30.6
							[4.2]	[2.8]
All	9.5	9.8	38.2	59.4	212.2	330.0	59.8	72.1
			[4.0]	[6.1]	[22.3]	[33.7]	[6.3]	[7.4]

Note: the prices are the region average of the community level prices. The community level prices are the median of prices reported by respondents at the household level. Nitrogen price is calculated by DAP price divided by 0.18 based on the fact that the nitrogen content in 100 kg of DAP is 18 kg

References

Adesina AA (1996) Factors affecting the adoption of fertilizers by rice farmers in Cote d'Ivoire. Nutr Cycl Agroecosyst 46(1):29–39

Bationo A, Mokwunye AU (1991) Role of manures and crop residue in alleviating soil fertilizer constraints to crop production: with special reference to the Sahelian and Sudanian zones of West Africa. Fertil Res 29(1):117–125

Cozzolino D, Moron A (2003) The potential of near-infrared reflectance spectroscopy to analyze soil chemical and physical characteristics. J Agric Sci 140(1):65–71

Gregory DI, Bumb BL (2006) Factors affecting supply of fertilizer in Sub-Saharan Africa, vol 24. D iscussion paper. Agricultural and Rural Development, World Bank, Washington, DC.

Hassan RM, Karanja DD (1997) Increasing maize production in Kenya: technology, institutions, and policy. In: Byerlee D, Eicher CK (eds) Africa's emerging maize revolution. Lynne Rienner Publishers, Boulder

Henao J, Baanante C (2006) Agricultural production and soil nutrient mining in Africa: implications for resource conservation and policy development. Background paper for African fertilizer summit, Abja

International Fertilizer Development Center (IFDC) (2006) Africa fertilizer summit proceedings, Abuja, June 9–13, 2006, An International Center for Soil Fertility and Agricultural Development, Muscle Shoals, AL

Jayne TS, Govereh J, Wanzala M, Demeke M (2003) Fertilizer market development: a comparative analysis of Ethiopia, Kenya, and Zambia. Food Policy 28(4):293–326

Kelly VA (2006) Factors affecting demand for fertilizer in Sub-Saharan Africa, vol 23, Agricultural and rural development discussion paper. World Bank, Washington, DC

Koning N, Smaling E (2005) Environmental crisis or 'lie of the land'? The debate on soil degradation in Africa. Land Use Policy 22(1):3–11

Kyriazidou E (1997) Estimation of a panel data sample selection model. Econometrica 65(6):1335–1364

Manlay R, Feller C, Swift MJ (2007) Historical evolution of soil organic matter concepts and their relationships with the fertility and sustainability of cropping systems. Agric Ecosyst Environ 119(3 and 4):217–233

Marenya PP, Barrett CB (2009) Soil quality and fertilizer use rates among smallholder farmers in Western Kenya. Agric Econ 40(5):561–572

Mbata JN (1997) Factors influencing fertilizer adoption and rates of use among small-scale food crop farmers in the rift valley area of Kenya. Q J Int Agric 36(3):285–289

Morris ML, Kelly VA, Kopicki RJ, Byerlee D (2007) Promoting increased fertilizer use in Africa: lessons learned and good practice guidelines. World Bank, Washington, DC

Ngugi DN, Karau PK, Nguyo W (1990) East African agriculture, 3rd edn. Macmillan, London

Nyoro J, Kirimi L, Jayne TS (2004) Competitiveness of Kenyan and Ugandan maize production: challenges for the future, vol 10, Working paper. Egerton University/Tegemeo Institute, Nairobi

Palm CA, Gachengo CN, Delve RJ, Cadisch G, Giller KE (2001) Organic inputs for soil fertility management in tropical agroecosystems: application of an organic resources database. Agric Ecosyst Environ 83(1):27–42

Pender J, Place F, Ehui S (2006) Strategies for sustainable land management in the East African highland. International Food Policy Research Institute, Washington, DC

Shepherd KD, Walsh MG (2002) Development of reflectance spectral libraries for characterization of soil properties. Soil Sci Soc Am J 66(3):988–998

Smale M, Jayne TS (2003) Maize in Eastern and Southern Africa: 'seeds' of success in retrospect, vol 97, EPTD discussion paper. International Food Policy Research Institute, Washington, DC

Smaling EMA, Nandwa SM, Janssen BH (1997) Soil fertility in Africa is at stake. In: Buresh RJ, Sanchez PA, Calhoun F (eds) Replenishing soil fertility in Africa. Soil Sci Soc Am, Madison

Sserunkuuma D (2005) The adoption and impact of improved maize and land management technologies in Uganda. Electron J Agric Dev Econ 2(1):67–84

Tiessen H, Cuevas E, Chacon P (1994) The role of soil organic matter in sustaining soil fertility. Nature 371(27):783–785

Yanggen D, Kelley VA, Reardon T, Naseem A (1998) Incentives for fertilizer use in Sub-Saharan Africa: a review of empirical evidence on fertilizer response and profitability, vol 70, MSU international development working paper. Michigan State University, East Lansing

Part III
The Role of Fertilizer Markets and Fertilizer Application

Chapter 10
Chemical Fertilizer, Organic Fertilizer, and Cereal Yields in India

Kei Kajisa and N. Venkatesa Palanichamy

Abstract Concern about the sustainability of food production has been leading to a revival in the use of organic fertilizer in modern agriculture as this is seen as an appropriate way to maintain soil health. The purpose of this study is to shed light on the role of the demand, supply, and market of farmyard manure (FYM) in maintaining and improving cereal yields by analyzing a data set of farm households from 1993 to 2003 in Tamil Nadu, India. We find the dairy sector development and associated increase in FYM supply contribute to the productivity improvement of upland cereals. However, due to high transportation costs of FYM, the impact is spatially constrained within the small area where dairy sector development has taken place.

Keywords Chemical fertilizer • Organic fertilizer • Organic matter • Farmyard manure • Cereal yield • India • White revolution

10.1 Introduction

Concern about the sustainability of food production has been leading to a revival in the use of organic inputs in modern agriculture as this is seen as an appropriate way to maintain soil health by providing soil organic matter and micronutrients

K. Kajisa (✉)
School of International Politics, Economics, and Communication, Aoyama Gakuin University, 4-4-25 Shibuya, Shibuya-ku, Tokyo 150-8366, Japan
e-mail: k.kajisa@sipeb.aoyama.ac.jp

N.V. Palanichamy
Department of Agricultural and Rural Management, Centre for Agricultural and Rural Development Studies, Tamil Nadu Agricultural University, Coimbatore, Tamil Nadu 641-003, India
e-mail: nvpchamy@yahoo.com

K. Otsuka and D.F. Larson (eds.), *An African Green Revolution: Finding Ways to Boost Productivity on Small Farms*, DOI 10.1007/978-94-007-5760-8_10,

(Rasmussen et al. 1998). More recently, a steep rise in international prices of chemical fertilizer has been further turning people's attention to organic fertilizer. Furthermore, a few studies show that the improvement of soil organic matter makes chemical fertilizer more effective (Marenya and Barrett 2009; Tiessen et al. 1994). On top of that, growing concern about poverty and hunger in sub-Saharan Africa (SSA) leads to reexamination of the role of locally produced organic fertilizer for boosting crop productivity because expensive inorganic fertilizer on international markets becomes even more expensive at the farm gate in Africa due to poorly developed internal transportation systems (Otsuka and Yamano 2005; Chap. 9). The summary of empirical studies indicates the considerable contribution of organic matters to crop yield increase in African agricultural systems (Place et al. 2003). Among many kinds of organic fertilizer, farmyard manure (FYM) is the most common form of organic fertilizer in developing countries and thus the attention to the potential of FYM in agricultural development has been increasing.

Meanwhile, several studies indicate the limitations of the roles of FYM in agricultural development. First, experiments in agronomy show that, for lowland rice cultivation, the use of organic fertilizer including FYM has little impact on its productivity, even though it has a discernible impact on upland cereals such as maize and wheat (Hati et al. 2007; Dawe et al. 2003; Edmeades 2003; Rasmussen et al. 1998). This means that organic fertilizer is effective but not a panacea. Secondly, the availability of FYM could be limited because FYM is bulky and untradeable except in a small area (Dawe et al. 2003; Place et al. 2003). To write a right prescription for agricultural development, the potential and limitation of FYM-based development strategy must be clearly understood.

Existing studies contribute to this end partially but not fully. For comprehensive understanding, we have to consider three interlinked aspects: (1) demand side, (2) supply side, and (3) the market of FYM which connects suppliers and users. Regarding the demand side, a number of studies in agronomy and soil science on the impact of FYM use on crop productivity are available (Balwinder et al. 2008; Pampolino et al. 2008; Sahrawat 2005; Dawe et al. 2003; Hati et al. 2007; Rangaraj, et al. 2007; Yadav and Chhipa 2007; Mussgnug et al. 2006; Somasundaram et al. 2004). Independently from this group of studies, there also exist some studies showing an association between livestock sector development and increased availability of FYM (Delgado et al. 2008; Motavalli et al. 1994). However, literature has rarely assessed the FYM markets. Although many of them are informal, FYM markets do exist (Ghosh 2004). For example, our data indicate that 34% of FYM is purchased from markets (otherwise, self production) for paddy and 41% for upland cereals. The lack of the study on FYM markets is critical because the potential of FYM-based development depends on functioning markets (Place et al. 2003). To authors' knowledge, no empirical study has examined the three related aspects simultaneously under a cohesive framework.

The purpose of this study is to fill the above-mentioned research gap by analyzing a data set of farming households from 1993 to 2003 in Tamil Nadu, India. This data set has several advantages for our analyses. First, it has a panel structure, which allows us to circumvent some econometric problems. Second, it has detailed

FYM records. Third, the different level of the progress in dairy sector development, which was induced by a so-called "White Revolution" in India, generates wide variations in the changes in FYM availability. This provides us with a good opportunity to explore the causal impact of dairy sector development on crop production through the changes in FYM availability at the markets.

This chapter proceeds as follows. Section 10.2 describes the features of our data set. Section 10.3 reviews literature in agronomy and soil science on the impact of FYM and explains our approach to an impact assessment. Our estimation models and results are shown in Sects. 10.4 and 10.5. Section 10.6 summarizes the results and presents implications of the study.

10.2 Data

Our analysis relies on the data set collected by Tamil Nadu Agricultural University under the Cost of Cultivation of Principal Crops (hereafter, CCPC) scheme with the aim of surveying the cultivation costs of principal crops. In the CCPC scheme, enumerators live close to survey villages and keep daily records of sample farmers' farm and non-farm activities throughout a year. This intensive data collection work provides us with very detailed information on farming practices, including production of, application of, and transactions with FYM.

The data has been collected annually since 1971. The latest round available for analyses is the data collected in 2003. The scheme expanded the questionnaire from the round starting in 1993 and has been collecting more detailed data on FYM as well as data on soil condition. Thus, we use the data from 1993 to 2003 in this study. The scheme has been carrying out the sampling of 60 villages and 10 farming households from each village and the same households were surveyed for three consecutive years until the next round of sampling, generating 3-year rotating panels.[1] The raw data contain plot level identification by household, which could have allowed us to construct plot-level panel data. However, unfortunately, the available data set does not have that information anymore. Hence, our analyses rely on household-level panel data. Since the CCPC scheme has been sampling

[1] The sampling involves three stages. First, districts are classified into six agro-climatic zones, and then districts in each zone are selected so that the crop area of sample districts becomes proportional to the crop area of the zone. Second, villages in each district are selected so that the crop area of sample villages becomes proportional to the crop area of the district. Third, in each selected village, farming households are selected in accordance with the size of landholdings. The five size classes are operational holdings with areas less than 1 ha, between 1 and 2 ha, between 2 and 4 ha, between 4 and 6 ha, and greater than 6 ha. In each size class, two households are selected by simple random sampling, generating a sample of ten farmers in each village. If in any village a particular size class does not contain even two households, more households are selected from the adjacent size group to make up for the deficit. The data for 2004 of the 2002–2004 round panel are not available yet. We treat this set also as a 2-year panel.

two farmers from every operational holding group regardless of the actual distribution of operational holdings, we weight the sample with the population distribution of operational holdings taken from the nearest agricultural census (Government of Tamil Nadu 2002).[2]

10.3 Literature Review and Our Approach

Field experiments show that FYM has little direct impact on paddy yield as the release of nitrogen from FYM is slow in flooded conditions and irrigation water helps to maintain soil health (Balwinder et al. 2008; Pampolino et al. 2008; Sahrawat 2005; Dawe et al. 2003). Meanwhile, in most cases, the impact on upland crops, including upland cereals, is expected to be high, firstly because the release of nitrogen is fast under aerobic conditions, and secondly because the degradation of soil organic matter and the deficiency of micronutrients in soil are usually problematic under aerobic conditions (Hati et al. 2007; Rangaraj et al. 2007; Yadav and Chhipa 2007; Mussgnug et al. 2006; Somasundaram et al. 2004). Another group of studies shows that FYM or organic matters in soil indirectly increases yield by making external nutrient more absorbable to crops (Marenya and Barrett 2009; Tiessen et al. 1994). This sheds light on the role of FYM as a complement to chemical fertilizer. At the same time, we should note that the significance of the influence varies considerably under different agroecological conditions (Edmeades 2003; Yadav 2003). Therefore, the empirical section of this study starts by confirming these established features in our data set by estimating yield functions by crop.

After identifying the differential impacts of FYM, we examine what kind of limitations exit in achieving the potential impacts. Dawe et al. (2003)'s benefit cost simulation implies that the positive benefit of FYM use is achieved only in the area where FYM is locally produced because FYM's long-distance transportation cost is prohibitively high. FYM used to be available anywhere over Tamil Nadu but it is no so anymore because the distribution of cattle has been becoming more uneven geographically. The progress of mechanization resulted in the disappearance of draught animals in many parts of the state. At the same time, however, the development of dairy sector, which is induced by a so-called "White Revolution" started around 1970 and massively supported by the government in the 1980s, has increased the number of milking cattle (cows and buffaloes). This process has been associated with the introduction of improved cows and buffaloes which have higher milk (and manure) productivity (Sharfuddin 1984). As a result, cattle have been concentrating in the areas agro-ecologically and economically suitable for dairy activities.

We can observe such feature mentioned above in Table 10.1 that shows different paths taken by typical four dairy developed (DD) districts and four less dairy

[2] The agricultural census was available in 1976–1977, 1985–1986, and 1995–1996. We rely on the 1995–1996 census as it is the nearest to the period of our analyses (1993–2003).

Table 10.1 Summary statistics of dairy sector development in Tamil Nadu from 1993–2003

		1993–1995	1996–1998	1999–2001	2002–2003
No. of improved cows and buffaloes per household	LDD[a]	0.12	0.00	0.00	0.00
	DD[b]	0.14	0.11	0.11	0.18
FYM use (t ha^{-1})	LDD	1.1	1.1	0.2	0.0
	DD	1.9	2.1	1.4	1.1
FYM price (Rs. per t)	LDD	112.5	120.8	151.6	na
	DD	95.1	113.9	134.6	127.4
NPK price (Rs. per kg)	LDD	11.4	11.2	11.7	13.1
	DD	10.1	11.1	11.7	11.8

Source: CCPC data set
[a] Less dairy-developed districts which consist of Dharmapuri, Thiruvallur, Vellore, Thiruvannamalai
[b] Dairy-developed districts which consist of Villupuram, Cuddalore, Salem, and Thirunelveli

developed (LDD) districts. A clear contrast between DD and LDD is that the improved ones disappeared in LDD in the period of 1996–1998, while they have been raised constantly, albeit not increasingly, in DD throughout the periods.[3] Accordingly, the FYM application per ha has drastically declined over time in LDD and became even zero in the last period, while in DD the decline has proceed at a much slower pace. The FYM market shows a consistent feature with this; the FYM price seems to have an upward trend in both DD and LDD, but the level of price is always higher in LDD and the market eventually disappeared in the last period when FYM was not applied at all in LDD. For comparison, we show also the price of NPK fertilizer and find no discernible price difference between LDD and DD except in the last period, indicating a spatially integrated chemical fertilizer market. In summary, these observations imply that the dairy sector development has deterred the thinning of FYM market and, thus, the associated decline in FMY use, but that such an effect is limited in a relatively small area where the dairy-sector development has taken place. This is basically consistent with our conjectures based on literature review. We statistically examine our conjectures in the empirical section.

For a better understanding of the analytical results, we make a few remarks on the situation of the FMY use. First, according to the figures in Table 10.1, FYM price relative to NPK fertilizer has been increasing and the use of FYM has been declining in both DD and LDD. In the following sections, a positive association between the dairy sector development and FYM use is implied, but it is misleading to interpret that result as the progress of organic farming; rather, the result indicates the determent of the decline in FYM use. Second, the comparison of FYM price with NPK fertilizer price in Table 10.1 indicates that 100 kg of FYM is about as expensive as 1 kg of NPK fertilizer. A typical FYM contains about 1.6–3.9% of NPK nutrients (or 1.6 kg or 3.9 kg of NPK in 100 kg of FMY) together with other micro nutrients, indicating that FYM is

[3] The existence of improved cows and buffaloes in LDD in the first period seems to be related with the persisting impact of the strong supports by the government for the achievement of White Revolution.

a cheaper source of nutrients.[4] However, we have to note that the release of nutrients from FYM is slower and thus the effect may be discounted by the farmers. In addition, application of FYM is more laborious than that of chemical fertilizer. Hence, it is better to refrain from comparing FYM and chemical fertilizer just based on their nutrients and prices. Third, related to this, the lagged impact of FYM could be captured by examining productivity change of the same plot over time. However, since we do not have plot identification, what we can observe is household-level productivity by crop in which different plots are included in each year as farmer may engage in crop rotation. Hence, one of the limitations of this study is that we give up estimating the long term impact of FYM application on crop productivity.

10.4 Estimation Models

Two kinds of econometric problems could arise in the analysis of FYM: (1) reverse causality and (2) self-selection. The former could occur when productive (thus wealthy) farmers tend to own more cattle and apply more FYM (resulting in an upward bias), and the latter could occur when farmers with soil degradation (thus currently low productivity) tend to apply more FYM to restore soil health (resulting in a downward bias). In setting up our models, we explicitly consider how to circumvent these possible problems with our panel data.

We define a yield function of farming household i in village j at time t as

$$y_{ijt} = f(l_{ijt} n_{ijt} m_{ijt}; \phi_{ij}),$$

where y is the yield of either paddy or upland cereals per hectare, l is the hours of labor input per ha, n is the amount of NPK fertilizer applied per ha, m is the amount of FYM applied per ha, and ϕ is aggregated influence from technology, farm management ability, access to irrigation, soil condition, and agro-ecological environment, which is time-invariant at least in the short run (a household-level fixed effect). For econometric estimation, we consider a second-order local approximation to this general form, which gives a quadratic yield function defined as

$$\begin{aligned} y_{ijt} &= \alpha_0 + \alpha_1 l_{ijt} + \alpha_2 l_{ijt}^2 + \alpha_3 n_{ijt} + \alpha_4 n_{ijt}^2 + \alpha_5 m_{ijt} + \alpha_6 m_{ijt}^2 \\ &\quad + \alpha_7 l_{ijt} n_{ijt} + \alpha_8 l_{ijt} m_{ijt} + \alpha_9 n_{ijt} m_{ijt} + \phi_{ij} + \varepsilon_{ijt}, \end{aligned} \tag{10.1}$$

where ε is the error term for random productivity shocks. Technical interdependencies are captured by the interaction terms, and α_9 becomes positive if FYM and NPK are complementary factors. One advantage of this functional form is that we can include observations with zero input values without any manipulation for log transformation. This is appealing to us as some farmers do not apply FYM.

[4] According to a local specialist on FYM, the nutritional composition of FYM is as follows. Nitrogen (0.15–0.75%), Phosphorus (0.1–0.6%), Potash (0.75–1.8%), Calcium (<0.1%), Magnesium (<0.1%), Ferrous (200 PPM), Born and Zink (<5PPM), Organic carbon (0.5–1.5%).

Econometrically, a self-selection bias due to the relationship between soil degradation and manure application is expressed as a possible negative correlation between m and ϕ. Estimation with household fixed effects purges the influence of ϕ.

From the quadratic function, the derived factor demand for FYM can be expressed as a linear function of input prices normalized by the output price. It is important to note that the nature of the FYM market is different from that of other goods. Since the nutrient content of 1 kg of FYM is lower than that of chemical fertilizer, a large amount of FYM must be applied to improve soil fertility and thus the transportation cost becomes high.[5] If one purchases FYM outside of one's own yard, its effective price including the transportation cost could be prohibitively high. Meanwhile, transportation costs would be minimal if one uses one's own FYM. Since effective price of these two kinds of FYM could be different, we must treat them separately. To capture the price effect of FYM produced *outside* of one's own yard, we use village average FYM price. As a proxy of price of own FYM, we use the number of ordinary cattle and the number of improved cattle owned by the household, assuming that the more cattle the household owns, the cheaper the own FYM is. We distinguish between ordinary and improved cattle as improved one has higher manure productivity. Under this set up, the factor demand function we estimate will be

$$m_{ijt} = \beta_0 + \beta_1 \frac{\overline{p}^m_{jt}}{\overline{p}^y_{ijt}} + \beta_2 \frac{c^{ord}_{ijt}}{\overline{p}^y_{ijt}} + \beta_3 \frac{c^{imp}_{ijt}}{\overline{p}^y_{ijt}} + \beta_4 \frac{\overline{p}^n_{jt}}{\overline{p}^y_{ijt}} + \beta_5 \frac{\overline{p}^l_{jt}}{\overline{p}^y_{ijt}} + \phi'_{ij} + \varepsilon'_{ijt} \qquad (10.2)$$

where p^f is the price of factor f on which an upper bar, if any, indicates averaging over the households in village j, and c^{ord} and c^{imp} are the number of ordinary and improved cattle owned by household i. Since this is a derived demand, it is possible to calculate βs from the parameters of the yield function. However, we prefer to estimate βs directly from the data and statistically check farmers' behavioral consistency between production practice and factor demand.

If c_{ijt}(c^{ord}_{ijt} and c^{imp}_{ijt}) is a function of y_{ijt}, estimation bias due to the reverse causality would arise. However, note that the number of cattle recorded in our data set is the number at the beginning of the season, and that the number of cattle is a stock.[6] Thus, the level of c_{ijt} is a function of the past series of yields (as proxy of household income) and cattle prices in the following manner:

$$c_{ijt} = \gamma_0 + \sum_{h=1}^{N} \gamma\, \overline{p}^c_{jt-h} + \sum_{h=1}^{N} \gamma\, y_{ijt-h} + \phi''_{ij} + \varepsilon''_{ijt} \qquad (10.3)$$

Since (10.3) does not contain y_{ijt}, we treat the system of equations from (10.1), (10.2), and (10.3) as a recursive one that has no influence from a reverse causality.

[5] For those who applied FYM (i.e., FYM input >0), 7.62 t ha^{-1} is applied to paddy and 6.17 t ha^{-1} for upland cereals on average.

[6] In the CCPC scheme, a change in any kind of stock during the survey year is recorded in a separate survey module.

Hence, our econometric strategy is to estimate (10.1) by household fixed effects models with time-varying village dummies. Such dummies are introduced to capture the influence of village-level covariate shocks such as weather shocks. For instance, in our Eq. (10.1), village-level shocks could affect individual yield and they could also be correlated with the amount of each input because village-level covariate shocks, rather than idiosyncratic shocks, could influence factor prices in the village. Therefore, unless we control for them, our results would suffer from omitted variable bias. In the estimation of the Eq. (10.2) with household fixed effect models, we use time-varying district dummies because, otherwise, village-level price variations are completely captured by time-varying village dummies.

Our next step is to estimate price transmission functions at household and at village levels in order to analyze spatial limitations of FYM markets. Since FYM is de facto a non-tradable goods because of its high transportation costs, we hypothesize that FYM markets are not integrated beyond the village where FYM is produced. For comparative analysis, we estimate the price transmission function for paddy as an example of tradable goods. The household-level transmission function is defined as

$$\ln p_{ijt} = \delta_0 + \delta_1 \ln \overline{p}_{jt}^{-i} + \delta_2 \ln \overline{\overline{p}}_{t}^{-j} + \phi_{ij}''' + \varepsilon_{ijt}''', \tag{10.4}$$

where *ln* indicates log transformation, p_{ijt} is the price of either FYM or paddy for household i in village j at time t, $\overline{p}_{jt}^{-i}$ is the average price in village j excluding household i, and $\overline{\overline{p}}_{t}^{-j}$ is the district average price excluding village j. We expect that only δ_1 becomes positively significant for FYM and both δ_1 and δ_2 for paddy. We estimate the functions by household fixed effects models with year dummy. Neither village nor district-level time-varying dummies are included as we have village and district-level explanatory variables.

The above function, however, does not indicate the transmission relationship between villages and districts. Hence, we estimate village-level function defined as

$$\ln \overline{p}_{jt} = \eta_0 + \eta_1 \ln \overline{\overline{p}}_{t}^{-j} + \varphi_j + \upsilon_{jt} \tag{10.5}$$

where φ is the village fixed effects and v is the village-level random error. We estimate this by village fixed effects models with year dummy, expecting insignificant η_1 for FYM and significant and positive η_1 for paddy.

Last, in order to understand the association between dairy sector development and FYM price, we try to identify the determinants of village-level FYM price. Since the FYM price in village j at year t ($\overline{p}_{jt}$) is determined by supply of and demand for FYM of the village, we include, as supply side factors, the number of ordinary cattle (c^{ord}), the number of improved cattle (c^{imp}), and the area for fodder crops (a^{fodder}) (as a proxy for cheaper feed for cattle) in village j, while the area for paddy (a^{paddy}) and the area for upland cereals (a^{upland}) in village j are included as demand side factors. We employ the first three variables to capture the level of dairy sector development. However, these variables can take large values simply because the size of village is large. Hence, we use village averages as our explanatory variables and thus the estimation function is defined as follows.

$$\ln \overline{p}_{jt} = \kappa_0 + \kappa_1 \overline{c}_{jt}^{ord} + \kappa_2 \overline{c}_{jt}^{imp} + \kappa_3 \overline{a}_{jt}^{fodder} + \kappa_4 \overline{a}_{jt}^{paddy} + \kappa_5 \overline{a}_{jt}^{upland} + \varphi_j' + v_{jt}' \quad (10.6)$$

where φ' is the village fixed effects and v' is the village-level random error. We expect that an increase in supply reduces the price (i.e. negative coefficients from κ_1 to κ_3), while increase in demand increases the price (i.e. positive coefficients of κ_4 and κ_5). Since improved cattle produce more manure than ordinary cattle, we expect that increase in the number of improved cattle through dairy development greatly contribute to the reduction in FYM price.

The summary statistics, description, and unit of all the variables are reported in Appendix Tables 10.A.1, 10.A.2, 10.A.3, and 10.A.4.

10.5 Regression Results

10.5.1 Yield Functions

Table 10.2 shows the estimation results of paddy, upland cereals, and sorghum yield functions. The upland cereals consist of sorghum, finger millet, pearl millet, and maize. Among them, we separately show the result of sorghum as it is by far the most common in Tamil Nadu.

The results of the first model for paddy show that the coefficients of labor, NPK fertilizer, and the squared term of each have expected signs, although the squared term of labor is not statistically significant. Meanwhile, the coefficients of FYM and FYM squared are not significant at any acceptable level of significance, indicating that no direct impact exists, which is consistent with the findings from field experiments. Hence, FYM by itself will not contribute for paddy yield increase. However, it is worth noting that the interaction term between FYM and NPK is positive and significant. Thus, our data indicate that FYM application still has a complementary impact together with chemical fertilizer use.[7,8]

The results for upland cereals and sorghum in the same table indicate that the coefficients of FYM and its squared term are in clear contrast to those for paddy; they are significant with conventional signs in the both models. This is consistent with the findings from field experiments. This indicates that the FYM use contributes to productivity improvement even when chemical fertilizer is completely

[7] Ghosh (2004) find the same complementarities in his estimation of quadratic paddy yield function, although it is a cross-section analysis.

[8] A puzzling result is that although the impact of FYM arises only indirectly, farmers apply a substantial amount of FYM to paddy (see Table 10.A.1). A possible reason we find during our field interview is that farmers apply FYM on paddy plot as nutrients will be mixed better in soil during the plowing under flooded condition. On the other hand, when farmers apply FYM on dry land, dried FYM could be brown away. Note that nutrient release is slow under flooded condition. Hence, in this case, farmers expect that the impact of FYM on flooded paddy filed appears in the subsequent crop when it is cultivated under rainfed condition. To examine this point accurately, we need plot level panel data, which are unfortunately unavailable.

Table 10.2 Estimation results of paddy and upland cereal yield functions in Tamil Nadu

	Paddy yield ($t\ ha^{-1}$)	Upland cereal yield ($t\ ha^{-1}$)	Sorghum yield ($t\ ha^{-1}$)
Labor (1,000 $h\ ha^{-1}$)	2.990	1.960	3.060
	(3.34)***	(2.79)***	(4.51)***
$Labor^2$	−0.301	−1.137	−2.644
	(0.52)	(1.86)*	(3.69)***
NPK ($t\ ha^{-1}$)	10.759	7.261	3.185
	(4.34)***	(1.77)*	(2.82)***
NPK^2	−12.442	−18.797	−5.955
	(2.02)**	(0.80)	(2.79)***
FYM ($t\ ha^{-1}$)	−0.034	0.193	0.450
	(0.81)	(4.14)***	(5.48)***
FYM^2	−0.001	−0.009	−0.038
	(0.89)	(1.77)*	(5.11)***
Labor*NPK	−5.935	17.988	3.804
	(2.69)***	(2.49)**	(2.04)**
Labor*FYM	0.009	−0.245	−0.759
	(0.32)	(2.98)***	(5.01)***
NPK*FYM	0.237	−0.070	2.032
	(1.88)*	(0.07)	(5.03)***
Constant	1.628	0.412	0.169
	(3.62)***	(2.42)**	(1.02)
Time-varying dummies	Year*Village	Year*Village	Year*Village
Fixed effects	Household	Household	Household
Observations	2,445	555	345

Robust *t* statistics in parentheses
Use the census in 1995–1996 as weights
*Significant at 10%; **Significant at 5%; ***Significant at 1%

inaccessible. In addition, if we limit our sample only to sorghum, a complementary impact (the coefficient of NPK*FYM) becomes significant. This is consistent with the study by Motavalli et al. (1994) in ICRISAT villages.

10.5.2 Factor Demand Functions

The estimation results of factor demand function are presented in Table 10.3.[9] Regarding the results for paddy, we do not find any statistically significant determinants. Since the impact of FYM on paddy is marginal, farmers do not seem to react to the price changes

[9] To compute the village average FYM price, at least one farmer in a village must have record of FYM value. Hence, for this analysis, we excluded the observations in the villages where no one uses FYM at all for any purposes. The sample size is reduced from 2,445 to 1,294 for paddy, from 555 to 199 for upland cereals, and from 345 to 110 for sorghum.

Table 10.3 Estimation results of FYM demand functions for paddy and upland cereals in Tamil Nadu

	FYM for paddy (t ha^{-1})	FYM for upland cereals (t ha^{-1})	FYM for sorghum (t ha^{-1})
FYM price ($\overline{p}^m / p^y$)	0.229	−0.404	−1.102
	(1.11)	(1.51)$^+$	(1.79)*
No. of ordinary cattle (c^{ord} / p^y)	0.655	0.391	0.390
	(1.18)	(0.51)	(0.47)
No. of improved cattle (c^{imp} / p^y)	−0.266	0.105	−0.002
	(0.44)	(0.33)	(0.01)
NPK price ($\overline{p}^n / p^y$)	−0.390	−0.826	−0.920
	(0.76)	(3.02)***	(3.81)***
Wage rate ($\overline{p}^l / p^y$)	−0.114	0.349	−0.694
	(0.22)	(0.21)	(0.57)
Constant	6.266	13.576	29.045
	(0.87)	(0.90)	(2.09)**
Time-varying dummies	Year*District	Year*District	Year*District
Fixed effects	Household	Household	Household
Observations	1,294	197	110

Robust t statistics in parentheses
Use the census in 1995–1996 as weights
The census in 1995–1996 is used for weights for the computation of village average
*Significant at 10%; **Significant at 5%; ***Significant at 1%; +Significant at 15%

actively. Meanwhile, the FYM price is negative and weakly significant for upland cereals and significant at 10% for sorghum. The negative coefficient of NPK fertilizer price is consistent with the existence of complementary effect.

Although we include the number of ordinary cattle and that of improved cattle, they are not statistically significant in any models. As we will see in the next sub-section, the FYM markets are fairly integrated *within* a village. Hence, the result indicates that as long as someone else own the cattle, farmers can purchase FYM at reasonable price.

Summarizing these results, a key finding of the analyses up to this sub-section is that in case of upland cereals, the reduction in FYM price induces farmers to apply more FYM through transaction in its markets, and the increase in FYM application contributes to the productivity improvement through the direct and complementary impacts. Meanwhile, such clear features are not observed in case of paddy. The next-subsection examines whether farmers at distant locations (for example, outside of the village) can benefit from the market transaction of FYM.

10.5.3 Price Transmission Functions

We exhibit the results of the FYM market analysis in Table 10.4. A household-level analysis shows that when the FYM price increases by 1% among the other households

Table 10.4 Results of manure and rice price transmission analyses in Tamil Nadu

	Household-level price $\ln p$		Village-level price $\ln \overline{p}$	
	FYM	Paddy	FYM	Paddy
Av. price of the other HHs ($\ln \overline{p}^{-i}$)	0.784	0.743		
	(3.81)***	(17.89)***		
Av. price of the other villages ($\ln \overline{\overline{p}}^{-j}$)	0.097	0.156	0.032	0.275
	(1.07)	(3.04)***	(0.20)	(2.87)***
Constant	0.278	0.606	2.213	4.463
	(0.59)	(1.82)*	(5.86)***	(7.56)***
Time-varying dummies	Year	Year	Year	Year
Fixed effects	Household	Household	Village	Village
Observations	1,070	2,787	233	502

Robust *t* statistics in parentheses
Use the census in 1995–1996 as weights for household-level regressions
The census in 1995–1996 is used for weights for the computation of village average
*Significant at 10%; ***Significant at 1%

in the same village ($\ln \overline{p}^{-i}$), the price of own FYM corresponds to this by 0.784%, implying that the FYM market is fairly integrated within a village. However, the results show that the price of own FYM receives no influence from a price change in other villages (the coefficient of $\ln \overline{\overline{p}}^{-j}$, 0.097, is not statistically significant). The reason for this is that a price change of FYM in other villages is not transmitted to the village-level FYM price, as statistically indicated by the village-level analysis in the same table (the coefficient of $\ln \overline{\overline{p}}^{-j}$, 0.032, is not statistically significant). This result does not necessarily mean that the FYM market is not integrated beyond the village at all. If the other village locates near the village we could find the integration to some extent. However, we do not have data on the distance between sample villages, and thus our analysis only allow for a binary "within" versus "beyond" village effect. Hence, we would like to interpret this result such that the FYM market is integrated only within a small area.

In contrast to FYM, paddy markets are fairly integrated both within and beyond the village. The household-level regression results show a positive and significant influence not only from the price within a village ($\ln \overline{p}^{-i}$) but also from the price in other villages ($\ln \overline{\overline{p}}^{-j}$) with a smaller influence from other villages. The village-level analysis also shows a significant influence from the price in other villages ($\ln \overline{\overline{p}}^{-j}$) on the village-level price. The elasticity of household-level price against village-level price for paddy (0.743) is as high as that for FYM (0.784), indicating at least within a village the FYM market is integrated as well as the paddy market.

10.5.4 Determinants of Village-Level FYM Price

The determinants of village-level FYM price are presented in Table 10.5. A key finding from this result is that increase in the improved cattle is significantly associated with the reduction in village-level FYM price, indicating the contribution of

Table 10.5 The determinants of village-level farm-yard manure (FYM) price

	$\ln \overline{p}$
Av. no. of ordinary cattle ($\overline{c}^{ord}$)	0.026
	(0.91)
Av. no. of improved cattle ($\overline{c}^{imp}$)	–0.253
	(1.67)*
Av. area of fodder ($\overline{a}^{fodder}$)	–0.612
	(0.85)
Av. area of paddy ($\overline{a}^{paddy}$)	0.089
	(1.61)
Av. area of upland cereals ($\overline{a}^{upland}$)	0.597
	(2.18)**
Constant	2.115
	(23.88)***
Time-varying dummies	Year
Fixed effects	Village
Observations	262

Robust *t* statistics in parentheses
The census in 1995–1996 is used for weights for the computation of village average
*Significant at 10%; **Significant at 5%; ***Significant at 1%

dairy sector development. Meanwhile, the number of ordinary cattle has no significant impact on price, presumably because they are fed often by grazing, rather than stall fed, and thus the collection of dung for manure is not an easy task. Among the demand side factors, the cultivated area of upland cereals has highly significant coefficient, which is consistent with the importance of FYM for upland cereal production. Putting these results together with the results of price transmission analysis, we conclude that the dairy sector development reduces FYM price in a small area but, because of the non-integrated feature of FYM markets, the benefit of the reduction is spatially limited within the small area where the development has taken place.

10.6 Conclusion and Policy Implications

Through our analyses, four main findings have emerged. First, a direct impact of FYM application exists only for upland cereals but not for paddy. Second, a complementary impact through an increase in the marginal product of chemical fertilizer is observed for both paddy and upland cereals (sorghum). Third, reflecting the existence of the benefit of FYM application, farmers react to FYM price changes actively. However, our fourth finding indicates that the price reduction of FYM due to dairy sector development does not spread beyond the small area where the dairy development has taken place.

These findings reveal that the potential of FYM application is high for upland cereal cultivation. Since the direct impact exists, FYM-based development would be effective for upland cereals even when chemical fertilizer is completely inaccessible, which may be the case in many regions in SSA. In addition, if chemical fertilizer is accessible to some extent, through the indirect complementary impact, FYM use would increase effect of chemical fertilizer, resulting in further improvement of crop productivity. A moderate potential exists also in paddy cultivation through the indirect impact. Hence, the dairy sector development, which brings about the price reduction in FYM, contributes mainly to a productivity improvement of upland cereals. This is the additional but significant benefit of the dairy sector development beyond its direct impact of milk income increase.

In Tamil Nadu's context, the aggregated impact, nevertheless, may be moderate because upland cereals are minor crops. In addition, the impact in Tamil Nadu has been realized as the determent of decline in FYM use, rather than the progress of organic farming. However, for other developing regions and countries where upland cereals are major staples, the potential impact could be large. Such regions and countries include not only India's semi-arid tropics but also many parts of SSA (McIntire et al. 1992). In fact, the data from ICRISAT's villages in semi-arid tropics show a greater increase in sorghum yield when chemical fertilizer and FYM are combined (Motavalli et al. 1994). Marenya and Barrett (2009b) find in western Kenya that the impact of chemical fertilizer on maize yield is low when the level of soil organic matter is lower than a threshold level, which is consistent with our results on the complementarities between FYM and NPK. Based on the results, they argue that, in addition to high chemical fertilizer price and limited credit access, low soil fertility and resulting low marginal return of chemical fertilizer can be a reason for unduly low application of chemical fertilizer (8 kg ha^{-1} $year^{-1}$ against 102 kg ha^{-1} $year^{-1}$ in developing countries) and low productivity in SSA (Morris et al. 2007). Our study indicates that the dairy sector development can be a solution to this problem through the complementary effect of FYM application.

In summary, an important lesson from our findings is that dairy sector development can be a key component of a development strategy when the targeted commodities are upland cereals. However, at the same time, as a limitation of this strategy, we should note that, due to a non-tradable attribute of FYM, the impact of dairy sector development is limited within a small area where the development has taken place. Although the spatially-wider achievement of dairy sector development is important to overcome this limit, this does not necessary mean that we have to achieve it all over because not all areas are agro-ecologically or economically suitable for dairy activities. For such unsuitable upland areas, designing alternative development strategies such as the introduction of other forms of organic fertilizer and integrated soil fertility management for productivity improvement or non-farm sector development for poverty reduction may be needed, which is beyond the scope of this study but an important future research agenda for us.

Appendix

Table 10.A.1 Description, measure, and summary statistics of the variables for yield functions in Tamil Nadu

			Paddy		Upland cereals		Sorghum	
Variable	Description	Measurement	Mean	S.D.	Mean	S.D.	Mean	S.D.
Y	Yield	t ha^{-1}	4.407	1.216	1.159	1.014	0.979	0.794
Labor	Labor input	1,000 h ha^{-1}	1.183	0.409	0.494	0.278	0.344	0.161
NPK	NPK input	t ha^{-1}	0.193	0.064	0.021	0.040	0.084	0.180
FYM	FYM input	t ha^{-1}	2.098	4.183	0.434	1.749	0.183	1.001

Table 10.A.2 Description, measure, and summary statistics of the variables for farm-yard manure (FYM) demand functions in Tamil Nadu

			Paddy		Upland cereals		Sorghum	
Variable	Description	Measurement	Mean	S.D.	Mean	S.D.	Mean	S.D.
M	FYM input	t ha^{-1}	4.420	5.176	1.535	3.070	0.724	1.980
$\overline{p}^m / p^y$	Village average FYM price (real price in terms of output)	kg of output per kg of FYM	10.54	3.510	11.231	3.597	11.585	2.870
c^{ord} / p^y	Number of ordinary cattle (divided by output price)	Number kg^{-1}	2.284	2.633	1.659	1.904	1.811	2.036
c^{imp} / p^y	Number of improved cattle (divided by output price)	Number kg^{-1}	0.139	0.684	0.3119	0.791	0.233	0.627
$\overline{p}^n / p^y$	Village average NPK price (real price in terms of output)	100 kg of output per kg of NPK	11.031	1.276	11.178	2.099	11.201	2.156
$\overline{p}^l / p^y$	Village average labor wage rate (real rate in terms of output)	100 kg of output per hour of labor	7.815	2.678	6.560	1.812	6.956	1.651

Table 10.A.3 Description, measure, and summary statistics of the variables for manure and rice price transmission functions in Tamil Nadu

			FYM		Paddy	
Variable	Description	Measurement	Mean	S.D.	Mean	S.D.
Household-level analysis						
p	Price for household i	Rs. per 100 kg	10.689	3.411	506.648	120.027
$\ln \overline{p}^{-i}$	Village-level average price for village j excluding household i	Rs. per 100 kg	10.576	3.366	508.436	112.313

(continued)

Table 10.A.3 (continued)

Variable	Description	Measurement	FYM Mean	FYM S.D.	Paddy Mean	Paddy S.D.
$\overline{\overline{p}}^{-j}$	District-level average price excluding village j	Rs. per 100 kg	10.439	2.651	510.283	103.913
Village-level analysis						
$\overline{p}$	Village-level average price for village j	Rs. per 100 kg	10.619	3.708	513.770	114.583
$\overline{\overline{p}}^{-j}$	District-level average price excluding village j	Rs. per 100 kg	10.894	2.731	509.202	101.309

Table 10.A.4 Description, measure, and summary statistics of the variables for the determinants of village-level farm-yard manure (FYM) price

Variable	Description	Measurement	Mean	S.D.
$\overline{p}$	Price of FYM (village average)	Rs. per 100 kg	10.636	3.549
$\overline{c}^{ord}$	Number of ordinary cattle (village average)	Number	2.000	1.656
$\overline{c}^{imp}$	Number of improved cattle (village average)	Number	0.159	0.399
$\overline{a}^{fodder}$	Planted area for fodder (village average)	ha	0.0155	0.0609
$\overline{a}^{paddy}$	Planted area for paddy (village average)	ha	0.798	0.811
$\overline{a}^{upland}$	Planted area for upland cereals (village average)	ha	0.0926	0.221

References

Balwinder KB, Guputa RK, Bhandari AL (2008) Soil fertility changes after long term application of organic manures and crop residues under rice-wheat system. J Indian Soc Soil Sci 56(1):80–85

Dawe D, Doberman A, Ladha JK, Yadav RL, Bao L, Guputa RK, Lal P, Panaullah G, Sariam O, Singh T, Swarup A, Zhen QX (2003) Do organic amendments improve yield trends and profitability in intensive rice systems? Field Crops Res 83(2):191–213

Delgado CL, Narrod CA, Tiongco M (2008) Determinants and implications of the growing scale of livestock farms in for fast-growing developing countries, vol 157, Research report. International Food Policy Research Institute, Washington, DC

Edmeades DC (2003) The long-term effects of manures and fertilizers on soil productivity and quality: a review. Nutr Cycl Agroecosyst 66(2):165–180

Ghosh N (2004) Reducing dependence on chemical fertilizers and its financial implications for farmers in India. Ecol Econ 49(2):149–162

Government of Tamil Nadu, Evaluation, Applied Research Department (2002) Economic appraisal: Tamil Nadu. Government of Tamil Nadu, Chennai

Hati KM, Swarup A, Dwivedi AK, Misra AK, Bandyopadhyay KK (2007) Changes in soil physical properties and organic carbon status at the topsoil horizon of a vertisol of Central India after 28 years of continuous cropping, fertilization and manuring. Agric Ecosyst Environ 119(1 and 2):127–134

Marenya PP, Barrett CB (2009) State-conditional fertilizer yield response on Western Kenyan Farms. Am J Agric Econ 91(4):991–1006

McIntire J, Bourzat D, Pingali PL (1992) Crop-livestock interaction in Sub-Saharan Africa. World Bank, Washington, DC

Morris ML, Kelly VA, Kopicki RJ, Byerlee D (2007) Promoting increased fertilizer use in Africa: lessons learned and good practice guidelines. World Bank, Washington, DC

Motavalli PP, Singh RP, Anders RP (1994) Perception and management of farmyard manure in the semi-arid tropics of India. Agric Syst 46(2):189–204

Mussgnug F, Becker M, Son TT, Buresh RJ, Vlek PLG (2006) Yield gaps and nutrient balances in intensive, rice-based cropping systems on degraded soils in the Red River Delta of Vietnam. Field Crops Res 98(2 and 3):127–140

Otsuka K, Yamano T (2005) The possibility of a Green Revolution in Sub-Saharan Africa: evidence from Kenya. J Agric Dev Econ 2(1):7–19

Pampolino M, Laureles EV, Gines HC, Buresh RJ (2008) Soil carbon and nitrogen changes in long-term continuous lowland rice cropping. Soil Sci Soc Am J 72(3):798–807

Place F, Barrett C, Freeman HA, Ramisch J, Vanlauwe B (2003) Prospects for integrated soil fertility management using organic and inorganic inputs: evidence from smallholder African agricultural systems. Food Policy 28(4):365–378

Rangaraj T, Somasudaram E, Mohamed AM, Thirumurugan V, Ramesh S, Ravi S (2007) Effect of agro-industrial wastes on soil properties and yield of irrigated finger millet in coastal soil. Res J Agric Biol Sci 3(3):153–156

Rasmussen P, Goulding KWT, Brown JR, Grace PR, Janzen HH, Korschens M (1998) Long-term agroecosystem experiments: assessing agricultural sustainability and global change. Science 282(30):893–896

Sahrawat KL (2005) Fertility and organic matter in submerged rice soils. Curr Sci 88(5):735–739

Sharfuddin RCR (1984) Dairy co-operative. Indian J Mark 15(2 and 3):9–11

Somasundaram E, Sankaran N, Thiyagarajan TM (2004) Efficacy of organic sources of nutrients and panchagavya spray on productivity of crops in maize based cropping system. J Agric Res Manage 3(1):46–53

Tiessen H, Cuevas E, Chacon P (1994) The role of soil organic matter in sustaining soil fertility. Nature 371(27):783–785

Yadav RL (2003) Assessing on-farm efficiency and economics of fertilizer N, P and K in rice-wheat systems in India. Field Crops Res 81(1):39–51

Yadav KK, Chhipa BR (2007) Effect of FYM, gypsum and iron pyrites on fertility status of soil and yield of wheat irrigated with high RSC water. J Indian Soc Soil Sci 55(3):324–329

Chapter 11
The Demand for Fertilizer When Markets Are Incomplete: Evidence from Ethiopia

Daniel Zerfu Gurara and Donald F. Larson

Abstract The under-utilization of fertilizers is viewed as a hurdle to the adoption of more productive and sustainable agricultural techniques in sub-Saharan Africa. In this chapter, we investigate the role incomplete markets play in determining the use of chemical fertilizers among Ethiopian farmers with the aim of identifying policies that would encourage the adoption of profitable and sustainable agricultural practices. The results of regression analysis show that high transport costs, illiteracy, adverse local climates, and limitation in risk and credit markets are major constraints on the functioning of fertilizer markets, suggesting that government actions to close knowledge gaps and lower transportation costs can increase fertilizer use among farmers.

Keywords Green Revolution • Chemical fertilizer • Organic fertilizer • Soil fertility • Incomplete markets • Credit markets • Insurance markets

11.1 Introduction

Average agricultural productivity in Africa is lower than in other regions and available technologies that show great promise in field studies are not always adopted by farmers. Consequently, the gains in rural income and the reduction in rural poverty that characterized Asia's Green Revolution have not reached Africa (Otsuka et al. 2009).

D.Z. Gurara
African Development Bank, 15 Avenue du Ghana, P.O. Box 323-1002, Tunis-Belvedère, Tunisia
e-mail: d.gurara@afdb.org

D.F. Larson (✉)
Development Research Group, World Bank, 1818 H Street, NW, Washington, DC 20433, USA
e-mail: dlarson@worldbank.org

K. Otsuka and D.F. Larson (eds.), *An African Green Revolution: Finding Ways to Boost Productivity on Small Farms*, DOI 10.1007/978-94-007-5760-8_11,

Several reasons are given for the slow pace of productivity growth in African agriculture, but many researchers and policy makers view the limited take up of fertilizers as a key cause. At present, most productive technologies depend on abundant soil nutrients. In Africa, soils are depleted in many places and practices aimed at boosting soil fertility, including the application of chemical and organic fertilizers, are not commonplace. Moreover, studies suggest that because of this, nutrients are continually extracted from African soils to feed current crops.[1] For these reasons, the under-utilization of fertilizers is viewed as a hurdle to the adoption of more productive and sustainable agricultural techniques.

Several arguments are advanced to explain why fertilizer use is low in Africa, including some that suggest failed markets. In turn, this has prompted calls for African governments to take a more active role in fertilizer markets and to offer targeted subsidies that promote the use of fertilizers, high-yielding seeds and other components of more productive techniques.[2]

In this chapter, we investigate the role incomplete markets play in determining the use of chemical fertilizers among Ethiopian farmers with the aim of identifying policies that would encourage the adoption of profitable and sustainable agricultural practices. We use panel data to estimate a selection model and this allows us to distinguish empirically between determinants that condition the use of fertilizer and those that affect the conditional quantities of fertilizer that farmers demand. The panel data that we employ is well suited to our purpose since it contains a relatively even mix of farmers that use fertilizers and those that do not. The survey also contains information that we use to proxy the effects of incomplete credit and insurance markets, and information hurdles that are related to how well fertilizer markets work. The survey also contains information about characteristics that potentially help households cope with poorly functioning markets.

Empirical results from the panel selection model suggest that a variety of factors stand in the way of the widespread adoption of fertilizer-using production technologies in Ethiopia. Chief among these are high transport costs, illiteracy, adverse local climates, and limitation in risk and credit markets. Households that are wealthier, better educated, or have greater political authority are better able to overcome obstacles to adoption; however, this, in turn, suggests a reinforcing effect on rural poverty as less-advantaged households continue to rely on less productive farming techniques that continue to deplete soils. At the same time, the results suggest that fertilizer markets are not all together missing in rural Ethiopia, and that government actions to close knowledge gaps and lower transportation costs can increase fertilizer use among farmers.

With this as background, the rest of this chapter is organized as follows Sect. 11.2 provides additional background information about Ethiopian agriculture and related studies of fertilizer demand. Section 11.3 develops the empirical model, whereas

[1] Henao and Baanante (2006), reported in Morris et al. (2007), estimate that 85% of African farmland suffer soil nutrient loses at a rate of 30 kg per year or greater.

[2] See, for example, the African Union (2006), issued during the 2006 Africa Fertilizer Summit.

Sect. 11.4 presents the descriptive statistics followed by the empirical results reported in Sect. 11.5. Section 11.6 concludes.

11.2 Background

Agriculture is the mainstay of the Ethiopian economy with its hefty contribution of about half of the GDP. Yet its growth rate over the past four decades had been quite low. Per capita agricultural output in 2003 is not so much different from its level in 1961. This is coupled with the dominance of smallholder farmers with land holding size of 0.96 ha per household, and yield of 1,167–2,122 kg/ha for the main cereal crops (CSA 2008). As a consequence, few farming households produce significant surpluses to market and many are unable to consistently feed themselves.[3] Yet results from research stations and pilot programs suggest that the economic returns to fertilizer, when combined with improved seeds and better farming techniques, can be high (Bekele and Höfner 1993; Howard et al. 2003). This has prompted the Government of Ethiopia to emphasize agriculture in its economic growth strategy, Agriculture Development Led Industrialization (ADLI). And in particular, the government and donors have supported a number of productivity-enhancing efforts, such as the provision of extension services, the introduction of high yield variety seeds and the timely provision of chemical fertilizers.[4]

Between 1991 and 1995, fertilizer use and intensity in Ethiopia rose dramatically from 110,000 mt (21 kg/ha) to 300,000 (35 kg/ha) in 1999, levels that compare favorably with many countries in Africa (World Bank 2006). Nevertheless, in subsequent years, fertilizer consumption and intensification fluctuated considerably and intensification has only recently resumed a steady upward trend. Consequently, like many countries in Africa, the penetration of productivity enhancing inputs such as chemical fertilizer, improved seeds, and pesticides is very low. As a proportion of the total crop area at the national level, around 45% is treated with chemical fertilizer (with the average application of 81 kg/ha) and only 15.2% is treated with by pesticides. Only 3.5% of the land is planted with improved seeds and extension agents only reach 11% of total cropland (CSA 2008).

Several arguments are offered in the economics literature about why fertilizer use is low in Ethiopia and among African farmers generally.[5] Not all of the arguments imply failed markets, though several do. In particular, some economists argue that because natural and man-made transportation systems are poorly developed in Africa, high transport costs work to raise the price of inputs and lower the farm-gate price of

[3] Demeke et al. (1998) noted that 790 kg of grain is needed to meet the minimum calorie requirement assuming 156 kg/person/year grain requirement for a household of five.

[4] See Byerlee et al. (2007) for a discussion of government policy aimed at promoting fertilizer use in Ethiopia.

[5] For a related discussion in the context of Ethiopia, see Croppenstedt and Demeke (2003).

goods destined for distant urban centers. This discourages the use of chemical fertilizers in particular, which, in Africa, are mostly imported. Consequently, techniques that prove profitable in Asia, where populations and transport systems are denser, are uneconomic in Africa.[6] Other policies can exacerbate this problem. For example, in a comparative study, Jayne et al. (2003) gives examples of policy-driven charges, including port charges and fuel taxes, that add to already high input costs.

That a supply of fertilizer is unavailable locally when all farmers find fertilizer use uneconomical would not be a market failure as such, since it does not imply that resources are misallocated. However, farms and farmers are heterogeneous and it may be the case some farmers who would purchase fertilizer find supplies difficult to obtain when aggregate local demand is not sufficient to attract traders to remote locations. This is especially true when demand is spatially dispersed and highly unpredictable due to variable output prices and weather conditions. Consequently, the locations where fertilizers can be purchased may be distant, leading to higher transaction costs.[7]

A much discussed obstacle to fertilizer use and the take-up of new technologies has to do with risk and missing insurance markets. High-yielding farming methods often require greater up-front investments in inputs and labor that are at risk should crops fail. Consequently, farmers may rationally choose low-productivity approaches that lower investment risks and allow households to diversify limited labor resources (Dercon 1996; Dercon and Christiaensen 2007). For similar reasons, farmers who are willing to take on riskier technologies may be unable to finance up-front input purchases since the same risks faced by farmers may also make formal and informal lenders hesitant to provide credit to farmers and to input suppliers (Jayne et al. 2004; Gregory and Bumb 2006). Potentially, production risks to farmers and lenders could be addressed by insurance or other contractual arrangements. However formal insurance markets are hampered by asymmetric information problems and often poorly supported by weak institutions. Informal approaches for sharing risk can work well when risks are idiosyncratic, but fail in the face of systemic risks like drought (Larson et al. 2004; Skees et al. 2005).

There are additional reasons for low fertilizer use tied to knowledge and information hurdles. Several authors note that levels of education are especially low in Africa which slows the dissemination of new ideas and techniques (Sachs et al. 2004). A related argument is that the institutions charged with dissemination are also weak, partly because of low stocks of human capital and partly due to governance. A second set of arguments are more subtle and focus not on the capacity to learn but on difficulties in discovering appropriate techniques. This argument goes that, because soil and growing conditions are heterogeneous in Africa, an identical technology identically applied can have different outcomes for neighboring households.

[6] See the general discussion in Sachs et al. (2004). Morris et al. (2007) provide estimates based on case studies from Africa.

[7] Gregory and Bumb (2006) provide examples. Xu et al. (2009b) discuss how heterogeneity in local conditions affects the profitability of fertilizer use.

Consequently, advice offered by extension agents gleaned from nearby field stations can be inappropriate for many farmers. For the same reason, the efficacy of learning by example is limited; instead, farmers must learn through experimentation, which is risky and comes with added costs.

As discussed, the empirical model we develop in the next section is motivated by the presumption that Ethiopian farmers decide on fertilizer input levels in a constrained environment. This distinguishes our model from earlier efforts to measure the determinants of fertilizer demand in Ethiopia. Most empirical studies to date, by relying on tobit-model estimates, implicitly characterize zero-valued observations of fertilizer demand as corner solutions – that is, zero-value outcomes are modeled as though they emerged from a self-truncating process.[8] Alternatively, when farmers who would use fertilizer are additionally constrained by poorly performing markets, an additional process is implied. We return to this topic in the next section.

11.3 The Empirical Framework

The starting point for the empirical model is an input demand function, derived from the standard farm household profit maximization conditions via Hotelling's lemma. As discussed, the subsequent choice of an appropriate empirical model depends critically on assumptions about the determinants of zero-value outcomes. In our case, we assume that, in some instances, farmers forgo fertilizer-using techniques because the opportunity costs of doing so are high. But we also allow for instances in which market imperfections further truncate implicit demand. In particular, we include in the estimation of fertilizer demand, an additional step, and we use a set of determinants to control for the potential effects of incomplete markets.[9] The result is panel selection model, which provides estimates of the demand for fertilizer that are adjusted for selection bias.[10] In particular, the demand for chemical fertilizer (f_{it}) is specified as:

$$f_{it} = \beta' x_{it} + \varepsilon_{it} + c_i \geq 0,$$

[8] For example, see studies by Bacha et al. (2001), Croppenstedt and Demeke (2003), Fufa and Hassan (2006), Wubeneh and Sanders (2006), and Alem et al. (2010) have looked at different variants of this issue in Ethiopia.

[9] The alternative is that zero-use outcomes are voluntary corner-solutions that are optimal from the household's perspective.

[10] A related double-hurdle model is often used in cross-section studies, including the study by Croppenstedt and Demeke's (2003) mentioned earlier. The applied component is a mixed-model that combines a probit model of adoption with a truncated model of intensity. The focus of the model is on the conditional demand for fertilizer once the adoption hurdle has been cleared (Duan et al. 1984) As Ahn (2004) points out, the distinction between the hurdle and selection models blurs in the case of panel data when the non-zero population varies over time.

where f_{it} represents chemical fertilizer used in kilograms per hectare, x_{it} is a vector of variables affecting demand for chemical fertilizer, where c_i is a random effect associated with farm *i*, and $\varepsilon_{it} \sim N(0,\sigma^2)$. In practice, the demand equation is only observed when farmers are not additionally constrained by weak markets and the corresponding selection mechanism is given by

$$z_{it}^* = \boldsymbol{\alpha}' 2\mathbf{w}_{2it} + u_{it} + d_i; \, z_{it} = 1\ (z_{it}^* > 0)$$

and w_{it} is a vector of variables that describe the completeness of local markets and the abilities of farmers to cope with incomplete markets, where d_i is a random effect associated with farm *i*, and $u_{it} \sim N(0,1)$. The two random errors are potentially correlated with $\text{corr}(e,u) = \rho$... The random effects, c_i and d_i, are assumed to be bivariate-normally distributed with zero means and standard deviations, σ_c and, σ_d respectively. The random effects are also potentially correlated with $\text{corr}(e,u) = \rho$. A simulated maximum likelihood method is used to estimate the model.[11]

Our estimation strategy follows the parametric approach suggested by Greene (2007), based on efforts by Verbeek (1990), Zabel (1992), Verbeek and Nijman (1992), which we prefer to nonparametric approaches. See, for example, Kyriazidou (1997), and Honore and Kyriazidou (2000). While there is an advantage to the nonparametric estimators, as they are robust to distributional misspecifications of the correlation between the unobserved effects and observed variables, from a practical perspective, the approach is less useful for policy-motivated research. As Greene (2007: E30-2) emphasizes, the parameters of the selection model are not the slopes of the index function, the implicit behavioral component of the model. Consequently, additional and more detailed assumptions about the underlying functional form and parametric family are still needed to interpret the conditional means and partial effects from the model, which are often key to policy discussions.

Turning to the choice of determinants, the arguments in the selection equation are chosen to capture factors that determine access to chemical fertilizer through markets – that is, factors that increase transaction costs or discourage the entry of intermediaries and factors that allow households to compensate for poorly performing markets. As discussed, market risk, low-moisture climates and remoteness are expected to result in thin fertilizer markets and high transaction costs and we include average rainfall (climate), price variability, and the distance to fertilizer distribution centers to take this into account. While credit and insurance markets are weak in Ethiopia, some farmers do gain access to credit and farmers can take additional meliorative actions before abandoning fertilizer-using techniques. Therefore we include access to credit and a set of arguments (literacy, age and participation in extension programs) that indicate the capacity of the household to overcome weak credit and insurance markets. For the same reason, we also include a measure of wealth: the ownership of a house with a corrugated iron roof. Moreover, the dominant role of state actors in rural Ethiopian fertilizer and credit markets is an important

[11] LIMDEP 9 is used to estimate the model. See Greene (2007) for greater detail.

aspect that likely shapes outcomes since fertilizer distribution and credit provision are often handled at a local administrative level. In the face of weak governance and communication systems, holding a position in local government can result in better information or preferential treatment. To test for this, we include a dummy variable to indicate whether the head of the household is a member of the village council.

Input demand is measured in kilos per hectare and we include as determinants farmed area and the relative price of fertilizer. We also include an indicator of whether organic fertilizer is used. Soil moisture is import to the efficacy of fertilizer use and farmers have some scope for adjusting its use as the growing season progresses (Larson and Plessmann 2009). We therefore include weather outcomes measured relative to long-term averages. In the face of imperfect markets, consumption and production are inseparable and endogenous shadow prices enter into the input demand functions (de Janvry et al. 1991). Nevertheless, the input demand equations can be estimated indirectly by using instruments for the shadow prices (Fafchamps and Quisumbing 1999). For this reason, we also include a set of household characteristics: household size, wealth, age and education. To account for the cumulative effects of technology adoption, we include a time effect.[12]

11.4 Data Sources and Descriptive Statistics

The household data used in our analysis comes from a countrywide panel household survey conducted in 2004 and 2006 by the Ethiopian Economic Association (EEA) and the World Bank. The survey consists of about 2,300 randomly selected households in 115 villages (kebeles) stratified by agro-ecological zone and region to ensure coverage of all the agricultural production systems of the country. After adjusting for missing data, there are data on 2,140 matching households. The first round focused on extension services while the core section was the land certification program during the second round, so there are some differences in the questionnaires. Because of missing values, our sample is unbalanced and contains 4,126 observations on 2,104 households.[13]

The survey data was supplemented with district level data on output prices from Ethiopia's Central Statistical Authority and fertilizer prices from the Ministry of Agriculture and Rural Development, Agricultural Marketing Directorate. Data on rainfall for the survey years came from the Ethiopian Meteorological Agency.[14] The climate variable was constructed from average values (1960–1990) of historical spatial data prepared by the Climate Research Unit (Mitchell et al. 2002).

[12] For a discussion of how adoption rates are expected to spread with time, see Feder and Umali (1993).

[13] By design, the survey is meant to capture the diverse geography of Ethiopia, and may not be nationally representative of Ethiopian farmers.

[14] Survey sites were matched with data from the closest weather station.

Table 11.1 Sample statistics on output and fertilizer use in Ethiopia

	2004	2006
Value of output in Birr per ha	1,405.90	1,807.70
Chemical fertilizer in kg	42.00	55.70
Manure in kg	229.00	549.30
Chemical fertilizer in kg per fertilized hectare	116.00	167.00
Percentage of households using:		
Chemical fertilizer	68.50	63.30
Manure	28.10	58.00
Improved seeds	35.60	30.10

For each year of the panel, we construct a measure of weather by taking the ratio of average rainfall for the growing season over the 30-year average for the same period and place.

Consistent with the national figures, the 2006 wave survey data show that the Amhara and Oromia regions have higher proportions of fertilizer user households (61 and 58%, respectively) followed by SNNP and Tigray regions (57 and 55%, respectively). However, in terms of consumption per hectare of fertilized land, SNNP and Tigray regions exhibit higher fertilizer use with per hectare averages of 153 and 144 kg, respectively, followed by Amhara and Oromia regions with per hectare averages of 129 and 111 kg, respectively. The lower intensity of fertilizer use in Oromia may be related to the region's soil type.[15]

Table 11.1 indicates that around 68 and 63% of the households in the sample used one or multiple types of fertilizer (DAP, Urea or mix of the two) in 2004 and 2006, respectively. The average fertilizer consumption increased from 42 to 55.7 kg/ha in between the two rounds when both user and non-user households are considered. A similar trend is also observed when user households alone are considered as average consumption has increased from 61 to 88 kg/ha during the same period. At a plot level, average consumption per hectare of fertilized plot increased from 116 to 167 kg/ha. Though fertilizer use increased significantly, it remained below the recommended 200 kg/ha mark.

As discussed, several factors may explain the low fertilizer adoption rate in Ethiopia. However, even when farmers adopt chemical fertilizers, the application of fertilizer is below recommended rates. The problem is further aggravated by the poor timing of the fertilizer application; a shortage of well-trained agricultural extension workers; and the generally ineffective transmission of the government's research outputs. For instance, while the Ethiopian Agricultural Research Institute recommends the application of urea over two cycles, the 2004 wave of the panel data shows that farmers who apply fertilizer do so all at once and none of the

[15] For instance, the Ethiopian Ministry of Agriculture study recommends a lower dose of fertilizer per hectare (at 150 vs. the usual 200 kg/ha.) for vertisols soil type with improved Durhum wheat seed type.

surveyed households applied urea in two cycles. The combined result of low-dose application and non-optimal timing may explain the low response of output for chemical fertilizer, which in turn, further discourages potential adopters.

The structure of fertilizer markets in Ethiopia has constantly changed since the mid-1990s when, following the fall of the Derg regime, the state monopoly on the distribution of fertilizer was lifted. By 1996, 67 private wholesalers and about 2,300 retailers handled roughly two-thirds of the fertilizer market (World Bank 2006). By the study period, this had changed. By 2004, a combination of companies with potential political affiliation and a public enterprise, the Agricultural Input Supply Corporation, dominated the wholesale market, with cooperatives handling an increased share of the wholesale market by 2006 and fertilizer was distributed to farmers through a combination of extension agents, local governments and cooperatives and some private retailers.[16]

The changing structure of fertilizer markets came as a consequence of a government decision to promote "packets" of high-yielding seeds, fertilizers and extension services. Following a set of successful pilots, the approach was rapidly expanded under Ethiopia's Participatory Demonstration and Extension Training System (PADETS). For the most part, the packets are sold on credit after a 10–35% down payment (DSA 2006). Credit is extended through the Commercial Bank of Ethiopia through cooperatives, local authorities and micro-lending institutions, which also handle record keeping and the collection of interest and principal.

Timely and adequate supply of fertilizer is one of the major problems reported by a significant proportion of the households surveyed in the 2004 round. More than 70% of the households reported that fertilizer is often supplied late and around 40% of the households reported that supplies were inadequate. The survey results also point to high fertilizer price and tight credit repayment schedules as problems that constrain fertilizer use.

Yet, despite the public sector's dominant role, a study by Heisey and Norton (2007) suggests that, on average, the margins between farmgate and import prices for fertilizers in Ethiopia compare well to similarly calculated margins in South Africa and other African economies – although the margins are still high in comparison to Asia or Latin America. The authors speculate that this is because of the implicit subsidy of distributing fertilizer supplies through extension agents and other local institutions. Even so, Mezgebo (2005) notes large regional differences in fertilizer margins, suggesting a diversity in how local markets function.

Credit plays an important role in acquiring fertilizer. In the 2004 survey wave, 61.5% of households received fertilizer on credit, while cash purchases accounted for only 37.7% (Table 11.2). Moreover, the pattern of credit finance across the different regions is quite skewed. While credit finances more than 80% of purchases in Oromia, shares of credit-purchases drop to 35.7 and 40% in SNNP and Amhara regions, respectively. In Tigray, credit finances around 63% of fertilizer purchases.

[16] A 2006 report by the Ethiopian Economic Association/Ethiopia Policy Research Institute (EEA/EEPRI) estimated that the public sector handled about 70% of the retail market.

Table 11.2 Procurement channels for organic and chemical fertilizer in Ethiopia in 2004 (% of total)

	Organic fertilizer	Chemical fertilizer
Cash	10.3	37.7
Credit	10.2	61.5
Own (Left over)	–	0.1
Land owner	13.5	0.1
For free	9.7	0.6
Own animal dung	56.4	–

In terms of access to credit (from all type of sources such as friends, banks, microfinance and cooperatives), more households in Amhara and Oromia have access to credit than households in Tigray and SNNP. The average obtainable credit ceiling is also higher in Oromia and Amhara.

As mentioned, supplying fertilizer and facilitating credit for its acquisition has been an important component of Ethiopia's PADETS program. In the 2004 survey, households were asked if they participated in extension program and how productivity gains from the package compared with the traditional practices. More than 64% of the sampled households participated in the package and 95% of the households found the new technology more profitable than the traditional one. More than 50% of the households responded that production increased by half while 20% of households reported production increase of more than 50%. Only a small proportion of households (7%) felt that the extension package was riskier than the traditional practices and only 11% of the households opted out of the extension package.

Among these, 33% of households reported that a lack of credit was a major reason why they withdrew from the program. That credit would be constrained is perhaps understandable, as 62% of households ran into difficulty repaying the input loan on one or more occasion. The three major reasons for default on the input loans are low yield due to rain failure, low output price and timing of repayment (being forced to repay immediately after harvest when output prices are depressed). The efficacy of fertilizer also depends on land preparation and fertilizer application may be constrained by a lack of essential farm implements. The Ethiopian Agricultural Research Institute best practice guides suggest three to five cycle of pre-harvest land preparation to get the optimal results from fertilizer applications, but 57% of households report that they lack land-preparation tools.

In the 2006 wave of the survey, the share of chemical fertilizers is around 18% of the total value of outputs while the share of all purchased inputs (including expenditures on chemical fertilizer, improved and traditional seeds, hired labor, transportation, rented in oxen and tractor) is around 27%. The figures may be indicative of the profitability of chemical fertilizers and other purchased inputs. However, adoption seems to be lower as only 45% of the cropped area is covered by chemical fertilizer; and the intensity of fertilizer use is on average below the recommended dose of 200 kg/ha.

Table 11.3 Estimation results of fertilizer application function in Ethiopia

	Coefficient	Std. error
Selection equation parameters		
Household size	0.103***	0.012
Age of household head	–0.007***	0.002
Education of household head	1.132***	0.212
Credit	0.336***	0.062
Extension	0.881***	0.059
Fertilizer distribution center	0.253***	0.061
Climate	0.002***	0.000
Price risk	–0.081**	0.039
Wealth	0.122**	0.059
Village Council membership	0.412***	0.085
Selection-corrected regression parameters		
Relative price	–14.272*	7.816
Use of organic fertilizer	9.916*	5.867
Area	–0.897***	0.159
Household size	0.021	1.131
Age of household head	0.078	0.215
Education of household head	24.532**	11.313
Wealth	9.736*	5.486
Weather	–14.246	9.399
Time effect	39.847***	7.210
Error structure		
Disturbance standard deviation, σ	0.008	0.641
Correlation between regression and probit, €	–0.160***	0.043

Note: The model was estimated using LIMDEP version 9.0
*Significant at 10%; **Significant at 5%; ***Significant at 1%

11.5 Empirical Results

Estimation results from the panel selection model are given in Table 11.3. In general, the model fits well and the parametric values and related statistical test are consistent with the notion that fertilizer markets in Ethiopia are incomplete.[17] At the same time, the model suggests that, when circumstances are favorable, the demand for fertilizer by farmers is sensitive to relative prices and behaves as expected.

[17] On a technical note, errors associated with two components of the model are correlated, as expected. See the lower panel of Table 11.3.

11.5.1 Selection Equation

The first three determinants in the selection equation – household size, the age and state of literacy for the household head – relate to the household's ability and capacity to implement farming techniques that use chemical fertilizers. The results indicate that larger households and households headed by younger and literate farmers are more likely to use fertilizers. The negative effect of age may be due to a reluctance to accept new technologies and lack of technical capability to use chemical fertilizers effectively. In addition, for chemical fertilizers to give the maximum return, intensive pre-harvest land preparation and keeping the optimal application time are required. Advancing age may limit the ability of farmers to fulfill these essential and physically demanding requirements in the face of weak labor market. In a related way, the positive effect of household size may be the consequence of a more abundant supply of family labor. In addition, large families may adopt highly productive inputs to fulfill higher food requirement in cases where expanding land holding is restricted by imperfect or missing land markets. (We return to this topic below.) The negative effect of illiteracy may also be the combination of different effects. One channel is that illiterate farmers may not have the minimum required knowhow for effective application of chemical fertilizer. The other possible channel, suggested by Pitt and Sumodiningrat (1991) is that illiterate farmers are more risk averse and, consequently, less willing to take on the additional risks associated with fertilizer use (Table 11.2).

The next set of determinants indicates that better access to fertilizer markets and supporting services increases the probability of adopting chemical fertilizer. Access to credit, participation in the extension program and proximity to fertilizer distribution centers all increase the probability of adopting chemical fertilizer in the estimated model. As discussed, fertilizer markets are more likely to be incomplete in dryer climates that disfavor fertilizer use and in areas where relative prices are more volatile. The estimation results are consistent with this view as the probability of adoption rises with average rainfall during the growing season and falls with price volatility.

Results associated with the last two determinants are consistent with the notion that wealth and social capital or authority help farmers mitigate incomplete credit and insurance markets. Greater wealth is expected to increase the credit-worthiness of farmers and also allow farmers to self-finance. Wealthier farmers are also better able to self-insure against temporary shortfalls in income.[18] Our results are consistent with these expectations and suggest that wealthier households are more likely to adopt chemical fertilizers. The selection equation also indicates that being in a position of local authority increases the probability of adopting fertilizer-using techniques. Because fertilizer and credit supplies are channeled through a variety of local public and cooperative agencies, holding political power at the local administrative unit may facilitate access to credit and fertilizer. For one, politically active households may have better information on state-sponsored programs. As discussed,

[18] There is some empirical evidence that wealthier households tend to avoid riskier inputs (Pitt and Sumodiningrat 1991). In our study, this does not appear to be a dominant effect.

to obtain fertilizer credit, farmers are required to pay a certain percentage of the fertilizer cost as down payment, but the size of the down payment is not uniform. For example, in the Oromiyo region, the down payment is generally 25–30%, but can run as high as 60% and some farmers receive loans without a down payment. Discretion is given to local agencies, so the system is potentially open to local influence (DSA 2006). At the same time, farmers that are politically active may also have unobservable entrepreneurial characteristics that make them more likely to adopt new technologies and also adopt leadership roles. Our result may capture any or all of these channels.

11.5.2 Demand for Chemical Fertilizer

The middle panel of Table 11.3 gives the parameters of the fertilizer demand equation, adjusted for selection. The first result is that Ethiopian farmers in the sample were sensitive to the price of fertilizer relative to output prices. This is consistent with theory and also consistent with literature suggesting that, by raising fertilizer costs and reducing farm-gate prices for surplus production, high transport costs dampen the use of fertilizers by African farmers. The results also indicate that farmers who use manure also use more chemical fertilizer. To some extent, organic and chemical fertilizers are substitutes, since both are sources of plant nutrients. However, it is also true that the efficacy of chemical fertilizers depends on the organic content of the soils. In Africa, where soils are often depleted, consistent practices of fallowing or using manure are required to build up the soils carbon content, resulting in a complementary relationship between the two.[19] Our analysis suggests this later relationship is dominant in our sample.

The results also suggest a statistically significant and negative association between the intensity of fertilizer and farm size. This is contrary to the positive correlation between land size and fertilizer use in Feder et al. (1985), but consistent with the results reported in Croppenstedt and Demeke (2003) in the case of Ethiopia and Nkonya et al. (1997) in the case of Tanzania. This may have to do with a willingness by farmers with smaller holdings to invest greater time and effort into their limited holdings. In addition, Nkonya et al. (1997) suggest that farmers with larger holding sometimes hedge risks by applying on a smaller proportion of their land, thereby reducing average application rates.

Parameter estimates associated with age and the literacy of the household head, which were significant in the adoption of fertilizer-use, were not statistically significant determinants of the intensity of fertilizer use.[20] In the case of household size, this suggests that the channel through which household size affects adoption

[19] Marenya and Barrett (2009a) and Matsumoto and Yamano (2009) provide evidence that the organic content of soils, which can be built up through the application of organic fertilizer, increases the efficacy of chemical fertilizers.

[20] Nkonya et al. (1997) report a similar finding.

has more to do with food demand that weak labor markets. From a technical perspective, excluding these variable from the model does not materially affect the parameters of the remaining determinants (see Appendix Table 11.A.1). In contrast, literacy does have a positive impact on fertilizer use.

As discussed, wealth can mitigate weaknesses in credit and insurance markets. Separate from the capacity to take-on risk, wealth may also be associated with a greater preference for risk. Consistent with these arguments, the parameter on wealth in the demand equation is positive and statistically significant, as it was in the adoption equation. In general, fertilizer applications are made before weather outcomes are known; although farmers can make some adjustments when fertilizer is applied in a series of doses. As discussed, this latter practice is not common in Ethiopia and the results suggest that, although climate is an important determinant of whether fertilizer-using practices are adopted, contemporaneous weather outcomes do not affect fertilizer demand in a statistically measureable way. And finally, the time effect estimate suggests that fertilizer demand increased between the first and second waves, even after adjusting for changes in other determinants. This provides indirect evidence of a positive trend in the adoption of fertilizer-using technologies that is consistent with government efforts to promote more productive technologies.

11.6 Conclusion

Cereal yields and fertilizer use are low in Ethiopia relative to other regions and there is a widely expressed view that a significant change in Ethiopian agriculture is impossible without a significant increase in the use of fertilizers. Moreover, because supplies of organic fertilizer and the scope for increased fallowing are limited in Ethiopia, any significant growth in fertilizer use will depend on an increase in the application of chemical fertilizers. Even though research from field studies and pilot programs in Ethiopia shows that the economic returns to chemical fertilizer are sometimes large relative to its cost, the application rates are nonetheless low. Based on a panel household data from 2004 to 2006, this chapter examines the role of missing and imperfect markets that affects farmer choices about chemical fertilizer applications in Ethiopia.

Taken together, the results suggest that fertilizer markets are not altogether absent in rural Ethiopia and, for some farmers, work as expected. At the same time, the modeling results provide evidence that high transport costs, limitations in complementary markets for credit and insurance, adverse climate and illiteracy all conspire to limit the adoption of chemical fertilizers in Ethiopia. Moreover, the combination of factors that promote or impede effective fertilizer markets differs among locations, making it difficult to find a single production technology that is uniformly profitable. One implication is that inconsistencies can be expected between field studies and pilots finding large returns to the use of fertilizers in Ethiopia and studies finding that fertilizer applications are uneconomical since market conditions are crop and place dependent. This in turn has consequences for incomes, food security and the sustainability of soils.

The results suggest that households with greater wealth, human capital and greater authority can apply those assets to overcome hurdles that stand in the way of making greater use of fertilizers. The finding offers some encouragement, but it also implies a self-enforcing link between low agricultural productivity and poverty, since low-asset households are less able to compensate for weaknesses in Ethiopian fertilizer markets. The study suggests that the provision of extension services can be effective in helping households participate in fertilizer markets and that lowering transport costs can improve the intensity of fertilizer use in Ethiopia by lowering the cost of fertilizer and boosting the value of outputs at the farmgate.

Appendix

Table 11.A.1 Estimation results of fertilizer application function in Ethiopia from the base model and alternative model

	Base model		Alternative model	
	Coefficient	Std. error	Coefficient	Std. error
Selection equation parameters				
Household size	0.103***	0.012	0.101***	0.012
Age of household head	−0.007***	0.002	−0.007***	0.002
Literacy of household head	1.132***	0.212	1.049***	0.210
Credit	0.336***	0.062	0.338***	0.062
Extension	0.881***	0.059	0.866***	0.059
Fertilizer distribution center	0.253***	0.061	0.268***	0.061
Climate	0.002***	0.000	0.002***	0.000
Price risk	−0.081**	0.039	−0.084**	0.032
Wealth	0.122**	0.059	0.104*	0.059
Village council membership	0.412***	0.085	0.422***	0.085
Selection-corrected regression parameters				
Relative price	−14.272*	7.816	−14.693**	7.182
Use of organic fertilizer	9.916*	5.867	9.948*	5.875
Area	−0.897***	0.159	−0.895***	0.158
Household size	0.021	1.131		
Age of household head	0.078	0.215		
Literacy of household head	24.532**	11.313	23.473**	11.242
Wealth	9.736*	5.486	9.944*	5.366
Weather	−14.246	9.399	−13.889	9.320
Time effect	39.847***	7.210	40.377***	6.936
Error structure				
Disturbance standard deviation, σ	0.008	0.641	0.008	0.566
Correlation between regression and probit, €	−0.160***	0.043	−0.170***	0.041

Note: The model was estimated using LIMDEP version 9.0
*Significant at 10%; **Significant at 5%; ***Significant at 1%

References

Africa Union (2006) Abuja declaration on fertilizer for the African Green Revolution. Paper presented at the Africa fertilizer summit, Abuja, 13 June 2006

Ahn J (2004) Panel data sample selection model: an application to employee choice of health plan type and medical cost estimation. Paper presented at econometric society 2004 far eastern meetings, Econometric Society

Alem Y, Bezabib M, Kassie M, Zikhali P (2010) Does fertilizer use respond to rainfall variability? Panel data evidence from Ethiopia. Agric Econ 41(2):165–175

Bacha D, Aboma G, Gemeda A, Groote HD (2001) The determinants of fertilizer and manure use in maize production in Western Oromiya, Ethiopia. In: Friesen DK, Palmer AFE (eds) Proceedings of the seventh Eastern and Southern Africa regional maize conference. CIMMYT, Nairobi Kenya

Bekele T, Höfner W (1993) Effects of different phosphate fertilizers on yield of barley and rape seed on reddish brown soils of the Ethiopian highlands. Fertilizer Res 34(3):243–350

Byerlee D, Spielman DJ, Alemu D, Gautam M (2007) Policies to promote cereal intensification in Ethiopia: a review of evidences, vol 00707, Discussion paper. International Food Policy Research Institute, Washington, DC

Croppenstedt A, Demeke M (2003) Technology adoption in the presence of constraints: the case of fertilizer demand in Ethiopia. Rev Dev Econ 7(1):58–70

CSA (2008) Agricultural sample survey 2007/2008: farm management practices, vol 3. The Federal Democratic Republic of Ethiopia Central Statistical Agency, Addis Ababa

de Janvry A, Fafchamps M, Sadoulet E (1991) Peasant household behaviour with missing markets: some paradoxes explained. Econ J 101(409):1400–1417

Demeke M, Kelly V, Jayne TS, Said A, Vallee JCL, Chen H (1998) Agricultural market performance and determinants of fertilizer use in Ethiopia, vol 10, Grain market research project working paper. Ministry of Economic Development and Cooperation, Addis Ababa

Dercon S (1996) Risk, crop choice, and savings: evidence from Tanzania. Econ Dev Cult Change 44(3):485–513

Dercon S, Christiaensen L (2007) Consumption risk, technology adoption and poverty traps: evidence from Ethiopia, vol 265, CSAE working paper series. Centre for the Study of African Economies, Oxford

Development Studies Associates (DSA) (2006) Study on improving the efficiency of input market, vol 1. Development Studies Association, Addis Ababa

Duan N, Manning W, Morris C, Newhouse J (1984) Choosing between the sample selection model and the multi-part model. J Bus Econ Stat 2(3):283–289

Fafchamps M, Quisumbing AR (1999) Human capital, productivity, and labor allocation in rural Pakistan. J Hum Res 34(2):369–406

Feder G, Umali DL (1993) The adoption of agricultural innovations: a review. Technol Forecast Soc Change 43(3 and 4):215–239

Feder G, Just RE, Zilberman D (1985) Adoption of agricultural innovations in developing countries: a survey. Econ Dev Cult Change 33(2):255–298

Fufa B, Hassan R (2006) Determinants of fertilizer use on maize in eastern Ethiopia: a weighted endogenous sampling analysis of the extent and intensity of adoption. Agrekon 45(1):38–49

Greene WH (2007) Limdep version 9.0 econometric modeling guide. Econometric Software, Bellport

Gregory DI, Bumb BL (2006) Factors affecting supply of fertilizer in Sub-Saharan Africa, ARD Working Paper 24, Discussion paper. Agricultural and Rural Development, World Bank, Washington, DC

Heisey PW, Norton G (2007) Fertilizers and other farm chemicals. In: Evenson RE, Pingali PL (eds) Handbook of agricultural economics, vol 3. Elsevier, Amsterdam

Henao J, Baanante C (2006) Agricultural production and soil nutrient mining in Africa: implications for resource conservation and policy development. Background paper for African fertilizer summit, Abja

Honore BE, Kyriazidou E (2000) Panel data discrete choice models with lagged dependent variables. Econometrica 68(4):839–874

Howard J, Crawford E, Kelly V, Demeke M, Jeje JJ (2003) Promoting high-input maize technologies in Africa: the Sasakawa-Global 2000 experience in Ethiopia and Mozambique. Food Policy 28(4):335–348

Jayne TS, Govereh J, Wanzala M, Demeke M (2003) Fertilizer market development: a comparative analysis of Ethiopia, Kenya, and Zambia. Food Policy 28(4):293–326

Jayne TS, Yamano T, Nyoro J (2004) Interlinked credit and farm intensification: evidence from Kenya. Agric Econ 31(2 and 3):209–218

Kyriazidou E (1997) Estimation of a panel data sample selection model. Econometrica 65(6):1335–1364

Larson DF, Plessmann F (2009) Do farmers choose to be inefficient? Evidence from bicol. J Dev Econ 90(1):24–32

Larson DF, Anderson JR, Varangis P (2004) Policies on managing risk in agricultural markets. World Bank Res Obs 19(2):199–230

Marenya PP, Barrett CB (2009) Soil quality and fertilizer use rates among smallholder farmers in Western Kenya. Agric Econ 40(5):561–572

Matsumoto T, Yamano T (2009) Soil fertility, fertilizer, and the maize Green Revolution in East Africa, vol 5158, World Bank policy research paper. World Bank, Washington, DC

Mezgebo T (2005) Structure, conduct and performance of fertilizer marketing industry of Ethiopia, Draft paper. Ethiopian Development Research Institute, Addis Ababa

Mitchell TD, Hulme M, New M (2002) Climate data for political areas. Area 34(1):109–112

Morris ML, Kelly VA, Kopicki RJ, Byerlee D (2007) Promoting increased fertilizer use in Africa: lessons learned and good practice guidelines. World Bank, Washington, DC

Nkonya E, Schroeder T, Norman D (1997) Factors affecting adoption of improved maize seed and fertilizer in Northern Tanzania. J Agric Econ 48(1):1–12

Otsuka K, Estudillo JP, Sawada Y (2009) Rural poverty and income dynamics in Asia and Africa. Routledge, London

Pitt MM, Sumodiningrat G (1991) Risk, schooling and the choice of seed technology in developing countries: a meta-profit function approach. Int Econ Rev 32(2):457–473

Sachs J, Mcarthur J, Schmidt-Traub G, Kruk M, Bahadur C, Faye M, Mccord G (2004) Ending Africa's poverty trap. In: Brainard WC, Perry GL (eds) Brookings papers on economic activity 1. Brookings, Washington, DC, Spring 2004

Skees J, Varangis P, Larson DF, Siegel P (2005) Can financial markets be trapped to help poor people cope with weather risks? In: Dercon S (ed) Insurance against poverty. Oxford University Press, Oxford

Verbeek M (1990) On the estimation of a fixed effects model with selectivity bias. Econ Lett 34(3):267–270

Verbeek M, Nijman T (1992) Testing for selectivity bias in panel data models. Int Econ Rev 33(3):681–703

World Bank (2006) National fertilizer sector project: implementation completion report. World Bank, Africa Region, Washington, DC

Wubeneh NG, Sanders JH (2006) Farm-level adoption of sorghum technologies in Tigray, Ethiopia. Agric Syst 91(1 and 2):122–134

Xu Z, Guan Z, Jayne TS, Black R (2009) Factors influencing the profitability of fertilizer use on maize in Zambia. Agric Econ 40(4):437–446

Zabel JE (1992) Estimating fixed and random effects models with selectivity. Econ Lett 40(3):269–272

Chapter 12
Technology Adoption in Agriculture: Evidence from Experimental Intervention in Maize Production in Uganda

Tomoya Matsumoto, Takashi Yamano, and Dick Sserunkuuma

Abstract To investigate the impact of a policy intervention on technology adoption by small scale farmers, we conducted sequential field experiments on maize production in Uganda in 2009, in which we provided a free maize start-up package to sample farmers in randomly selected villages. Subsequently, we conducted a sales experiment in each of the treatment and control villages involving all the sample households and their randomly selected neighbors in the treatment villages. The findings of this study suggest that the distribution of modern agricultural inputs has a significantly positive effect on their adoption by farmers who have little experience in their use. We also find a large impact of the credit intervention and significant spillover effects. In short, a small-scale intervention could have a large impact on farmers' demand for modern inputs and maize yield.

Keywords Technology adoption • Experimental intervention • Maize production • Uganda • Green Revolution • High yielding varieties • Chemical fertilizer • Input subsidy • Small scale farmers

---

T. Matsumoto (✉)
Division of Policy Research, National Graduate Institute for Policy Studies, 7-22-1 Roppongi, Minato-ku, Tokyo 106-8677, Japan
e-mail: tmatsumo@grips.ac.jp

T. Yamano
Division of Social Science and Economics, International Rice Research Institute, CG Block, NASC Complex, DPS Marg, Pusa, New Delhi, India
e-mail: t.yamano@cgiar.org

D. Sserunkuuma
Department of Agricultural Economics and Agribusiness, Makerere University, P.O. Box 7062, Kampala, Uganda
e-mail: Sserunkuuma@agric.mak.ac.ug

K. Otsuka and D.F. Larson (eds.), *An African Green Revolution: Finding Ways to Boost Productivity on Small Farms*, DOI 10.1007/978-94-007-5760-8_12,

12.1 Introduction

Successful technology adoption in agriculture is the key to the success of the Green Revolution. In Asian and Latin American countries, the dissemination of modern agricultural technology in the form of chemical fertilizers and high-yielding varieties has boosted crop yield drastically since the 1960s (Kikuchi and Hayami 1985; Evenson and Gollin 2003). In contrast, agricultural productivity in most Sub-Saharan African countries has been stagnant. Researchers and policy makers agree that, to realize a Green Revolution in Africa, an increase in the use of fertilizers and improved seed technologies is inevitable (Morris et al. 2007). Indeed, input subsidy programs to boost agricultural productivity are being reconsidered by many African countries (Denning et al. 2009; Minot and Benson 2009), which abolished subsidies and state monopolies on input distribution as part of the structural adjustment programs in the late 1980s, due to high fiscal cost and ineffective implementation (Kheralah et al. 2002). To avoid repeating the past mistakes, there is need to carefully examine the effectiveness of input dissemination programs and to find efficient ways to implement them, with due consideration of country-specific factors that could affect the fate of these programs.

To investigate the impact of a possible policy intervention on technology adoption by small scale farmers, we conducted sequential field experiments on maize production in Uganda in 2009. First, prior to the first cropping season of 2009, we conducted a randomized experiment that involved distribution of a free maize start-up package to each sample farmer in villages that were randomly selected from the sites where we conducted panel surveys in 2003 and 2005. In addition to the maize package, sample households in the treatment villages received a 2-h training session on the use of the provided inputs, unlike their cohorts in control villages. Subsequently, in the intermediate period between the 1st and 2nd cropping seasons, we conducted a sales experiment in each of the treatment and control villages involving all the sample households and their randomly selected their neighbors in the treatment villages. The purpose of the sales experiment was to collect information on input demand of the households for hybrid seeds and fertilizer and how it differs among farmers in control and treatment villages. The neighbors of the treatment households were included to measure knowledge spill-over effects.[1] In the sales experiment, we offer three different price levels with and without a credit option.

Using the information gathered at the sales meeting, we estimated the demand curves for each input for the different types of households with and without the credit option. The results show that, first, the average purchase quantity for the treatment households is much higher than that of the control households, while that

[1] Because of the reflection problem in the estimation of spill-over effects (Manski 1993), the identification of such effects using survey data is not an easy exercise (Conley and Udry 2001; Munshi 2004). However, our approach is experimental and hence less susceptible to the reflection problem.

of the neighboring households lies in between. For instance, the average quantity of hybrid seed purchased by the treatment households at the market price was 2.1 kg/ha; that by the control households was 1.1 kg/ha; and that by the neighbor households was 1.3 kg/ha. We observed a similar pattern for fertilizers. Second, the results indicate low price elasticity: the average purchase quantity for hybrid seed increased by 5–9%, following a 10% price discount. Third, the credit option had a large impact on the purchase quantities of all inputs and for all types of households. For example, the average quantity of hybrid seed purchased at the market price by the control households increased by 68% when the credit option was made available; and by 59 and 70% for the treatment households and their neighbors, respectively.

We also simulate the yield gain from using modern inputs purchased at the sales experiment. The results show that discounting the input price would have very minor impact on yield, while credit would have large impact. The yield would more than double if farmers switched from the local variety to hybrid seed and applied chemical fertilizers at the level purchased by treatment households when credit is made available.

The findings of this study suggest that the distribution of modern agricultural inputs has a significantly positive effect on their adoption by farmers who have little experience in their use. The intervention had a spillover effect on the neighbors' adoption, too. We also find a large impact of the credit intervention, which suggests that farmers would drastically increase the use of inputs if credit was offered. The impact of credit was largest among treatment households who obtained the free trial packages in the previous season because of the acquired knowledge on usage and profitability of the modern inputs through the intervention. This shows that a small-scale intervention could have a large impact on farmers' demand for modern inputs.

The rest of this chapter is organized as follows. Section 12.2 describes the current farming system in Uganda. Section 12.3 discusses a series of the interventions that we have conducted in Uganda since January 2009. Section 12.4 discusses the village level and household level data by type of household in the sample. Section 12.5 reports the key results of the sales experiment and the yield prediction based on the quantities of the modern agricultural inputs purchased at the sales experiment, and Sect. 12.6 concludes the chapter.

12.2 Maize Production in Uganda

In Africa, the level of chemical fertilizer use and the adoption rate of high-yielding maize varieties are generally much lower than in most Asian countries (see Chap. 8). However, there is also large variation across African countries. One example is the interesting contrast in the use of modern inputs on maize production between two neighboring countries, Kenya and Uganda (Chap. 9). Table 12.1 compares input use on maize production between Kenya and Uganda using the data from the RePEAT

Table 12.1 Comparison of input use in maize production between Kenya and Uganda

Plot level summary statistics	Kenya 2004/2007	Uganda 2003/2005
Hybrid seed use: (%)	59.0	4.9[a]
	(49.2)	(21.6)
Average inorganic fertilizer application (kg/ha)	94.7	2.4
	(124.5)	(18.9)
Average organic fertilizer application (kg/ha)	1,935	86
	(4,835)	(768)

Source: Matsumoto and Yamano (2009). Standard deviations are in parentheses
[a] This number is recalculated in this study because Matsumoto and Yamano (2009) did not differentiate the types of the improved seed. It is obtained as the proportion of maize plots where seeds with the price being more than or equal to 3,000 Ush were planted. That is, we assumed that the seeds whose price is more than or equal to 3,000 Ush were hybrid

survey in Kenya and Uganda.[2] Only 6% of farmers in Uganda planted hybrid maize seed and applied negligible amount of chemical and organic fertilizers on the maize plots in the survey years. In contrast, about 60% of Kenyan farmers planted hybrid seeds and used 94 kg/ha of chemical fertilizers and more than 1 t/ha of organic fertilizers on the maize plots. They have been using such inputs for a decade or even longer.[3] As a consequence, the average maize yield is higher in Kenya than in Uganda (see Chap. 9).

Uganda is a land-locked country and imports most of the modern inputs for crop production from overseas through Kenya. Due to the high transportation costs, the market price of those inputs is higher in Uganda than in Kenya (Omamo 2003), and the converse is true for profitability. The low profitability of using modern inputs is one of the major reasons for their low adoption rate and application level among Ugandan farmers (Chap. 9). In addition, in the past, the issue of land scarcity was less severe in Uganda than Kenya, owing to the presence of uncultivated land and favorable climatic conditions for crop production in wider areas in Uganda than Kenya. Thus, Ugandan farmers had little incentive to use modern inputs for intensive farming. In addition, because of the low potential demand for these inputs, the supply network is less developed in Uganda than Kenya.

However, the conditions have of late changed significantly in Uganda. First, because of high population pressure and the limitation of expansion of arable land through land clearing, arable land is increasingly becoming scarce and the average

[2] RePEAT stands for Research on Poverty, Environment, and Agricultural Technologies which is a research project by a research team of GRIPS and Foundation for Advanced Studies on International Development (FASID, Japan) aiming to identify constraints and effective technologies to reduce poverty in East African countries, especially, Kenya, Uganda, and Ethiopia, through empirical analyses based on field data on agricultural production collected from farm households. RePEAT also indicates our intention to repeat data collection to construct panel data over a long time (See Yamano et al. 2004 for more details).

[3] The RePEAT surveys in Kenya mainly cover areas in Central, Rift Valley, Nyaza, and Western province where population density is relatively high and crop production is relatively suitable.

land size per household has been decreasing.[4] Second, recent hikes in crop prices are prompting farmers to change their perception of crop production. Farmers have started to consider crop production as a business enterprise rather than purely for subsistence. These factors have created high potential demand for intensive farming methods among crop farmers in Uganda. Since these modern approaches require a different set of skills, a lack of knowledge on their usage and profitability might be a large deterrent to their adoption by farmers with little experience and knowledge. Thus, we expect that a small intervention involving one-time material support and training on the usage of such modern inputs may have a large impact on the adoption of modern agricultural inputs among Ugandan farmers.[5]

12.3 Experimental Design

This experimental intervention was carried out as a part of the Global Center of Excellence (GCOE) Project of National Graduate Institute for Policy Studies (GRIPS), Japan in collaboration with Makerere University, Uganda.[6] The farmers selected for this intervention were chosen from the sample households of a panel survey called the Research on Poverty, Environment, and Agricultural Technology (RePEAT) survey. The RePEAT survey in Uganda covers the Eastern, Central, and Western regions and consisted of 940 households in 2003 and 894 households in 2005 from 94 villages (Fig. 12.1).[7] For convenience, we refer to these households as "RePEAT households" hereafter.

12.3.1 Randomized Experiment: Maize Package Distributions

The intervention was a sequential randomized-controlled trial. In the first exercise, in February and March 2009, prior to the first cropping season, we distributed free maize inputs to 504 RePEAT households. These households reside in 61 villages (26 in Eastern, 20 in Central, and 15 in Western region) that were randomly chosen from

[4] The estimate of annual population growth rate in 2005 in Uganda was 3.58% (world rank 11th) while that of Kenya was 2.36% (world rank 42nd).

[5] Duflo et al. (2008,2009) focus on the self-control problem of farmers in terms of ability to save for the purchase of inputs in subsequent planting season in order to explain the low application rate of chemical fertilizer on maize production in Western Kenya. In the context of farming in Uganda, however, it may not be a major reason to explain the low adoption and application rate of modern inputs because only few farmers have had experience of using such inputs. Those who do not know about the inputs would not struggle with a decision whether they save for inputs or not.

[6] The GCOE project of GRIPS was financially supported by Ministry of Education, Culture, Sports, Science and Technology, Japan.

[7] The smallest local administrative unit in Uganda is LC1. We call the LC1 "village" in this chapter.

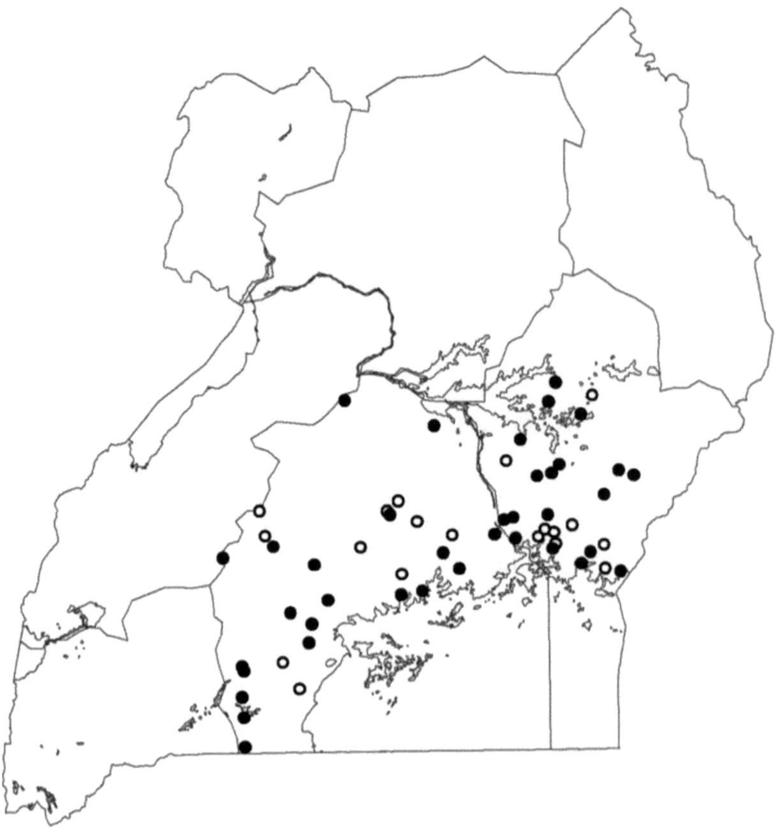

Fig. 12.1 Survey villages in Uganda (**Black circles*: treatment village/*White circles*: control village)

the villages covered by the RePEAT survey. For convenience, we call the 61 villages "treatment villages" to distinguish them from the remaining 30 villages (13 in Eastern, 10 in Central, and 7 in Western region) that are referred to as "control villages."[8] The free inputs distributed to the farmer in the treatment villages comprised of 2.5 kg of hybrid seed, 12.5 kg of base fertilizer, and 10 kg of top-dressing fertilizer, which are the recommended input levels for growing a quarter acre of maize.[9] In addition, a

[8] Three out of the 94 RePEAT survey villages are excluded from this experimental intervention. Two of them are located in Kapchowa district closed to the Kenyan border. Their application rate of chemical fertilizers and the adoption rate of hybrid maize seed were exceptionally high in the 2005 RePEAT survey. The other village has been involved in Millennium Village Project by United Nations since 2008. These villages are very different from others in terms of experience of the use of modern inputs.

[9] The market value of these inputs was 52,500 USH (26.8USD) in Febraury 2009.

2-h training session on the use of the modern inputs was given by an extension worker to the sample households in the treatment villages.[10]

We chose larger numbers of the treatment villages than the control villages because we expected that the effect of the experimental intervention on the adoption behavior in the subsequent seasons would differ across the treatment villages depending on the yield performance of the free inputs due to regional factors such as climate, soil conditions, etc. Thus, it was preferred to have wider variation across the treatment villages and, hence, the choice of more villages as the treatment villages.

12.3.2 Sales Experiment

The second exercise occurred during the intermediate period between the first cropping season and the subsequent season, in which we revisited 46 treatment and 23 control villages in the Eastern and Central regions[11] to sell the same inputs that were previously provided for free to the sample farmers. We held a sales event in each of the treatment and control villages and invited all the sample households as well as randomly selected neighbors of the sample households in the treatment villages (called the neighbor households hereafter).[12] The purpose of the sales experiment was to gather information on input demand of the participating households for hybrid seeds and fertilizer and to make comparison across the three groups—the control, treatment, and neighbor households.

To obtain information of their demand in response to changes in price, we used a "price contingent order form" which asked farmers to fill out how much of each input they would buy at different discount levels (see Appendix). Three discount rates from the market price were offered, namely, 0, 10 and 20%.[13] Which discount rate would be applied to the actual sales was not determined until they filled out the order form completely, although the participants were informed at the beginning of the sales experiment that one of the discount rates would be randomly chosen later and that they would have to pay for the amounts indicated on the form at the chosen discounted price.

---

[10] Figure 12.A.1 shows the time line and the number of sample households involved in each project for the RePEAT study. In the initial RePEAT household survey in 2003, there were ten households in each village. Because of attrition, 106 households were dropped out in the 61 treatment villages.

[11] The villages in Western region were excluded for the second exercise because of time and budget constraint. Thus, in this study, we use the samples from Eastern and Central regions only.

[12] To select the neighbor households in the treatment villages, we asked each of the target households to list 5–10 households as his/her neighbors, and then randomly selected one household from the list. We expect that this selection procedure of the neighbors mitigates the selection bias issue which would occur if the target households in the treatment villages invite households to which they think our exercise would be useful or beneficial.

[13] We were interested in collecting the information on the purchase quantities at wider range of discount rates. However, because of the possibility of the participants making large profit by reselling inputs to other residents or even input dealers, we decided not to offer higher discount rates.

We also used an order form for credit purchase, on which participants indicated how much of each input they would buy if credit was available. In the proposed credit scheme, the participants were requested to pay the balance, that is, the total payment with interest minus the initial payment, at the end of the subsequent season.[14]

After a group of the participants had filled out the forms, the group leader drew a ball from a bingo cage to randomly determine the discount rate; and a second ball to determine whether the credit option was actually available to the group or not. The chance of winning the credit option was one in ten. Finally, at the end of the sales experiment, the participants did, in fact, purchase inputs as indicated in the order forms at the discount level and with or without the credit option as determined by the bingo game.

12.4 Data

In the following analyses, we use information collected from the participating households in both the treatment and control villages in Eastern and Central regions. Table 12.2 shows the number of sample villages and households for each event by region and type of household. The geographic distribution of those villages is given in Fig. 12.1.

12.4.1 Village and Household Characteristics in 2005

Table 12.3 shows the characteristics of villages and households in the RePEAT 2005 survey by participation type. Due to the nature of the random assignment of free input distribution, there is presumably no systematic difference in the pre-intervention characteristics between these two groups. The test statistics of the difference in mean of the key variables shown in Column 3 confirms the presumption. There is no variable which is statistically significant different between these two groups. Our samples were small-scale farmers in rural areas who on average cultivated 1.2 ha of land, had slightly less than 8 family members, and earned 1.7 million Shillings in year 2005.[15] A quarter of the income came from sources other than farming. More than 80% of them grew maize, and

[14] We randomly assigned different minimum down payment and interest rate for credit sales across communities. The interest rates offered are 5, 10, or 15% per cropping season. The minimum down payment offered are 20, 30, or 40%.

[15] Exchange rate on August 15, 2005 was 1,811.23 UGX per US dollar.

Table 12.2 Number of households by participation group in experiments in Uganda

			Type of village and household			
Event	Region		Total	Control	Treatment	Neighbor
RePEAT 2005 survey						
	Eastern	Village	39	13	26	
		Household	372	125	247	
	Central	Village	30	10	20	
		Household	277	95	182	
Free input distribution						
	Eastern	Village	26	0	26	
		Household	242	0	242	
	Central	Village	20	0	20	
		Household	135	0	135	
Sales experiment						
	Eastern	Village	39	13	26	
		Household	513	110	210	193
			(0.12)	(0.04)	(0.09)	(0.18)
	Central	Village	30	10	20	
		Household	304	78	128	98
			(0.20)	(0.11)	(0.16)	(0.30)

Note: Sample attrition rates in the sales experiment are shown in parentheses

few farmers used modern inputs. The average use of fertilizer and the adoption rate of hybrid seed were negligible in both the control and treatment villages in 2005.

12.4.2 Demand for Inputs by Household Types

The simplest approach to observe the impact of free input distribution on the adoption behavior for modern inputs in the subsequent season is to compare the mean values of the purchase quantities at the sales experiment between the household types. For convenience, let us denote x_i as the purchase quantity of the *i*-th household. Let I_T, I_C, and I_N be the set of households who belong to the treatment, control, and neighbor households, respectively. Since the assignment of the treatment status was random, the average effect of the free input distribution on the purchase quantity is simply given by $E[x_i \mid i \in I_N] - E[x_i \mid i \in I_C]$. Also, its effect on the purchase quantity of the neighbor households is given as $E[x_i \mid i \in I_N] - E[x_i \mid i \in I_C]$. Since we collected the purchase quantity data with and without the credit option, we are also able to see the effect of the credit option on the purchase quantity by household type, i.e., $E[x_i \mid i \in I_O, CR = 1] - E[x_i \mid i \in I_O, CR = 0]$ for $O = T, C, N$, where CR is a binary variable taking the value of 1 if the credit option is available and 0 otherwise.

Table 12.3 Summary statistics of key variables in maize production in Uganda

	Village type					
Event	Control		Treatment			
RePEAT survey in Aug–Sep 2005	(1)		(2)		(3)	
Num. of villages	23		46			
	Mean[a]		Mean[a]		Difference[b]	
Village characteristics						
1 If public electricity is available	0.17	(0.39)	0.20	(0.40)	–0.02	(0.10)
1 If mobile network is available	0.91	(0.29)	0.89	(0.31)	0.02	(0.08)
1 If any primary school	0.65	(0.49)	0.67	(0.47)	–0.02	(0.13)
1 If any secondary school	0.13	(0.34)	0.11	(0.31)	0.02	(0.09)
1 If any health facility	0.83	(0.39)	0.67	(0.47)	0.15	(0.11)
Longitude (°)	33.03	(0.98)	32.97	(1.06)	0.06	(0.26)
Latitude (°)	0.60	(0.45)	0.59	(0.63)	0.01	(0.14)
Altitude (m)	1251.07	(181.8)	1204.68	(140.4)	46.39	(43.2)
Household characteristics						
Household size	7.94	(3.86)	7.80	(4.16)	0.14	(0.33)
1 If head is female	0.16	(0.37)	0.12	(0.32)	0.05	(0.03)
Head's age	46.86	(14.5)	46.27	(14.0)	0.59	(1.20)
Head's years of schooling	6.71	(3.42)	6.62	(3.16)	0.09	(0.30)
1 If having mobile phone	0.10	(0.29)	0.14	(0.34)	–0.04	(0.03)
Income (1,000sh)	1700.43	(1,165)	1691.60	(921)	8.83	(153.1)
Nonfarm income share	0.24	(0.29)	0.26	(0.29)	–0.02	(0.02)
Assets (1,000sh)	348.73	(1,117)	320.45	(763.6)	28.29	(83.9)
Cultivated land (ha)[c]	1.28	(1.03)	1.22	(1.12)	0.06	(0.09)
1 If planted maize	0.82	(0.38)	0.85	(0.35)	–0.03	(0.03)
Maize production among maize growers						
Yield (kg/ha)	1664.86	(1,460)	1436.13	(1,796)	228.73	(153.9)
Chemical fertilizer use (kg/ha)	2.77	(12.21)	1.29	(10.28)	1.48	(1.00)
1 If used hybrid seed[d]	0.06	(0.24)	0.06	(0.24)	0.00	(0.02)
Free input distribution in Feb–Mar 2009						
Participant characteristics						
1 If having mobile phone			0.35	(0.48)		
Cultivated land (ha)[c]			1.20	(0.87)		
1 If planted maize in 2008			0.87	(0.34)		
Maize production among maize growers						
Yield (kg/ha)			1534.05	(1,383)		
Chemical fertilizer use (kg/ha)			1.65	(11.47)		
1 If used hybrid seed			0.10	(0.30)		

**, *, + indicate 1%, 5%, 10% significance level, respectively

[a] Standard deviation in parentheses

[b] Standard error in parentheses

[c] Size of land cultivated (ha) in main cropping season

[d] Because of no direct information in the RePEAT survey in 2005 on whether the purchased seed was hybrid or other type, we assumed that the seed whose price per kg was more than 3,000 Ush was hybrid

12.5 Results

12.5.1 Average Purchase Quantity by Household Types

Table 12.4 shows the results of the average quantity purchased for each input at different discount rates by household type. The upper panels correspond to the results for cash purchase and the lower panels correspond to the results for credit purchase. Column 3 in Table 12.4 reports the difference in mean of the purchased quantities between the control and treatment households and the standard errors of the test statistics (in parentheses) corresponding to the null hypothesis in which the difference in mean is equal to zero. Similarly, Column 5 shows the difference between the control and neighbor households.

The difference in purchased quantities between the control and treatment households is statistically significant at the 1% level for all the inputs and at all the discount levels. This observation confirms the significant impact of free input distribution on the adoption and purchased quantity of modern inputs in the subsequent cropping season following free input distribution. The difference becomes larger with the availability of credit.

The purchased quantity of modern inputs by neighbor households is larger compared to the control households in all the cases. The difference is statistically significant for chemical fertilizers at all the discount levels, but is not significant for the hybrid seed as shown in the Table 12.4. The level of purchased quantities lies in between those for control and treatment households in all the cases.

The effect of credit is very large for all types of households, especially for the purchase of fertilizers. The credit option boosted the purchased quantities for fertilizers by more than threefold.

12.5.2 Prediction of Maize Yield with Purchased Inputs

From a policy perspective, we are also interested in knowing the level of yield gain corresponding to the use of modern inputs purchased at the sales experiment. Since we collected the purchase quantities at 3 different discount levels with and without credit from each household using the price-contingent order forms, we are able to estimate the yield gains in the six different arrangements (3 price levels times 2 credit arrangements) by household type.

However, average yield gains may not be properly estimated by simply calculating the mean yield at the different arrangements by household type because the number of observations may be insufficient for some arrangements, given that we actually sold the inputs under a single arrangement out of the six, based on the outcome of the bingo game in each village. Therefore, we instead first estimate the yield function first using maize production data in the 2nd cropping season of 2009, which was collected from the subsample of participants in the sales experiment.

Table 12.4 Purchase quantity of modern inputs at sales experiment held in August-September 2009 in Uganda

	Household type				
	Control	Treatment		Neighbor	
Discount	(1)	(2)	(3)	(4)	(5)
	Mean[a]	Mean[a]	Difference[b] vs. control	Mean[a]	Difference[b] vs. control
Cash purchase (kg)					
Input type					
Hybrid seed					
0%	1.06	2.06	−1.00**	1.26	−0.20
	(1.56)	(2.60)	(0.19)	(1.63)	(0.15)
10%	1.11	2.18	−1.07**	1.35	−0.24
	(1.63)	(2.77)	(0.20)	(1.78)	(0.16)
20%	1.19	2.37	−1.18**	1.47	−0.28
	(1.73)	(3.04)	(0.21)	(1.95)	(0.17)
Base fertilizer					
0%	0.63	2.33	−1.70**	1.01	−0.38+
	(1.86)	(5.11)	(0.32)	(2.30)	(0.19)
10%	0.76	2.54	−1.78**	1.14	−0.38+
	(2.16)	(5.32)	(0.34)	(2.53)	(0.22)
20%	0.87	2.82	−1.95**	1.39	−0.52*
	(2.35)	(5.69)	(0.36)	(3.20)	(0.26)
Top-dressing fertilizer					
0%	0.13	1.10	−0.98**	0.56	−0.43**
	(0.51)	(2.98)	(0.17)	(1.62)	(0.10)
10%	0.14	1.22	−1.08**	0.59	−0.45**
	(0.54)	(3.24)	(0.19)	(1.66)	(0.11)
20%	0.17	1.38	−1.21**	0.65	−0.48**
	(0.63)	(3.55)	(0.21)	(1.82)	(0.12)
Credit purchase (kg)					
Input type					
Hybrid seed					
0%	1.78	3.25	−1.47**	2.19	−0.41
	(2.72)	(3.75)	(0.29)	(2.93)	(0.27)
10%	1.84	3.37	−1.53**	2.24	−0.40
	(2.84)	(3.98)	(0.31)	(2.95)	(0.28)
20%	1.93	3.56	−1.63**	2.32	−0.39
	(2.94)	(4.29)	(0.33)	(3.05)	(0.29)
Base fertilizer					
0%	2.98	7.29	−4.30**	4.40	−1.42*
	(6.46)	(11.36)	(0.80)	(7.15)	(0.64)
10%	3.32	7.87	−4.55**	4.81	−1.50*
	(7.11)	(11.87)	(0.85)	(7.33)	(0.69)
20%	3.61	8.44	−4.83**	5.12	−1.51*
	(7.60)	(12.41)	(0.90)	(7.63)	(0.73)

(continued)

Table 12.4 (continued)

	Household type				
	Control	Treatment		Neighbor	
Discount	(1)	(2)	(3)	(4)	(5)
	Mean[a]	Mean[a]	Difference[b] vs. control	Mean[a]	Difference[b] vs. control
Top-dressing fertilizer					
0%	1.13	4.40	−3.27**	2.59	−1.46**
	(3.35)	(7.50)	(0.49)	(4.98)	(0.39)
10%	1.35	4.72	−3.37**	2.80	−1.45**
	(3.59)	(7.84)	(0.52)	(5.24)	(0.41)
20%	1.56	5.17	−3.61**	3.00	−1.44**
	(3.93)	(8.32)	(0.55)	(5.41)	(0.44)

**, *, + indicate 1, 5, 10% significance level, respectively
[a] Standard deviation in parentheses
[b] Standard error in parentheses

We then predict the yields, given average input levels by arrangement and household type using the sales data collected in the sales experiment.

We consider a simple yield function by following Matsumoto and Yamano (2009). The yield in kilograms per ha, denoted by *Y*, of the *p*th maize plot of the ith household living in the kth village is given as follows:

$$Y_{pik} = A_k \cdot F(S_{pik}, B_{pik}, T_{pik}) \cdot e^{\omega_{pik}} \tag{12.1}$$

where *A* is the Hicks neutral technology parameter or a total factor productivity given the village *k*, *F*(.) is an unknown function of inputs of *S*, *B*, and *T*. *S* is seed quantity planted (kg/ha), *B* is base fertilizer quantity (kg/ha), and *T* is top-dressing fertilizer quantity (kg/ha).[16] ω is an individual-level idiosyncratic shock. Taking logs of the yield function and using a second-order approximation of log of the unknown function of *F* (.), we have

$$\ln Y_{pik} = \ln A_k + \sum\nolimits_x \delta_x x_{pik} + \sum\nolimits_x \delta_{xx'} x_{pik} x'_{pik} + \omega_{pik} \tag{12.2}$$

for $x,\ x' \in \{S, B, T\}$. δ_x and $\delta_{xx'}$ are the parameters to be estimated.

We also consider the differential yield response to these inputs for different seed types (local vs. hybrid). Thus, we use an econometric specification which allows the parameters to be different depending on whether hybrid or local seeds are planted. These differential parameters can be estimated in a single regression model by introducing interaction terms of the binary indicator representing hybrid seed application with all the regressors.

[16] Organic fertilizers are ignored because they were applied to only few plots.

Table 12.5 Determinants of maize yield in 2nd cropping season of year 2009 in Uganda

	Dependent variable Log of maize yield (kg/ha)	
Seed (kg/ha)	0.0453***	(0.0092)
Seed squared	−0.0004**	(0.0001)
Base fertilizer (kg/ha) (Base)	0.0164	(0.0096)
Base squared	0.0002	(0.0004)
Top-dressing fertilizer (kg/ha) (Top)	−0.1263***	(0.0158)
Top squared	0.0158***	(0.0009)
Seed × Base	−0.0007	(0.0012)
Seed × Top	−0.0010	(0.0007)
Base × Top	−0.0114***	(0.0006)
1 If hybrid seed used (dHYB)	0.2945	(0.1876)
dHYB × Seed	−0.0154	(0.0142)
dHYB × Seed squared	0.0004	(0.0002)
dHYB × Base	−0.0040	(0.0129)
dHYB × Base squared	−0.0002	(0.0004)
dHYB × Top	0.1174***	(0.0192)
dHYB × Top squared	−0.0158***	(0.0009)
dHYB × Seed c Base	0.0001	(0.0013)
dHYB × Seed × Top	0.0019*	(0.0009)
dHYB × Base × Top	0.0114***	(0.0006)
Constant	6.3131***	(0.1356)
Village dummies	Included	
Number of observations	667	
Number of villages	54	
R-squared	0.16	

Robust standard errors in parentheses
*$p<0.05$; **$p<0.01$; ***$p<0.001$

Table 12.5 reports the estimates of the parameters. The model is estimated by the fixed effects regression method at the village level. We expect that the village fixed effects control for unobservable village factors such as weather shocks, soil qualities, and topographies which would affect the yield level.

Using the estimates of the parameters, we calculate the predicted maize yields in the cases where farmers plant hybrid seed and apply chemical fertilizers at their average purchase quantities in the sales experiment by household type, discount level and credit option. To calculate per ha input use from the information reported in the price contingent order forms, we assume a seed use of 25 kg/ha, which is the recommended level that was proposed to the participants in the sales experiment. Secondly, we calculate the plot size allocated to maize production based on the purchased quantity of hybrid seed, that is, (plot size in ha) = (purchased quantity of seed in kg)/25. Finally, the per ha use of fertilizers is obtained by dividing the purchased quantities by plot size. Plugging these numbers into the regression equation,

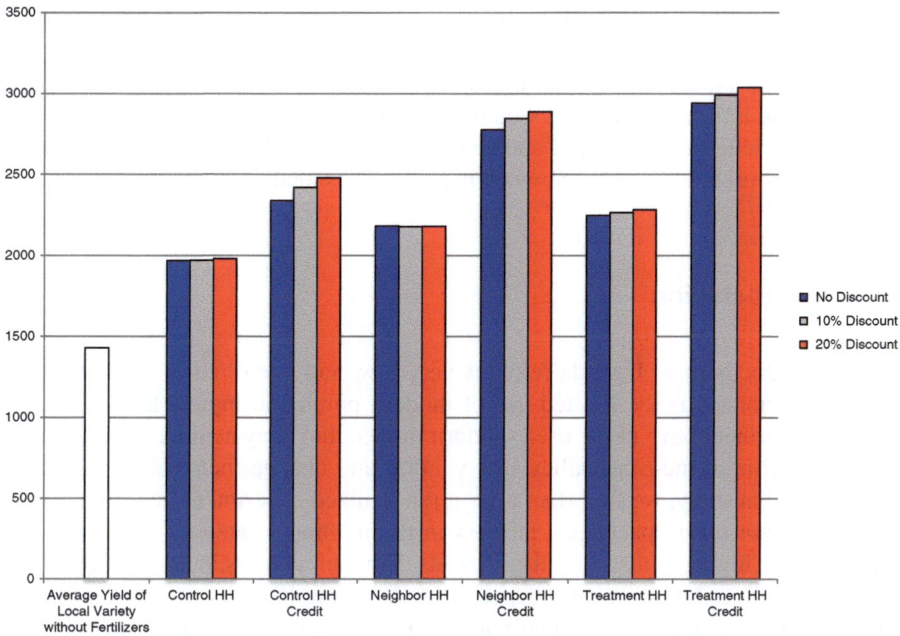

Fig. 12.2 Predicted maize yield with the use of purchased inputs in 2nd cropping season of 2009 in Uganda. The maize yields in the graph are the predicted values given the use of 25 kg/ha of the hybrid seed and the average purchase quantitiesof chemical fertilizersreported at the sales experiment in August and September 2009 by household type and by sales arrangement (in terms of discount level and credit avariability). As a referecnce, the average yield of local variety without fertilizers is also given

we obtain the predictions of maize yields by household type, discount level and credit option.[17]

The results of the simulation are summarized in Fig. 12.2. As a reference, the average maize yield for a local variety without fertilizers in the 2nd cropping season of 2009 is also reported.[18] First of all, we observe the predicted yields with hybrid seed and chemical fertilizers to be much higher than the average yield of local varieties without fertilizers. Secondly, as the discount level increases, the predicted yield slightly increases but the impact is very minimal. Thirdly, the predicted yield

[17] We estimate the semilog model, that is, $\ln Y_j = \alpha + x_j'\beta + \omega_j$, where α is the village fixed effect in our model and β is a slope coefficient vector. Hence, in order to obtain the prediction of the yield level, we convert the prediction of its log value to its level using the formula as follows: $E[Y_j \mid x_j = \bar{x}] = \exp(\bar{x}'\hat{\beta})E[\exp(\alpha + \omega_j)]$, where $\bar{x}$ is a vector of regressors having particular values such as average input levels. The estimate of $E[\exp(\alpha + \omega_j)]$ is obtained by $1/N \cdot \sum \exp(u_j)$, where $u_j = \ln Y_j - x_j'\beta$.

[18] We recorded 343 maize plots planted to local varieties in the 2nd cropping season 2009. No chemical fertilizers were applied in 334 out of 343 plots.

for the treatment households is the highest, while that for the control households is the lowest, and that for the neighbor households lies in-between. Fourthly, the impact of the credit option on predicted yield is large—being largest among the treatment households. The yield would more than double if farmers switched from the local varieties to hybrid seed and applied chemical fertilizers at the level that the treatment households purchased when the credit was made available.

12.6 Conclusion

Maize productivity in Uganda remains very low, and one obvious reason for the poor performance is the limited use of modern purchased inputs. Because many Ugandan farmers have never used modern inputs, they may acquire knowledge of their use from a one-time policy intervention and change their behavior permanently. In this study, we find that, after a randomized experiment involving distribution of free maize packages, farmers in the treatment category were found to have a much higher demand for these inputs than their cohorts in the control group, revealed in the subsequent sales experiment. Thus, our findings suggest that even a one-time policy intervention involving distribution of a free maize package will have a long-term impact on input demand because knowledge of the use and profitability of new inputs is acquired through the intervention. In addition, we find that neighbors of households in the treatment group have a higher demand for modern inputs than their cohorts in the control group. This likely to reflect the information spillover effect of the randomized experiment, which suggests that wider dissemination of new technology is possible even with a one-time, small-scale policy intervention. The major policy implication of this study is that in order to increase maize productivity by means of new technology, we must strengthen our extension system.

The findings of this study, however, show that Ugandan farmers face severe credit constraints because their demand for inputs increased significantly when they were given a credit option. During the sales experiment, we asked farmers to express their demand for the modern inputs with a credit option. Because they were told upfront that they had to buy the amounts of inputs entered in the order form if they had won a credit option, we consider the stated demand to be reliable. However, there is a possibility that participants over-stated their demand under the credit option with intentions of defaulting, if they doubted our ability to enforce repayment for the inputs received on credit. Further analysis is needed and will be conducted to account for the effect of opportunistic behavior (if any) by the participants; therefore, the credit results should be interpreted with caution. Nevertheless, these results suggest that the provision of affordable financial services in rural areas could prompt Ugandan farmers to change their farming methods, boost productivity, and improve their welfare. Such interventions, coupled with improvement in the distribution network for modern inputs, can increase farmers' knowledge about their usage and profitability, thereby spurring the demand for these inputs even without subsidies.

Appendix

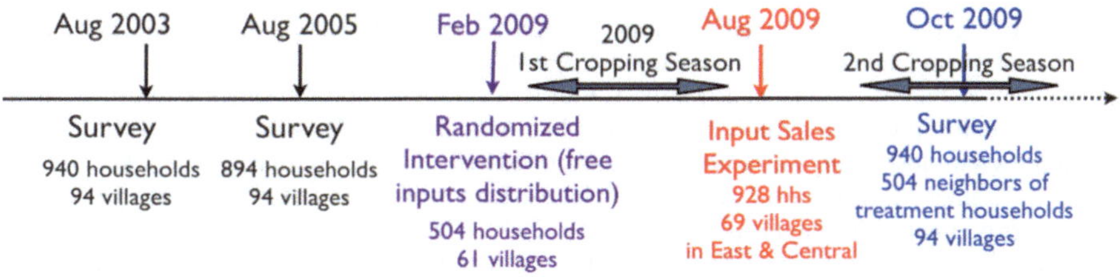

Fig. 12.A.1 Time line of surveys and field experiments in Uganda

Appendix 12.A.1. Price-Contingent Order Form Used in the Sales Experiment

Q1. Did you know the purpose of us coming is to sell the agricultural inputs? 1. Yes 2. No

Q1b. How many days ago did you know this sales experiment?

Q2. In the case of cash sales, how many kilograms of inputs do you buy?

	DK (Hybrid seed) (kg)	DAP (Base fertilizer) (kg)	UREA (Top-dressing fertilizer) (kg)	(Coordinator will help to calculate. Round-down the last two digits) **Total Amount** you would pay **today**
0% discount	(3,600)	(2,100)	(1,700)	Ush
10% discount	(3,240)	(1,890)	(1,530)	Ush
20% discount	(2,880)	(1,680)	(1,360)	Ush

Discount prices per kg (Ush) are given in the parentheses

Q3. In the case of credit sales, how many kilograms of inputs do you buy?

				(Coordinator will help to calculate. Round-down the last two digits in total amount)				
	DK (kg)	DAP (kg)	UREA (kg)	**Subtotal**	**Down payment** (above xx% of subtotal)[a]	**Balance** (Subtotal minus down payment)	**Interest** (zz% of balance)[a]	**Total amount** you pay **after harvest**
0% discount	(3,600)	(2,100)	(1,700)	Ush	Ush	Ush	Ush	Ush
10% discount	(3,240)	(1,890)	(1,530)	Ush	Ush	Ush	Ush	Ush
20% discount	(2,880)	(1,680)	(1,360)	Ush	Ush	Ush	Ush	Ush

[a] The numbers for xx and zz are preprinted and different across villages

Q4. If you decided to buy inputs, how did you finance the cost?
1. Own saving 2. Borrowing from relatives 3. Borrowing from friends 4. Other (specify)

References

Conley T, Udry C (2001) Social learning through network: the adoption of new agricultural technology in Ghana. Am J Agric Econ 83(3):668–673

Denning G, Kabambe P, Sanchez P, Malik A, Flor R et al (2009) Input subsidies to improve smallholder maize productivity in Malawi: toward an African Green Revolution. PLoS Biol 7(1) at http://www.plosbiology.org/article/info%3Adoi%2F10.1371%2Fjournal.pbio.1000023

Duflo E, Kremer M, Robinson J (2008) How high are rates of return to fertilizer? Evidence from field experiments in Kenya. Am Econ Rev 98(2):482–488

Duflo E, Kremer M, Robinson J (2009) Nudging farmers to use fertilizer: evidence from Kenya, vol 15131, NBER working paper. National Bureau of Economic Research, Cambridge, MA

Evenson RE, Gollin D (2003) Crop variety improvement and its effect on productivity: the impact of international agricultural research. CAB International, Wallingford, Oxford

Kheralah M, Delgado C, Gabre-Madhin E, Minot N, Johnson M (2002) Reforming agricultural markets in Africa, vol 38, IFPRI food policy statement. International Food Policy Research Institute, Washington, DC

Kikuchi M, Hayami Y (1985) Agricultural growth against a land resource constraint: Japan, Taiwan, Korea, and the Philippines. In: Ohkawa K, Ranis G (eds) Japan and the developing countries: comparative analysis. Blackwell, Oxford

Manski C (1993) Identification of endogenous social effects: the reflection problem. Rev Econ Stud 60(3):531–542

Matsumoto T, Yamano T (2009) Soil fertility, fertilizer, and the maize Green Revolution in East Africa, vol 5158, World Bank policy research paper. World Bank, Washington, DC

Minot N, Benson T (2009) Fertilizer subsidies in Africa: are vouchers the answer? vol 60, July, IFPRI Issues Brief. International Food Policy Research Institute, Washington, DC

Morris ML, Kelly VA, Kopicki RJ, Byerlee D (2007) Promoting increased fertilizer use in Africa: lessons learned and good practice guidelines. World Bank, Washington, DC

Munshi K (2004) Social learning in a heterogeneous population: technology diffusion in the Indian Green Revolution. J Dev Econ 73(1):185–213

Omamo SW (2003) Fertilizer trade and pricing in Uganda. Agrekon 42(4):310–324

Yamano T, Sserunkuuma D, Otsuka K, Omiat G, Ainembabazi JH, Shimamura Y (2004) The 2003 REPEAT survey in Uganda: results. http://www3.grips.ac.jp/~globalcoe/j/data/repeat/REPEATinUgandaReport.pdf

Part IV
Conclusion

Chapter 13
Towards a Green Revolution in Sub-Saharan Africa

Keijiro Otsuka and Donald F. Larson

Abstract In this volume we have seen evidence that existing newer varieties of rice and maize can be and have been successful, and that there is great scope for additional transfers and adaptations of technology from Asia, particularly in lowland rice production. Even though improved technologies are required to drive any green revolution, lessons from the chapters also indicate that technologies alone are not enough. They must be supported by working systems of input and product markets as well as credit markets that provide both the means for farmers to exploit new technologies and also the economic incentive to do so. Also needed is the effective extension system that brings new productive knowledge to farmers. These findings provide useful contexts for identifying important constraints and testing why they are not being overcome through induced innovation or government intervention in sub-Saharan Africa.

Keywords Green Revolution • Sub-Saharan Africa • Smallholder • Staple crops • Improved seeds • Inorganic fertilizer • Technology transfer

13.1 Introduction

The genesis of Asia's Green Revolution was scientific and entailed the successful creation and adoption of new varieties. It was remarkable and transformational because of the consequences of the new agronomy: farm incomes grew,

K. Otsuka (✉)
National Graduate Institute for Policy Studies, 7-22-1 Roppongi,
Minato-ku, Tokyo 106-8677, Japan
e-mail: otsuka@grips.ac.jp

D.F. Larson
Development Research Group, World Bank,
1818 H Street, NW, Washington, DC 20433, USA
e-mail: dlarson@worldbank.org

K. Otsuka and D.F. Larson (eds.), *An African Green Revolution: Finding Ways to Boost Productivity on Small Farms*, DOI 10.1007/978-94-007-5760-8_13,

regardless of scale; the price of food staples fell; and as a result, both rural and urban poverty declined regionally and globally. As described in Chap. 1, the successful agricultural development strategy was characterized by a focus on smallholder staple crops, private markets and new varieties that, in turn, depend on higher water and nutrient inputs. This approach also lies at the core of most agricultural strategies in Africa. Therefore, progress towards an-African Green Revolution requires an improved understanding about the appropriate roles of small farms, food staples, modern inputs (e.g., improved seeds and inorganic fertilizer), and the government in an African context, and how all these elements can be integrated into a viable development strategy. The studies in this book seek to help fill that gap – especially in light of the relationships between agro-climate, technology adoption and farm productivity, as the agro-climate in Sub-Saharan Africa is widely considered to be less favorable for a Green Revolution. On the whole, the findings are supportive of an Asian-style approach. This is encouraging, since most agricultural land in Africa is farmed by smallholders who grow staple crops, and it is hard to imagine how the sector as a whole can grow without improving smallholder productivity.

At the outset, it is important to emphasize that the availability of productive and profitable technologies is a prerequisite for a Green Revolution in Africa. As pointed out by Hayami and Ruttan (1985), the essence of the Green Revolution in Asia was the transfer of technology from temperate zones, such as Japan, to tropical areas in Asia, which required massive research and educational investments for adaptation.[1] Compared with such a north-to-south transfer of agricultural technology, technology transfers from tropical Asia to Africa are easier because of the similarity of the climates. Moreover, as we have seen in this volume, there is evidence that existing newer varieties of rice and maize can be and have been successful, and there is great scope for additional transfers and adaptations of technology from Asia, particularly in lowland rice production. There is scope for intra-African transfers as well, from places where green revolutions have already begun.[2]

Even though improved technologies are required to drive any green revolution, lessons from the chapters indicate that technologies alone are not enough. They must be supported by working systems of input and product markets as well as credit markets that provide both the means for farmers to exploit new technologies and also the economic incentive to do so. Also needed is the effective extension system that brings new productive knowledge to farmers.

Several case studies in this volume are undertaken in regions of Sub-Saharan Africa that already experienced some progress towards a green revolution, or are at the cusp of promising breakthroughs. In some cases, many of the key ingredients for a green revolution are in place (especially the technology), but take-off is proving slower than expected or is not yet scaling up to other similar areas. These

[1] Also see the collection of papers by Ruttan and Hayami edited by Otsuka and Runge (2011).

[2] In addition to relevant chapters in this volume, see Larson et al. (2010).

provide useful contexts for identifying important constraints and testing why they are not being overcome through induced innovation or government intervention. Moreover, the problems faced in Africa are not exactly those facing Asia when its green revolution began. This is partly because of the structural change that has occurred in global markets since that time and partly because conditions in Africa are more heterogeneous. This in turn suggests that a broader pallet of technologies and approaches will be required than in Asia in order to transform agriculture in Africa.

In Sect. 13.2, we briefly summarize key differences between current conditions in Africa and in Asia when the Green Revolution got under way: (1) a greater diversity of staple foods in African diets and production in combination with greater regional differences; (2) the trend of higher fertilizer prices relative to grain output prices during the last 40 years; and (3) a greater diversity of initial land conditions related to irrigation, urbanization and available agricultural lands. Next in Sect. 13.3, we summarize the major findings in this study in the light of four major issues identified in Chap. 1: (1) small vs. large farms; (2) food staple vs. high-value products; (3) low-input vs. high-input agriculture; and (4) role of market vs. government. Then we provide policy implications of this study towards a Green Revolution in Sub-Saharan Africa in Sect. 13.4.

13.2 Different Starting Points

13.2.1 A Broader Portfolio of Crops

As already discussed, African agriculture is characterized by smallholder farms devoted to staple crops. This was also true in Southeast Asia when the Asian Green Revolution began and remains broadly true today. And, as Evenson (2004) points out, the Green Revolution should be viewed as an on-going process that led to the production of more than 9,000 modern varieties for 11 food crops and based on the work of 7 international research centers and more than 500 national agricultural research systems between 1965 and 2000. Still, because of the dominant roles of rice and wheat in Asian diets and on Asian farms, breakthroughs that boosted the productivity of these two crops alone were sufficient to significantly affect productivity sector-wide and to bring down regional poverty rates. For example, the Food and Agricultural Organization estimates that rice alone accounted for 57% of the average daily intake of calories in Southeast Asia and rice and wheat in combination accounted for more than half of the calories in South Asia. Because the share of food budgets spent on staples is higher among the poor, rice and wheat played an even larger role in the diets of the poor. At the time, more than 80% of the population was rural in both regions, and the combined populations of the two regions accounted for more than 27% of the world's population. As a result, improved productivity in rice and wheat brought

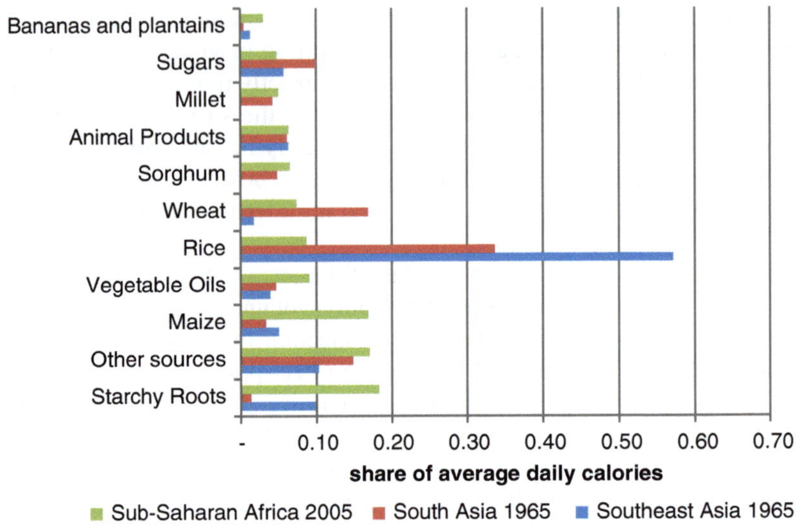

Fig. 13.1 Composition of diet in South and Southeast Asia in 1965 compared to Africa 2005 (Note: Based on calories derived from apparent consumption, Source: FAOSTAT (2010))

transformational changes to millions of people and disproportionate benefits to the rural poor, with global consequences.

Currently and historically, diets and farm output are more diverse in Africa. In Fig. 13.1, the composition of average diets in South and Southeast Asia in 1965 is compared to the composition of current diets in Africa.[3] The outsized role played by rice in Asia is apparent as is diverse nature of diets in Africa; maize and rice are important, but together they account for only 26% of caloric intake continent-wide. Root crops, including cassava and yams, are important as well, but including this additional class of staple crops still only accounts for 44% of calories in Sub-Saharan Africa. Moreover, as can be seen in Table 13.1, there are regional differences as well, with maize playing an especially important role in Eastern and Southern Africa while rice is most important in Western Africa.

In terms of policies aimed at boosting productivity, the diversity of African agriculture opens multiple fronts for the innovation of new technologies, but it also means that the research, dissemination and extensions systems in Africa must cover a broader portfolio of crops to bring about the same level of impact as the varietal innovations in rice and wheat achieved in Asia.

[3] Asian countries include Afghanistan, Bangladesh, Bhutan, India, Iran, Maldives, Nepal, Pakistan and Sri Lanka comprise South Asia. Southeast Asia includes Brunei Darussalam, Cambodia, Indonesia, Lao People's Democratic Republic, Malaysia, Myanmar, Philippines, Singapore, Thailand, Timor-Leste, and Viet Nam.

Table 13.1 Regional composition of diet in Africa, 2005

	Eastern Africa	Middle Africa	Southern Africa	Western Africa	Sub-Saharan Africa
Total calories per capita per day	2,010	1,823	2,884	2,580	2,260
Share of total					
Starchy roots	0.16	0.35	0.03	0.19	0.18
Other sources	0.21	0.17	0.09	0.15	0.17
Maize	0.24	0.14	0.31	0.09	0.17
Vegetable oils	0.06	0.09	0.11	0.12	0.09
Rice	0.07	0.05	0.05	0.12	0.09
Wheat	0.08	0.06	0.16	0.05	0.07
Sorghum	0.05	0.03	0.01	0.10	0.06
Animal products	0.07	0.05	0.13	0.05	0.06
Millet	0.01	0.02	0.00	0.10	0.05
Sugars	0.05	0.04	0.10	0.04	0.05
Bananas and plantains	0.04	0.03	0.01	0.03	0.03

Source: FAOSTAT (2010)
Note: Based on calories derived from apparent consumption

13.2.2 Fertilizer Use

An important component of Asia's Green Revolution was an increase in fertilizer use associated with the adoption of high-yielding varieties of grain. As several of the chapters in this volume point out, the willingness and ability of farmers to supplement soil nutrients is especially important for the adoption of new varieties of maize, rice and other crops since soils and soil management practices are frequently poor in Africa.

As Table 13.2 illustrates, lower yields in Africa go hand in hand with lower applications of fertilizers. The table provides average yields for maize and rice for regions in Sub-Saharan Africa and for South and Southeast Asia. Yields have been largely stagnant in Africa, while yields have more than doubled in Asia. Comparable aggregate data on fertilizer use is unavailable for 1965; however the table suggests that differences in fertilizer consumption are associated with yield differences between Asia and Africa as a whole and also among regions in Africa. Moreover, the one region where maize yield increases has kept pace with Asia is Southern African, a region with relatively high fertilizer consumption.

Because chemical fertilizers are mostly imported into Sub-Saharan Africa, the poor performance of transportation networks and the inefficiencies of markets are concerns, since these tend to raise the cost of inputs and lower the price of outputs at the farm-gate. The same concerns were present in Asia 40 years ago and governments sometimes intervened to maintain product prices or lower fertilizer prices. Still, the effects of the interventions on relative prices may have been less important than the cumulative effect of global trends.

Table 13.2 Average yields for rice and maize by region in Sub-Saharan Africa and Asia, 1965 and 2005

	Rice		Maize		Fertilizer consumption
	1965 (t/ha)	2005	1965	2005	2005-centered average (kg/ha)
Eastern Africa	1.81	2.25	0.95	1.19	11.46
Middle Africa	0.75	0.95	0.79	0.94	2.73
Southern Africa	2.27	2.32	1.05	3.49	43.66
Western Africa	1.08	1.60	0.77	1.59	5.07
Southeast Asia	1.66	3.88	0.93	3.12	100.95
South Asia	1.41	3.27	1.10	2.20	116.56

Source: FAOSTAT (2005)
Fertilizer consumption is measured as total nutrient weight of nitrogen, phosphate and potash fertilizer divided by arable land and area permanent crops. The valued is a 5-year average, centered on 2005

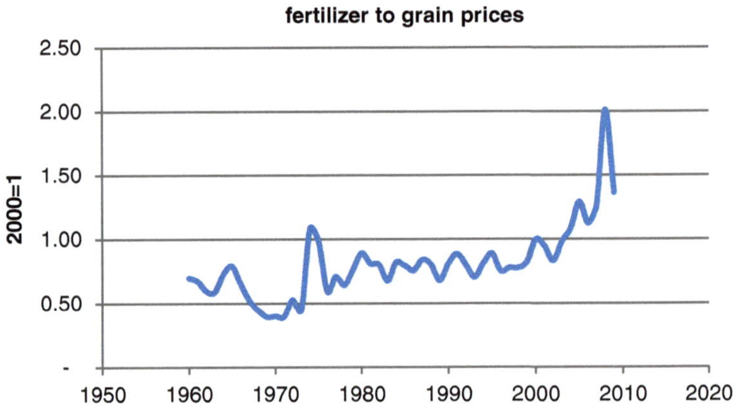

Fig. 13.2 The amount of grain needed to purchase fertilizers based on international price data, 1960–2009 Note: Based on the international grain and fertilizer prices expressed as indices (Source: World Bank Development Prospects Group 2010)

Figure 13.2 maps the international price of fertilizers relative to grains. Two indices comprise the ratio, based on data maintained by the World Bank (2010), and the ratio can be thought of as measuring the amount of grain required to buy a given amount of fertilizer. Though there has been significant year-to- year variation, the indices suggest that the amount of cereal required to purchase a given amount of fertilizer has risen by 78%, when comparing a 5 year average of prices around 2005 with a similar average centered on 1965. This means that, globally, the economic threshold for the use of chemical fertilizer has risen significantly since Asia's green revolution got underway. For policy, this means that efforts to reduce transport and transaction costs for input and output markets are crucial in Africa, as are efforts

Table 13.3 Installed irrigation as share of cropland by region in Sub-Saharan Africa and Asia, 1965 and 2005

	1965	2005
Sub-Saharan Africa	0.016	0.027
Eastern Africa	0.017	0.041
Middle Africa	0.005	0.006
Southern Africa	0.064	0.091
Western Africa	0.007	0.012
Southeast Asia	0.121	0.222
South Asia	0.212	0.430

Source: FAOSTAT (2010)
Note: Total area equipped for irrigation divided by arable and permanent crop land

aimed at finding productive and sustainable alternatives to farming methods that depend heavily on chemical fertilizer use.

13.2.3 Land and Demographics

When the Asia Green Revolution began, the continent's population was predominately rural and arable land was relatively scarce. Effectively, in many places in Asia, further significant growth in the production of staple crops required finding ways to improve land productivity. In turn, absent new technologies, this could only be accomplished through the build-up of capital investments – especially those related to land improvements – or by the use of increased labor or other inputs, which in turn added to production costs. Consequently, the arrival of high-yielding modern varieties directly addressed a key constraint in Asian agriculture.

Investing in irrigation is one way of improving land productivity and at the start of Asia's Green Revolution, the shares of cropland in South and Southeast Asia benefiting from investments in irrigation were higher than is the case for Sub-Saharan Africa today (Table 13.3) Moreover, irrigation efforts continued in Asia and the share of irrigated cropland doubled in both regions by 2005. During the same period, irrigation rates improved noticeably in Eastern and Southern Africa, but the overall gap between Sub-Saharan Africa and tropical Asia widened.

In general, the relationship between growth in production and constraints on land differs between Asia and Africa and is different among regions in Africa. The upper left quadrant of Table 13.4 shows the number of rural people per hectare of farm land in Sub-Saharan Africa and Asia. In 1965, there were 2.85 and 2.51 rural persons per hectare in Southeast and South Asia, respectively. In Sub-Saharan Africa, density levels were lower – averaging 1.47 persons per hectare for the region as a whole. This average masks wide differences, and regional rates ranged from less than one in Southern Africa to more than two in Eastern Africa. By 2005, population pressures had increased, and average levels for Sub-Saharan Africa as a whole approached levels similar to the 1965 values for Asia. Moreover, regional differ-

Table 13.4 Agricultural land relative to rural and urban populations by region in Sub-Saharan Africa and Asia, 1965 and 2005

	Arable land and permanent crops		Agricultural area	
	1965	2005	1965	2005
Rural population per hectare				
Sub-Saharan Africa	1.47	2.44	0.24	0.52
Eastern Africa	2.03	3.73	0.30	0.75
Middle Africa	1.31	2.64	0.18	0.43
Southern Africa	0.88	1.39	0.08	0.14
Western Africa	1.28	1.74	0.32	0.56
Southeast Asia	2.85	3.13	2.32	2.67
South Asia	2.51	4.81	1.74	3.60
Urban population per hectare				
Sub-Saharan Africa	0.31	1.31	0.05	0.28
Eastern Africa	0.19	1.07	0.03	0.21
Middle Africa	0.35	1.76	0.05	0.28
Southern Africa	0.66	1.80	0.06	0.18
Western Africa	0.29	1.24	0.07	0.40
Southeast Asia	0.71	2.46	0.58	2.10
South Asia	0.56	2.09	0.39	1.56

Source: FAOSTAT (2010) and authors' calculations

ences became more pronounced. Returning to the table, the 2005 rate of 2.64 in Middle Africa is comparable to the 1965 levels in Asia and density levels in Eastern Africa in 2005 is comparable to current rates in Southeast Asia.

Urbanization is a related aspect of the demography that sets the stage for agricultural transformations. In particular, the growth of large and nearby urban centers in Asia provided a natural market for agricultural surpluses as Asia's Green Revolution took root. In contrast, Africa's less populated landscape results in more dispersed marketing channels, which are in turn expected to result in higher transaction costs that exacerbate the growing expense of inputs relative to outputs. Still, while population densities are higher in Asia, Africa today is more urban than Asia 40 years ago. The lower left-hand quadrant of Table 13.4 shows the ratio of urban population to arable land. This provides a rough but scaled measure of the potential proximate demand for agricultural surpluses. African rates in 2005 are on average higher than in South or Southeast Asia in 1965. Moreover, unlike the case of rural rates, the urban rates are uniformly higher for all regions. As shown in Fig. 13.3, urbanization will continue, and by 2020, the urban shares of the population will be similar in Asia and Africa.

To this point, the discussion on land use has focused on land that is currently farmed for permanent or annual crops. Using the broader measure of "agriculture area" suggests that land remains much more abundant in Africa. This category of land includes cropland and land under permanent crops, but it also includes meadows and pastures used for grazing – land that could potentially be planted to crops. Density ratios based on this broader measure are reported in the two right-most

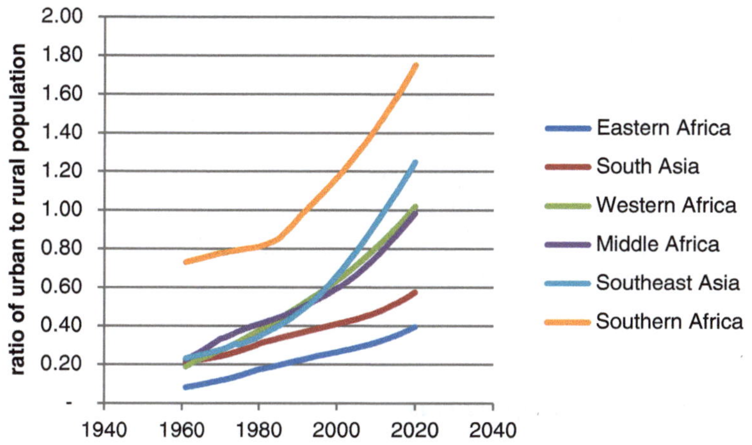

Fig. 13.3 Actual and predicted regional urbanization in Africa and Asia, 1961–2020 (Source: FAOSTAT 2010)

columns of Table 13.4. The inclusion of grazing lands reduces the density ratios considerable in Africa, but has considerably less effect on the ratios for Asia. Most likely, the difference in the two categories is made up of land that is remote or less well suited for crops, but there is potential for conversion from pastureland to cropland that is not present at the same level in South and Southeast Asia.

The stark differences between the two measures illustrate the complex and heterogeneous setting for agricultural policy in Africa. In general, growing urbanization has created and will continue to create a larger domestic market for surpluses from rural Africa. To compete against imports, increases in productivity are key and the pressures on currently farmed areas suggest that conditions are ripe for applying the same type of land-saving technologies that fueled Asia's Green Revolution, especially if improvements in transportation and distribution networks can be made. At the same time, there are places in Africa where land remains abundant that can be converted to cropland under the right circumstances. A global rise in food prices is one scenario leading to this, but the scenario is also associated with a worsening of poverty – especially among the urban poor. Making better use of land that is already farmed or bringing in new land at lower cost by the development of improved seeds and the adoption of improved farming practices that address the place-specific constraints on African staple crops is an alternative and more hopeful strategy.

13.3 Major Findings

In this section, we summarize some of the key findings from previous chapters in light of the core characteristics of an Asian-style policy, and interpret the results in light of key differences between Africa today and Asia at the start of its green revolution.

Table 13.5 Average farm size in India and Indonesia during Asia's Green Revolution

Country	Census year	Average farm size
India	1971	2.30
	1991	1.60
	1995–1996	1.40
Indonesia	1973	1.10
	1993	0.90

Source: Nagayets (2005)

13.3.1 Small vs. Large Farms

Smallholder agriculture was at the center of Asian agriculture when the green revolution began and improved productivity on small farms propelled the success of Asian agricultural policies. An encouraging lesson from the Asian experience is that rapid growth in agriculture is feasible without an accompanying change in farm structure. For example, census data shows that farms became smaller in India and Indonesia as Asia's Green Revolution progressed (Table 13.5). Small-holder agriculture is pervasive in Africa as well (Table 13.6).[4] And, as in Asia, there is no evidence that small farms in Africa are less productive than large farms or less innovative in the adoption of new technology. And in Asia, small farmers do not appear to apply smaller amount of external inputs, including inorganic fertilizer (Hayami and Kikuchi 1982; David and Otsuka 1994). Chapter 3 reports that small rice farmers achieved higher yields per hectare than large farmers in irrigated areas in Uganda and Mozambique. Furthermore, rice yields in these countries are comparable to those in Asia at the early stage of the Green Revolution. Farmers growing upland NERICA rice in Uganda are all small farmers, who achieved equally high yields regardless of the farm size (Chap. 6). According to the analysis of maize production in Kenya and Uganda in Chap. 9, farm size has no significant effect on the adoption of HYV maize and plot size has significantly negative effect on maize yield. Such inverse correlation between farm/plot size and productivity is widely observed in Sub-Saharan Africa (Holden et al. 2009). This can be explained by inefficient labor markets, high transaction costs due to asymmetric information, or inefficient land rental markets which fail to reallocate land from less to more productive uses (Carter 1984; Otsuka 2007; Lipton 2010). Furthermore, small farmers apply larger amount of inorganic fertilizer per hectare than large farmers in Ethiopia (Chap. 11). These findings clearly indicate that small farmers are no less productive than large farmers in Sub-Saharan Africa.

Thus, following Schultz (1964), we conclude that African small farmers are as entrepreneurial and innovative as large farmers and would be no different in this

[4] Interestingly, average farm size in both SSA and Asia was 1.6 ha in the early 2000s (von Braun 2005).

Table 13.6 The dominant role of smallholder farming in Asia and Africa

	Census year	Number of farms under 2 ha	Percentage of farms under 2 ha
Asia			
China	1997	189,394,000	98
India	1995–1996	92,822,000	80
Indonesia	1993	17,268,123	88
Bangladesh	1996	16,991,032	96
Viet Nam	2001	9,690,506	95
Africa			
Ethiopia	2001–2002	9,374,455	87
Nigeria	2000	6,252,235	74
DR Congo	1990	4,351,000	97
Tanzania	1994–1995	2,904,241	75
Egypt	1990	2,616,991	90

Source: Nagayets (2005)

respect from the small farmers who drove Asia's Green Revolution. This does not preclude the potential for successful large-scale agriculture in Africa. But it does cast doubt on policies that would abandon small-scale agriculture.

13.3.2 Food Staple vs. High-Value Products

As stated in Chap. 1, the choice between high-value products and food staples is false one and, especially going forward, urbanization and improvements in transport will undoubtedly create greater scope for high-value crops in Africa. However, as discussed in the previous section, agriculture in Africa is firmly set in staple crops as is the welfare of most Africans because of the dominant role of staples in African diets. Thus, in this study we focused on the possibility of dramatically improving the productivity of food staples.

Chapters 2 and 3 clearly demonstrated that lowland rice production in Sub-Saharan Africa is promising, particularly if irrigation is available. The chapters report that the rice technologies of Asia have been already been successfully adapted in some places in Africa and that further scope remains. In fact, irrigated rice yields in Sub-Saharan Africa, which use either modern high-yielding varieties developed in Asia or improved varieties using modern Asian varieties as parents, are not lower than irrigated rice yields in tropical Asia and are sometimes higher. Upland NERICA rice, which is cross-bred between Asian and African varieties, is also highly productive in some countries, such as Uganda and Benin where the yields are two to three times as high as those of traditional upland rice varieties (Chaps. 7 and 8). Maize yield can be as high as 2.2 t/ha, if purchased hybrid seeds are planted along with the application of inorganic fertilizer in Kenya (Chap. 9), whereas it can reach as much as 3.0 t in Uganda if credit is provided according to the result of experiments (Chap. 12).

Chapters 4 and 5, which compare the yields of rice, wheat, maize, sorghum, and millet between India and Sub-Saharan Africa, found large potential to transfer rice, maize, and wheat technology from Asia to Africa. This is not true for sorghum and millet as the yields of these two crops are comparable between India and Sub-Saharan Africa, despite the more favorable economic and climatic environments in India. Even if the development of hybrid millet and sorghum varieties contributed to the growth of yield and total factor productivity of these crops in India (Janaiah et al. 2005), the role of private sector seed suppliers is crucial in the development and dissemination of new varieties, as in the case of maize (Pray and Nagarajan 2009), which has been difficult to realize in SSA. As a result, no visible impacts of the new technologies, including use of the improved varieties developed in India, were found in SSA (Ndjeunga and Bantilan 2005). Moreover, although wheat is a Green Revolution crop and an important component of African diets, the scope for wheat as a focus of policy in Sub-Saharan Africa is limited, because it can be profitably grown only in limited low-temperature areas.

As discussed earlier, a successful African Green Revolution will require a broader portfolio of crop technologies than in Asia to achieve comparable levels of food security and poverty reduction. Though the topic lies outside the scope of the current volume, root crops as a group, including cassava and sweet potatoes, represent a potential third focus area for regional efforts. The crops are attractive since they grow year-round in tropical Africa and are an important component of smallholder livelihoods. Moreover, as with grains, achieved yields in Sub-Saharan Africa are lower than in Asia on average and also well below the potential yields indicated by the field tests of current varieties.[5] Still, some of the most promising technologies for root crops are based on trans-genetic modifications, which face special implementation hurdles (Cohen and Paarlberg 2004). For many reasons, Eicher et al. (2006) conclude that genetically modified varieties of cassava, sweet potatoes and four other staple crops will not reach small farms in Africa for 10–15 years.

Consequently, taken as a whole, the research reconfirms the reasoning behind current regional efforts that focus on the development and dissemination of new varieties for rice and maize along with related efforts to establish best smallholder farming practices. This should not preclude efforts to close the yield gap in other African staples, but it does suggest that the rice and maize are crops that merit special attention.

[5] For example, Fermont et al. (2009), Haggblade and Hazell (2010), and Johnson et al. (2006) discuss the potential for known cassava technologies in Africa and Qaim (2001) discusses the potential for sweet potatoes. See also Spielman and Pandya-Lorch (2009) for a collection of promising crops and technologies in SSA.

13.3.3 Low-Input vs. High-Input Agriculture Fertilizer Use

As discussed in the previous section, the starting point for an African Green Revolution differs in key ways and these differences affect most in this third component of an Asian style agricultural policy. Inorganic-fertilizer prices have risen relative to grain prices in international markets since Asia's Green Revolution began. High transport and transaction costs in Africa further exacerbate this problem (Chaps. 6, 7, 8, 9, 11, and 12). Moreover, the constraints on land are not as binding as in Asia, which opens the possibility for productivity improvements related to other inputs – especially from more efficient water use and improved resistance to pests and diseases. Consequently, an exact replication of the Asian-type "high-input farming" is not appropriate for Sub-Saharan Africa. Indeed, inorganic fertilizer is seldom or insufficiently applied in NERICA rice farming in both East and West Africa (Chaps. 6 and 7), and maize farming in Uganda (Chap. 9) and elsewhere in Sub-Saharan Africa (Chap. 8).

At the same time, however, it is obvious that high grain yields cannot be achieved without applying fertilizer, be it inorganic or organic. Furthermore, the use of organic fertilizer is known to be effective in restoring and improving the soil fertility, which, in turn, improves maize yield (Chap. 9). According to Chap. 11, inorganic and organic fertilizers are complements in Ethiopia. This point is also emphasized by the literature review on maize production in Chap. 8. In addition, the experience in India demonstrates the importance of using both inorganic and organic fertilizers in upland farming (Chap. 10).

Thus, developing and implementing farming systems based on the use of moderate amounts of both organic and inorganic fertilizer is especially appropriate for Sub-Saharan Africa at present. Needless to say, however, this does not justify prevailing unfavorable product price-input price ratios. In fact, we argue that the efficiency of marketing systems must be improved in order to intensify farming system further in Sub-Saharan Africa over time.

13.3.4 Water and Climate

In South Asia, the Green Revolution was associated with the expansion of irrigated areas. In Africa, the area under irrigation is quite small. Consequently, because modern varieties require greater amounts of water to complement larger applications of purchased inputs, climates in Sub-Saharan Africa where rainfall is low is considered a major constraint on the adoption of improved agricultural technologies.

Chapters 4 and 5 examine this premise, focusing on comparisons between Sub-Saharan Africa and India, because India's climate is not far different from that of Sub-Saharan Africa. These chapters suggest that the impacts of climate, i.e., temperature and rainfall, on crop yields have been lessened in the case of rice, wheat, and maize farming in India, key crops in Asia's Green Revolution. This is likely due to the fact that improved varieties are short-maturing and, hence, they can be harvested within a short rainy season. Another interesting finding is that weaker but similar changes, i.e., weakened impacts of climate on yields of rice, maize, and wheat, have been observed in Sub-Saharan Africa. This may be taken to imply that the Green Revolution technology has been adopted in some crop sectors in Sub-Saharan Africa, even though its impacts are still limited.

13.3.5 Role of Markets vs. Government

In this subsection we turn to lessons drawn concerning two areas of concern about the role of government in promoting an African Green Revolution. The first has to do with whether private markets can be relied upon to provide services essential to the production and marketing of modern varieties of rice and maize. The second area has to do with the efficacy of public research and extension networks.

By their nature, modern grain varieties require a greater coordination of markets. The land is used more intensively and this requires greater amounts of fertilizer. Production increases are intended for off-farm markets, creating a greater demand for agents to process, transport, store and market additional amounts of grain. For policy makers, a key question is whether this process is an endogenous one, in which intermediaries respond to market incentives set in motion by the adoption of new farming methods. The notion of endogenous market development is related to the induced innovation theory of Hayami and Ruttan (1985) that posits that increases in agricultural supply resulting from the adoption of high-yielding and fertilizer-using technologies induce their own market development.

While none of the chapters addresses the question of endogenous markets directly, several results reported in the chapters are consistent and perhaps indicative of the endogeneity of markets. For example, Kijima and Otsuka report in Chap. 6 that, in 2004, Ugandan farmers reported that the major constraint on the NERICA adoption was the lack of rice millers in nearby towns, but that the average distance from farm to mill shortened considerably by 2006 due to mushrooming of rice milers. In Chap. 9, Matsumoto and Yamano show that prices of both maize and inorganic fertilizers are fairly uniform in Kenya where many farmers apply inorganic fertilizer, whereas these markets are much less developed in Uganda where maize farming is still extensive without applying inorganic fertilizer. In Chap. 11, Zerfu and Larson find evidence that some of the same conditions that result in lower output prices and higher fertilizer prices at the farm-gate can also work to discourage intermediaries from entering as suppliers of fertilizers in remote parts of Ethiopia. In a similar way, according to the recent study by Yamano et al. (2011), milk markets in Kenya have been developed owing to the active participation of private traders and fresh banana markets in Uganda function more effectively due to the introduction of mobile phone networks. Although the evidence is not as concrete as we may wish, it seems that input and output markets are developing endogenously in many parts of Sub-Saharan Africa. Still, these results are not separate from broader government policies and investments and suggest that it is important for the government to support such market development by investing in the basic transportation and communication systems that facilitate the action of private agents.

As may be expected, collectively the studies suggest that credit markets do not work well in Sub-Saharan Africa. Though subsidies that lessen the need or cost of credit are difficult to sustain, they do give some indication of the benefits of finding sustainable solutions to credit hurdles. For example, Chap. 11 provides the evidence that inorganic fertilizer use increases when famers have access to subsidized credit in Ethiopia. The result of experiment in the provision of credit in Chap. 12 shows that the demands for hybrid seeds and inorganic fertilizer increase drastically if credit is provided. The absence of the use of inorganic fertilizer in the production of upland NERICA in many countries in Sub-Saharan Africa (Chaps. 6 and 7) as well as in lowland rice production in some countries (Chap. 3) is likely to be a consequence of the incomplete credit markets. According to Chap. 2, credit programs were introduced by the governments responding to increased demand for credit associated with increased adoption of the Green Revolution technologies in Asia. Such inducement effects may take place in Sub-Saharan Africa, once the Green Revolution takes off. However, Herdt (2010) warns that aid-supported credit schemes are rarely effective and nearly impossible to sustain. As an alternative, it is also important to mention that in Asia the common practice of fertilizer dealers is to provide fertilizer without immediate payment and to request payment with interest after harvesting while implicitly using standing crops as collateral (Hayami and Kikuchi 1982). Considering the immense difficulties in managing effective credit programs, how to provide credit at reasonable terms is a major challenge in realizing the Green Revolution in Sub-Saharan Africa.[6]

While none of the studies in this book looked directly at the performance of extensions services, several chapters report disappointing outcomes that can be attributed to inadequate information systems. In general, information about agricultural technology is a public good that will not be adequately supplied without public interventions. As discussed by Herdt (2010) in his review of agricultural aid, evaluations suggest that, overall, aid to research and extension is cost-effective and sustainable; however, aggregate support for agriculture and the support given to research and extensions in Africa have fluctuated with the changing priorities of donors and recipient governments. Moreover, there is little agreement on the best way of delivering research and extension services. Thus, it is possible that insufficient extension system is responsible for the failure to disseminate improved technologies in Sub-Saharan Africa.

This argument is supported by the finding in Chap. 6 that NERICA dissemination efforts were poorly targeted in Uganda and that an inadequate understanding by

[6] A joint project is under way between IRRI and BRAC (Bangladesh Rural Advance Committee) in which BRAC provides micro finance for the purchase of organic fertilizer for rice farmers in Tanzania under the condition that the repayment is made after harvest.

early adopters of how to replicate seeds contributed to a deterioration of the quality of self-produced rice seeds. In Asia rice seed suppliers are unimportant, as most farmers use carefully grown self-produced seeds. Another important finding is the increased use of hybrid maize seeds and inorganic fertilizer by those farmers who received free package of these inputs with instruction on their use during the previous season in Uganda (Chap. 12). This having been said, it is also the case that the delivery of adequate extension services is made more difficult by the diverse nature of Africa's agricultural sector. Chapter 7 illustrates how this is the case even for a single crop. The chapter chronicles how the heterogeneous conditions of farms and farmers in West Africa can lead to wide differences in the adoption and continued use of NERICA in West Africa, making it all the more difficult to anticipate how technologies that prove successful on experimental plots will fare on the farm. Still, what remains clear is that extension system must be strong and effective to realize Africa's Green Revolution.

13.4 Conclusions and Their Implications for Policy

Asia's Green Revolution centered on smallholder agriculture and modern varieties of staple crops, especially wheat and rice. The aim was to boost productivity on small farms throughout Asia with the expectation that food security and rural incomes would improve. An important consequence, not fully anticipated at the time, was that poverty would also fall as agricultural productivity improved regionally and globally and food became more affordable.

Current agricultural policies in Sub-Saharan Africa build on the Asian policy model and the collective research brought together in this volume largely lends support to this approach. Still, there are differences between Asia and Africa and this necessitates an African policy approach that draws selectively from the Asian experience. As discussed, at the start of Asia's Green Revolution, wheat and rice dominated the diets and farms of South and Southeast Asia. Moreover, growing conditions were fairly homogeneous for these crops across the region. Faced with growing demand, alternatives to improving land productivity were few, since land was often in relatively short supply. As a consequence, a narrow set of technical breakthroughs was sufficient to bring about large gains in farm productivity. Moreover, similarities among production systems in Asia lowered adoption hurdles. To start, homogeneous growing conditions meant that successful experiments were more easily scaled up and that successful farming strategies could spread through imitation. The smaller set of relevant technologies also made it easier for extension agents to know the appropriate technology for a given local setting and this simplified extension efforts. In addition, although the new varieties with their greater dependence on water and fertilizers meant significant adjustments to input markets, the already dominant role for rice and wheat in the food system meant that existing collection, storage, transport and marketing systems could be scaled-up rather than reinvented.

In Sub-Saharan Africa, conditions are more varied. Diets are more diverse, and agro-climatic conditions, soils, water resources, urbanization and land availability differ greatly across Africa. This means that Africa's Green Revolution will require a larger set of technologies and a wider set of adaptations. It also means that more experimentation and innovation is required to find successful farming practices since varied initial conditions limits replication and imitation as effective dissemination strategies.

Still, as in Asia, small farms growing staple crops are the backbone of African agriculture. And while there is greater scope for large-scale farming in the more land-abundant parts of the continent, significant productivity gains for African agriculture as a whole will require smallholder participation. Moreover, the chapters of this book suggest that this is feasible and that the process has already begun in some places.

Specifically, the studies suggest that lowland rice, upland NERICA rice, and maize can serve as core strategic crops, since improved technologies well suited for the production environment in Sub-Saharan Africa are available. Accumulated evidence points to high transferability of lowland rice technology from Asia to Sub-Saharan Africa, as was demonstrated in Chaps. 2 and 3. NERICA is high-yielding even with minimum application of chemical fertilizer in some areas in Sub-Saharan Africa, such as Uganda (see Chap. 6), and drought-tolerant in dry areas in West Africa (see Chap. 7). Also it seems clear that high-yielding hybrid maize varieties are available in Sub-Saharan Africa (Chaps. 8, 9, and 10), even though their yield potential needs to be enhanced. In contrast, millet and sorghum do not appear to be appropriate crops currently, partly because yield differences between Asia and Sub-Saharan Africa are not significant and partly because the development of new varieties of these crops requires institutional building for strengthening collaboration between public sector research institutions and private seed companies, indicating the limited opportunity to transfer technology from Asia to Sub-Saharan Africa (Chaps. 4 and 5). Indeed, responding to the increasing population pressure, we found evidence of a nascent Green Revolution occurring in rice and maize production in Sub-Saharan Africa, even though it is not yet as dramatic as it has been in Asia (Chap. 5).

Second, although the transferability of Asian lowland rice technology is potentially high, investments in irrigation and water management are much needed in Sub-Saharan Africa for the transfer to be successful (Chap. 3). Also further work is needed to develop modern rice varieties truly suitable for production environments in Sub-Saharan Africa, as modern varieties popular in Sub-Saharan Africa now were developed mostly in the 1970s in Asia or Sub-Saharan Africa based on the cross-breeding using Asian high-yielding varieties as parents. At present, much superior rice varieties have been developed in Asia and they can be used to develop better varieties uniquely suited for Sub-Saharan Africa. There is also need to develop a new generation of improved maize varieties that are drought-tolerant, pest-resistant, and nutrient-efficient (Chap. 8). It is important to note the conclusion from Chap. 8 that, as far as maize is concerned, "progress achieved in one tropical environment cannot be easily replicated in another" (page, xx), meaning that the direct transferability of Asian maize technology to Sub-Saharan Africa may be limited. Moreover, unlike rice Green Revolution in Asia, investments in maize research in SSA were not sustained and close long-term collaboration between international research centers and national agricultural research programs was missing.[7] Thus, increased investments in adaptive research on rice and maize as well as investments in irrigation are key, so as to facilitate the technology transfer from tropical Asia to Sub-Saharan Africa.

Third, while improving soil fertility is vital to improving African agriculture and to sustaining it, trends in global prices and high transaction costs limit the economic viability of chemical fertilizer use in some places in Africa. The review of literature on maize production in Sub-Saharan Africa (Chap. 8), the case study of maize Green Revolution in Kenya and Uganda (Chap. 9), the long-term analysis of cereal yields in India (Chap. 10), and the analysis of upland crop production in Ethiopia (Chap. 11) all point to the importance of soil fertility management by the use of organic fertilizer as well as its positive interaction with the use of inorganic fertilizer. Consequently, it is especially important for upland crops, particularly maize, to promote farming methods that use an optimum mix of manure, compost, and inorganic fertilizer; integrated dairy production and maize farming systems are a good example of this type of approach.[8]

Fourth, a strong and well directed extension system is vital for the dissemination of appropriate technologies. Various case studies, e.g., NERICA in Uganda in Chap. 6 and experiments on the distribution of hybrid maize seeds and chemical fertilizer in Uganda in Chap. 13, indicate the adverse consequences of an incomplete dissemination of important production knowledge to farmers. In fact, farmers in Uganda do not know how to produce high-quality rice seeds and how profitable the use of hybrid maize and inorganic fertilizer is. Supportive evidence for the weak extension system is also provided by the analysis of upland crop production in Ethiopia (Chap. 11) and the literature review of maize production (Chap. 8). Because of a greater heterogeneity in growing conditions, this task is more difficult in Africa than Asia; however, the possibility of realizing African Green Revolution will be enhanced significantly with a sufficiently improved network of extension systems capable of diffusing the Green Revolution technologies.

Finally, a number of obstacles stand in the way of effective input and output markets. Almost all the studies agree that investments in transport and communications infrastructure are a first critical step as this would reduce the price of chemical fertilizer and increase the prices of farm products, which is a prerequisite for large-scale Green Revolution in Sub-Saharan Africa. The studies also note credit constraints on fertilizer application in Ethiopia (Chap. 11) and Uganda (Chap. 12). In addition, the experimental research reported in Chap. 12 suggests that there are benefits of combining credit, inputs and extension to overcome learning hurdles as farmers may not know how to apply modern inputs optimally and, consequently, undervalue the

[7] According to Haggblade and Hazell (2010), dramatic success in improving productivity of cassava and cotton in SSA was based, among other thing, on the close and enduring collaboration between advanced agricultural research centers and national agricultural research systems.

[8] Our view is supported by classic studies of farming systems in Sub-Saharan Africa by Pingali et al. (1987) and McIntire et al. (1992).

benefits of fertilizer applications. These supporting programs must have high pay-offs when new technologies are being introduced. At the same time, past efforts to offer subsidized credit have not proved effective or sustainable in the medium term. Consequently, further research regarding the capacity of governments, banks, NGOs, cooperatives or fertilizer-dealers to facilitate credit is the critical remaining question.

In sum, the seeds for a Green Revolution are in place for many communities in Africa and substantial progress has been achieved in some places. The diversity of diets and growing conditions in Africa create additional hurdles that distinguish and will continue to distinguish the African and Asian experiences. In the short run, Africa's Green Revolution will be characterized as a mosaic of efforts that successfully adapt technologies that have proved successful in Asia and elsewhere in Africa, often based on existing varieties of rice and maize. Moreover, we believe that with adequate investments and appropriate policies, it is possible to launch and scale-up a Green Revolution in Sub-Saharan Africa.

References

Carter MR (1984) Identification of the inverse relationship between farm size productivity: an empirical analysis of peasant agricultural production. Oxf Econ Pap 36(1):131–145

Cohen JI, Paarlberg R (2004) Unlocking crop biotechnology in developing countries – a report from the field. World Dev 32(9):1563–1577

David CC, Otsuka K (eds) (1994) Modern rice technology and income distribution in Asia. Lynne Rienner, Boulder

Eicher CK, Maredia K, Sithole-Niang I (2006) Crop biotechnology and the African farmer. Food Policy 31(6):504–527

Evenson RE (2004) Food and population: D. Gale Johnson and the Green Revolution. Econ Dev Cult Change 52(3):543–569

Fermont AM, van Asten PJA, Tittonell P, van Wijk MT, Giller KE (2009) Closing the cassava yield gap: an analysis from smallholder farms in East Africa. Field Crops Res 112(1):24–36

Food and Agriculture Organization of the United Nations (FAO) (2005) FAOSTAT. FAO, Rome

Food and Agriculture Organization of the United Nations (FAO) (2010) FAOSTAT. FAO, Rome

Haggblade S, Hazell PBR (2010) Successes in African agriculture: lessons for the future. Johns Hopkins University Press, Baltimore

Hayami Y, Kikuchi M (1982) Asian village economy at the crossroads. Johns Hopkins University Press, Baltimore

Hayami Y, Ruttan VW (1985) Agricultural development: an international perspective. Johns Hopkins University Press, Baltimore

Herdt RW (2010) Development aid and agriculture. In: Evenson RE, Pingali PL (eds) Handbook of agricultural economics, vol 4. Elsevier, Amsterdam

Holden S, Otsuka K, Place F (2009) The emergence of land markets in Africa: assessing the impacts on poverty, equity, and efficiency. Resources for the Future, Baltimore

Janaiah A, Achoth L, Bantilan C (2005) Has the Green Revolution bypassed coarse cereals? The Indian experience. J Agric Dev Econ 2(1):20–31

Johnson ME, Masters WA, Preckel PV (2006) Diffusion and spillover of new technology: a heterogeneous-agent model for cassava in West Africa. Agric Econ 35(2):119–129

Larson DF, Otsuka K, Kajisa K, Estudillo J, Diagne A (2010) Fostering a Green Revolution in rice: can Africa replicate Asia's experience? In: Pandey S, Byerlee D, Dawe D, Dobermann A, Mohanty S, Rozelle S, Hardy B (eds) Rice in the global economy: strategic research and policy issues for food security. International Rice Research Institute, Los Banos

Lipton M (2010) From policy aims and small-farm characteristics to farm science needs. World Dev 38(10):1399–1412
McIntire J, Bourzat D, Pingali PL (1992) Crop-livestock interaction in Sub-Saharan Africa. World Bank, Washington, DC
Nagayets O (2005) Small farms: current status and key trends in the future of small farm. Paper in proceedings of a research workshop, Wye, UK, International Food Policy Research Institute, Washington, DC, 26–29 June 2005
Ndjeunga J, Bantilan MSC (2005) Uptake of improved technologies in the semi-arid tropics of West Africa: why is agricultural transformation lagging behind? J Agric Dev Econ 2(1):85–102
Otsuka K (2007) Efficiency and equity effects of land markets. In: Evenson RE, Pingali PL (eds) Handbook of agricultural economics, vol 3. Elsevier, Amsterdam
Otsuka K, Runge CF (2011) Can economic growth be sustained? The collected papers of Vernon W. Ruttan and Yujiro Hayami. Oxford University Press, Oxford
Pingali PL, Bigot Y, Binswanger HP (1987) Agricultural mechanization and the evolution of farming systems in Sub-Saharan Africa. Johns Hopkins University Press, Baltimore
Pray CE, Nagarajan L (2009) Improving crops for arid lands: pearl millet and sorghum in India. In: Spielman DJ, Pnadya-Lorch R (eds) Millions fed: proven successes in agricultural development. International Food Policy Research Institute, Washington, DC
Qaim M (2001) A prospective evaluation of biotechnology in semi-subsistence agriculture. Agric Econ 25(2–3):165–175
Schultz TW (1964) Transforming traditional agriculture. Yale University Press, New Haven
Spielman DJ, Pandya-Lorch J (2009) Highlights from millions fed: proven successes in agricultural development. International Food Policy Research Institute, Washington, DC
von Braun J (2005) Small scale farmers in a liberalized trade environment. In: Huvio T, Kola J, Lundström T (eds) Small-scale farmers in liberalised trade environment. Proceedings of the seminar, Haikko, Finland, 18–19 October, 2004. Department of Economics and Management Publications No. 38, Agricultural Policy, University of Helsinki, Helsinki
World Bank (2010) Rising global interests in farmland: can it yield sustainable and equitable benefits. Agricultural and Rural Development Department, World Bank, Washington, DC
Yamano T, Otsuka K, Place F (2011) Emerging development of agriculture in East Africa: markets, soil, and innovations. Springer, Amsterdam

ERRATUM TO

An African Green Revolution

Keijiro Otsuka and Donald F. Larson

DOI 10.1007/978-94-007-5760-8_14

MIX
Papier aus verantwortungsvollen Quellen
Paper from responsible sources
FSC® C105338

If you have any concerns about our products,
you can contact us on
ProductSafety@springernature.com

In case Publisher is established outside the EU,
the EU authorized representative is:
Springer Nature Customer Service Center GmbH
Europaplatz 3, 69115 Heidelberg, Germany

Printed by Libri Plureos GmbH
in Hamburg, Germany